Advances in Aquatic Ecology

Advances in Aquatic Ecology

Edited by Simon Oakenfold

MURPHY & MOORE
www.murphy-moorepublishing.com

Published by Murphy & Moore Publishing,
1 Rockefeller Plaza,
New York City, NY 10020, USA
www.murphy-moorepublishing.com

Advances in Aquatic Ecology
Edited by Simon Oakenfold

International Standard Book Number: 978-1-63987-015-8 (Hardback)

Cataloging-in-Publication Data

Advances in aquatic ecology / edited by Simon Oakenfold.
p. cm.
Includes bibliographical references and index.
ISBN 978-1-63987-015-8
1. Aquatic ecology. 2. Freshwater ecology. 3. Aquatic biology.
I. Oakenfold, Simon.
QH541.5.W3 A58 2022
577.6--dc23

TABLE OF CONTENTS

Preface VII

Chapter 1 **A technical guide to develop a statistically valid fish-based index in compliance with the water framework directive** **1**
Mehmet Borga Ergönül, Jan Breine and Erika Van den Bergh

Chapter 2 **Assessing functional diversity of macrobenthic assemblages in sewage-affected intertidal shores** **17**
G. V. Garaffo, M. L. Jaubet, E. N. Llanos, M. A. Saracho Bottero and R. Elías

Chapter 3 **A new insight into the influence of habitat on the biochemical properties of three commercial sea cucumber species** **32**
Mercedes González-Wangüemert, Christina C. Roggatz, Maria João Rodrigues, Luísa Barreira, Manuela Moreira da Silva and Luísa Custódio

Chapter 4 **Seasonal species diversity and abundance of phytoplankton** **45**
N. Heydari, S. M. R. Fatemi, A. Mashinchian, R. Musavi Nadushan and B. Raeisi

Chapter 5 **Horizontal and vertical variations in survival rates of juvenile bivalves, *Anadara kagoshimensis* (Tokunaga, 1906), *Ruditapes philippinarum* (Adams & Reeve, 1850) and *Atrina* sp., experimentally caged in the inner area of the Sea** **61**
Takeo Kurihara, Kazumasa Hashimoto, Shoji Nakano, Yukihiko Matsuyama, Atsushi Ito, Masaei Kanematsu and Kazumaro Okamura

Chapter 6 **An overview of Cladoceran studies conducted in mine water impacted lakes** **72**
Jaakko Johannes Leppänen

Chapter 7 **The East River tidal strait, New York City, New York, a high-nutrient, low chlorophyll coastal system** **87**
Yaqin Li, Shannon L. Meseck, Mark S. Dixon and Gary H. Wikfors

Chapter 8 **Difference in benthic invertebrate communities of headwater streams can be detected using a short elevation gradient** **100**
Armin Namayandeh, David V. Beresford, Keith M. Somers and Peter J. Dillon

Chapter 9 **Characterizing a novel predator–prey relationship between native *Diplonychus esakii* (Heteroptera: Belostomatidae) and invasive *Gambusia affinis* (Teleostei: Poeciliidae)** **112**
Xu Ouyang, Jiancao Gao, Bojian Chen, Zaizhao Wang, Hong Ji and Martin Plath

Chapter 10 **Surface water quality and deforestation of the Purus river basin** **123**
Eduardo Antonio Ríos-Villamizar, Maria T. F. Piedade, Wolfgang J. Junk and Andréa Viviana Waichman

Chapter 11 **Baseline survey of the fish fauna of a highly eutrophic estuary and evidence for its colonisation by Goldfish (*Carassius auratus*)** 131
James R. Tweedley, Chris S. Hallett and Stephen J. Beatty

Chapter 12 **Effects of temperature and salinity on larval survival and development of the western school prawn *Metapenaeus dalli*** 143
Jason A. Crisp, Gavin J. Partridge, Frances M. L. D'Souza, James R. Tweedley and Navid R. Moheimani

Chapter 13 **Saccate thallus of the red alga *Halosaccion glandiforme* harbor diverse invertebrate fauna** 153
Ritindra N. Bhaduri and Maya Wolf

Chapter 14 **An indoor pelagic mesocosm facility to simulate multiple water-column characteristics** 161
Ulf Båmstedt and Henrik Larsson

Chapter 15 **Inhibition of biofilm bacteria and adherent fungi from marine plankton cultures using an antimicrobial combination** 178
Vanessa Ochi Agostini, Alexandre José Macedo and Erik Muxagata

Chapter 16 **Yolk utilization and growth during the early larval life of the Silver Perch, *Bidyanus bidyanus* (Mitchell, 1838)** 191
Sulaeman and Ravi Fotedar

Chapter 17 **Capability of some agricultural wastes for removing some heavy metals from polluted water stocked in combination with Nile tilapia, *Oreochromis niloticus* (L.)** 201
Mohsen Abdel-Tawwab, Gamal O. El-Sayed and Sherien H. H. Shady

Chapter 18 **Diversity of denitrifying bacteria in the greenwater system of coastal aquaculture** 209
V. Kathiravan and K. K. Krishnani

Chapter 19 **Acute toxicity and risk assessment of three commercial detergents using the polychaete *Capitella* sp. C** 220
R. G. Uc-Peraza and V. H. Delgado-Blas

Permissions

List of Contributors

Index

PREFACE

The scientific study of the interactions of organisms with each other and with their environment is referred to as ecology. It also includes the relationships between individuals of the same species, between different species, and between organisms and their physical and chemical environments. Aquatic ecology focuses on the study of plants and animals as well as these relationships in the aquatic environments such as oceans, estuaries, lakes, ponds, wetlands, rivers and streams. The physical characteristics of an aquatic habitat affect the kind of organisms that are present there. Living organisms found in a particular ecosystem are directly affected by environmental characteristics such as nutrient concentrations, temperature, water flow and shelter. This book outlines the processes and applications of aquatic ecology in detail. There has been rapid progress in this field and its environmental applications are finding their way across multiple industries. Scientists and students actively engaged in aquatic ecology will find this book full of crucial and unexplored concepts.

This book is a comprehensive compilation of works of different researchers from varied parts of the world. It includes valuable experiences of the researchers with the sole objective of providing the readers (learners) with a proper knowledge of the concerned field. This book will be beneficial in evoking inspiration and enhancing the knowledge of the interested readers.

In the end, I would like to extend my heartiest thanks to the authors who worked with great determination on their chapters. I also appreciate the publisher's support in the course of the book. I would also like to deeply acknowledge my family who stood by me as a source of inspiration during the project.

Editor

1

A technical guide to develop a statistically valid fish-based index in compliance with the water framework directive: an evaluation for Turkish freshwaters

Mehmet Borga Ergönül · Jan Breine · Erika Van den Bergh

Abstract Fish are considered as reliable indicators of environmental quality and during the last decades several fish-based indices have been developed to assess the ecological status of water bodies in compliance with the water framework directive. The high number of fish species with a high degree of endemism, lack of knowledge on ecological requirements of certain fish species, absence of a complete dataset for reference conditions, and great variation among freshwater types complicate the development of fish-based indices for Turkish freshwaters. However, this is still possible using statistical approaches. This paper presents a practical guide which is mainly based on the recent approaches that have frequently been used in other countries worldwide. Data collection, metric selection and scoring, ecological quality ratio (EQR) calculation and validation steps are described and discussed with respect to the limitations and available data for Turkish freshwaters. The guide documented here, with minor adjustments, has a potential to be adopted as a fish-based assessment approach for the ecological status in freshwaters in Turkey and adjacent countries.

Keywords Ecological quality ratio · Fish-based index · Water framework directive · Modeling · Index of biotic integrity

Introduction

During the last decades, modification of natural flow regime, effluent discharge, excessive use of pesticides and fertilizers, deterioration of natural floodplain and river beds led to a series of alterations in the species composition and abundance of aquatic living organisms, and the physical and chemical features of the water bodies (Meybeck et al. 1996; Jekel 2005; Gallardo et al. 2016). Assessment of the ecological status of these ecosystems is important for managers and policy makers to design appropriate regulations and management strategies (Boulton 1999). Consequently, the European Union published the water framework directive (WFD) on 22 December 2000 which was entered into force on the same day. The main objective of WFD is to achieve

---

M. B. Ergönül (✉)
Department of Biology, Faculty of Science, Ankara University, Tandoğan, 06100 Ankara, Turkey
e-mail: ergonul@gmail.com

J. Breine
Instituut voor Natuur-en Bosonderzoek, Dwersbos 28, Linkebeek, 1630 Brussels, Belgium

E. Van den Bergh
Instituut voor Natuur-en Bosonderzoek, Havenlaan 88 bus 73, 1000 Brussels, Belgium

and maintain good ecological and chemical status in all water bodies to protect human health, water supply, natural ecosystems and biodiversity (EU WFD 2000).

Conventional methods based solely on chemical and/or physical water quality parameters are not considered as reliable approaches to evaluate the ecological status for surface waters (Meybeck et al. 1996; Hunting et al. 2017). Thus, WFD imposes the use of biological quality elements, including fish, to assess the ecological status in water bodies. In general, fish have representatives in a broad range of trophic levels, and a relatively high mobility and long life span, which makes them reliable indicators on a large spatial and long time scale (Karr 1981; Moore et al. 1997; Pont et al. 2007; Noble et al. 2007). Several studies emphasize the merits of fish data for environmental quality assessment (Fausch et al. 1990; Moore et al. 1997; Whitfield and Elliott 2002; Saravanan et al. 2003) and several statistical approaches have been developed to appraise the ecological status of water bodies by using fish-based indices (Fame Consortium 2004; Pont et al. 2007; EFI + Consortium 2009). Fish-based indices incorporating fish community data (species composition, abundance, biomass, etc.) and abiotic variables (pH, oxygen, altitude, slope, etc.) (Clavel et al. 2013; Aazami et al. 2015) are robust and powerful tools to reveal the effects of human activities (construction of dams, water abstraction, habitat fragmentation, etc.) on aquatic habitats (Lyons 2012; Sundermann et al. 2013; Mostafavi et al. 2015; Schnier et al. 2016).

The first fish-based index, the Index of Biotic Integrity (IBI), was developed for Midwestern USA streams (Karr 1981) and has been modified to be used all over the world. As such, several fish-based indices have been developed for lakes and rivers in America (Cunico et al. 2011), Canada (Stevens and Council 2008), New-Zealand (Joy and Death 2004), Africa (Hugueny et al. 1996) and Europe (Noble et al. 2007; Breine et al. 2010; 2015). Fish-based indices are available only from Pakistan (Qadir and Malik 2009), China (Wu et al. 2014), Taiwan (Hu et al. 2007) and Iran (Mostafavi et al. 2015) in Asia. Only one report on a fish-based index has been published recently for Akarçay and Küçük Menderes Basins in Turkey (Yerli et al. 2016). However, fish-based indices developed for specific regions and habitats with particular fish assemblages (e.g., EFI + for European freshwater fishes) cannot be used elsewhere without any adaptations and adjustments (Karr and Chu 1999; Pont et al. 2007; Wang et al. 2007). Thus, there is an increasing need for the development of fish-based national ecological evaluation tools to assist policy makers in Turkey.

Developing fish-based national ecological assessment tools for Turkish freshwaters has some drawbacks: the number of species is quite high with a considerable degree of endemism (Kuru et al. 2014), typology shows great variation among ecoregions and there are limited data on the distribution and species specific ecological requirements of certain fish (especially endemic fish) (Yerli et al. 2016). A total of 1814 riverine and 656 lacustrine freshwater habitats have been defined in 25 different catchment basins in Turkey (RBPAP 2013). These riverine and lacustrine habitats are classified into 56 and 23 different types, respectively, which is another challenging factor. The absence of a complete dataset on the reference conditions for riverine and lacustrine habitats also complicates the development of the indices. Fortunately, it is still possible to develop a valid fish-based index using appropriate statistical approaches. In this study, we present a statistical protocol for a fish-based index development process including data collection, metric selection and scoring, Ecological Quality Ratio (EQR) calculation and validation steps with respect to the limitations and drawbacks summarized above for Turkish freshwaters. Monitoring the biological quality elements without reliable and precise evaluation approaches is a waste of time and money (Irvine 2004). Thus, to be useful for the managers and policy makers, these indices must meet at least the following criteria: (1) they must be sensitive to human disturbance, (2) they should measure the variation in community patterns (i.e., species composition, abundance, biomass, etc.) and disturbance at a scale that is useful for management and (3) they should include interpretable metrics (Schoolmaster et al. 2013). The protocol explained here has the potential to be used in the development of a fish-based assessment approach for the freshwaters in Turkey and adjacent countries.

Index development

The development of a fish-based index is a complex procedure consisting of the following steps: data collection (fish sampling and abiotic data recording), selection of the candidate metrics and final metrics, scoring of the selected metrics, EQR calculation and determination of thresholds for integrity classes and

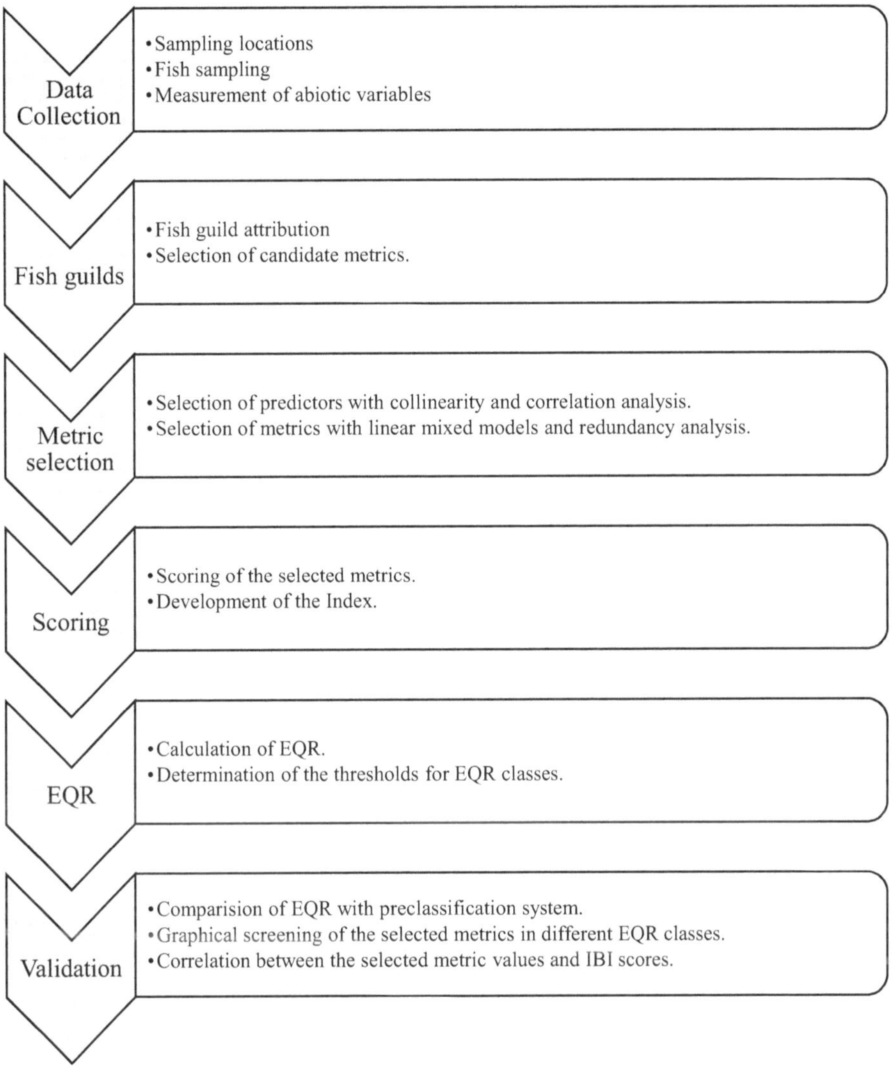

Fig. 1 Flowchart of a fish-based index development

validation. A brief workflow for the index development is given in Fig. 1. Each of these steps is summarized below.

Data collection

To develop a fish-based IBI for rivers, lakes or reservoirs, biotic and abiotic data should be collected from a wide range of sampling locations ranging from the least disturbed (i.e., having no or minor disturbances related to anthropogenic activities) to the most impacted locations, including a gradient of sites regarding human impacts (Gernes and Helgen 2002; Hering et al. 2006). The least impacted sampling locations for the different types can be selected utilizing pressure criteria (Anonymous 2003). Furthermore, the sampling locations selected on rivers should also represent an upstream–dow0010841nstream gradient along the whole river.

Fish sampling

For the assessment of ecological status in water bodies, the WFD proposes to use standardized fish sampling protocols as described by the European Committee for Standardization (CEN 2003a; CEN 2003b) with minor adjustments regarding the characteristics of water bodies and climatic conditions in the watershed. In general, electrofishing equipment and multi-mesh gill nets are recommended in rivers and lakes, respectively. The type and features of the electrofishing equipment chosen and the number of staff depend on the current velocity, width and depth. Backpack electrofishing devices are useful for wadable rivers. However, a boat may be required in large rivers. In each sampling location, electrofishing should include all habitats such as river banks, downed trees, and backwaters. Although the WFD has no recommendations for sampling season, late summer or early autumn seems to be suitable for Turkish rivers to avoid high water current velocity in spring. For lakes, the use of benthic and/or pelagic multi-mesh gill nets is recommended depending on the target fish and/or depth. The fishing effort (i.e., number of gill nets to be used) depends on the lake surface area, depth and habitat diversity. If available, a bathymetric map can be used to decide where to cast nets. Data collected by different sampling methods should not be combined or mixed to reduce the bias related to any single protocol (U.S. EPA 2002).

Fish caught in the field should be identified to the species level and the length (total length or standard length) and weight of each individual should be recorded. A few individuals can be retained and fixed in 10% formaldehyde if in situ identification is not possible. Photographs of live or freshly fixed specimens taken according to the tutorial described by Freyhof (2018) are useful for identification. To intercalibrate the fishing efforts, data should be transformed to catch per unit effort (number/biomass per m^2).

Abiotic variables

The quality and reliability of abiotic data play a key role for classification of sampling sites. Appropriate sampling equipment should be used and relevant sampling standards (e.g., ISO 5667-6 (2014) for rivers and ISO 5667-4 (2016) for lakes) should be followed. The sampling staff should be trained and experienced in such procedures. Further details can be found in the relevant standards (ISO Standard Series 5667; Part 1 (2010), 3 (2012), (2016) 4 and 6 (2014)).

Abiotic variables should include a series of both physico-chemical and hydromorphological parameters. Spot samples are suitable for rivers in most circumstances. In stratified lakes, samples from different depths may be needed to achieve a better interpretation. Additionally, monitoring the level of pollutants in samples from sediment and biota are useful to assess long-term human impacts.

The general objectives of this data collection are to rank the sampling sites according to a gradient of human disturbances in the preclassification phase and to understand why biological communities are damaged during the IBI interpretation phase (U.S. EPA 2002). Any information on the human activities which may affect fish communities and physical environment is crucial for the preclassification of sampling locations. A wide range of domestic, urban, agricultural, and industrial sources of pressures can be used for the preclassification. Pressure criteria can be based on waste water discharge, organic and inorganic pollution loads, land use, water withdrawal from lakes, rivers or groundwater, alteration of natural flow regime due to the presence of dams and regulators, channelization, presence of aquaculture facilities, deforestation, deterioration of riparian vegetation and flood plain, disturbance of river beds due to sand pits, etc. within a buffer area (e.g., 500 m) around the sampling point. A part of these data (e.g., presence of dams, deforestation, land use) can be extracted using GIS (Geographic Information Systems). Land use maps and low altitude aerial photographs taken from each sampling location are also useful.

The information on the human pressures including point and non-point sources of total nitrogen and phosphorus loads, organic and inorganic pollution loads is available via the technical reports of River Basin Protection Action Plans (RBPAP) and Sensitive Areas Project (SAP) for 25 different river basins in Turkey (RBPAP 2013; SAP 2016) and can assist in scoring the pressures to aid the preclassification of sampling locations. The categorical pressure variables (i.e., present or absent) are scored as 1 and 0, respectively. Other pressure variables which are counts or proportions (e.g., mean nutrient loads) are ranked and scored between 1 and 5: 1 indicating the best conditions and 5 the most unfavorable conditions among sampling locations. Thus, the sum of these scores can be used for preclassification; lowest corresponding to the least impaired and highest to the most impacted. A similar approach is described by Gernes and Helgen (2002). The preclassification system based on pressures caused by human activities is useful, particularly if the reference conditions are not known.

Selection of the candidate metrics

One of the most important steps for developing a statistically valid fish-based index is the metric selection using appropriate statistical tools. A metric is a variable assessing an ecological attribute of a fish community that is sensitive to human activity and reacts unambiguously to the variation of this human activity (Karr and Chu 1999).

Fish species utilizing and/or requiring similar resources (food, habitat, water quality, etc.) for feeding, reproduction and growth can be grouped into same guilds (Bain et al. 1988; Noble et al. 2007). Life history, residency status and tolerance to environmental disturbances of fish can also be used for guild attribution. The guild attribution of fish species should be the same among sampling locations and/or researchers (Schmutz et al. 2007). Candidate metrics include species richness indices, number of species, number and relative percentages of individuals and biomass of fish in each guild. Novel metrics which are potentially useful can be tested (Schmutz et al. 2007). A list of candidate metrics frequently used for fish-based indices and their predicted response to the environmental disturbances (see Sect. 2.1.3) (Fame Consortium 2004; Noble et al. 2007) is given in Table 1.

Selection of the final metrics

To fulfill the requirements of a statistically valid fish index, candidate metrics should be tested for their responsiveness to environmental disturbances. Selected metrics should cover the different ecological functions (Table 1) and redundant metrics should be avoided. Redundancy can be analyzed using correlation test. Different thresholds for redundancy are found in the literature; e.g., Hering et al. (2006) mention $c > 0.8$ but in other studies lower correlations ranging between 0.75 and 0.7 have also been accepted (Fame Consortium 2004; Breine et al. 2010).

Different selection approaches are available. Graphical screening of box-plots showing the metric values as a function of the preclassified sites can be used as a visual method for preliminary selection. Principal component analysis (PCA) and canonical correspondence analyses (CCA) were used frequently to monitor responsiveness (see review, ter Braak and Verdonschot 1995). Multimetric regression models including linear mixed effect models have been favored to select responding metrics in the recent fish-based index studies (Tableau et al. 2013; Breine et al. 2015; Smolinski and Calkiewicz 2015).

Whole data, including predictors (abiotic and pressure variables) and metrics (count, percentage), should be transformed to meet the specifications of linear models (Breine et al. 2015). The predictors included in modeling should be selected on the basis of redundancy (collinearity and correlation) tests to avoid circular thinking. Collinearity and correlation are tested with variance influence factor (VIF) and bivariate Pearson correlation tests, respectively. The full models should include initially all non-redundant predictors and a stepwise backward selection is applied until only statistically significant parameters remain.

The goodness-of-fit for each model can be defined by calculating the marginal and conditional R^2 values (Nakagawa and Schielzeth 2013). The metrics are selected on the basis of the significance of their response to predictors with a conditional R^2 value > 35%. An example for the full model is given below:

$$\text{model} < \text{lmer (Metric} \sim \text{predictor} - a + \text{predictor} - b + \text{predictor} - c + \text{predictor} - d + \text{predictor} - e)$$

Modeling of the fish data can be done using the statistical package program R (R Development Core Team 2012). To run linear modeling, the packages "lmerTest" and "lme4″ should be installed. The "MuMIn" package should be installed to calculate conditional R^2 values. Other packages are also available for same analyses. The excel data saved as.csv or.txt extension can be read and imported to R. Other statistical programs with different packages (e.g., SAS) can also be used. Since R is an open source program, it is more likely to get help from the scientific community.

Table 1 A list of candidate metrics frequently used in fish-based indices and their predicted response (up or down arrows) to increasing disturbances

Ecological function	Guilds and potential metrics	Explanation	Response	References
Species richness and composition	Shannon–Wiener	Diversity index	↓	Stirne et al. (1971)
	Pilou-Evenness Index		↓	Bellan-Santini (1980)
	Dominance Simpson's Index		↓	Simpson (1949)
	Camargo Index		↓	Camargo (1992)
	Number of species	Number of species caught	↓	Breine et al. (2004)
Condition and abundance	Number of individuals	Expressed as numbers/m^2	↓↑	Breine et al. (2004)
	Biomass	Expressed as g/m^2	↓↑	Breine et al. (2004)
	Age structure	Presence of different length classes	↓	Breine et al. (2004)
Habitat	Benthic	Bottom dwelling species	↓	Hughes et al. (2002)
	Rheophilic	Species preferring flowing water	↓	Breine et al. (2004)
	Eurytopic	Generalistic fish that are tolerant to a wide range of conditions and are found in a broad range of habitats	↑	Pont et al. (2006)
	Limnophilic	Fish preferring stagnant water	↑	Pont et al. (2006)
Residency status	Native	Species naturally occurring and self-sustaining population	↓	Schinegger et al. (2013)
	Invasive	Alien species having a negative effect on native species	↑	Lodge (1993)
	Exotic	Alien species	↑	Schinegger et al. (2013)
Tolerance/ resistance	Tolerant to oxygen deficiency	Fish species that can withstand periods of low oxygen concentrations	↑	Belpaire et al. (2000)
	Intolerant to oxygen deficiency	Fish species that always need high oxygen concentrations (> 6 mg/l)	↓	Didier (1997)
	Species that need shelter	Habitat should be diverse (pools and riffles, debris, etc.)	↓	Belpaire et al. (2000)
	Species that are habitat sensitive	Sensitive to channelisation, bank protection, dredging, etc.	↓	Mostafavi et al. (2015)

Table 1 continued

Ecological function	Guilds and potential metrics	Explanation	Response	References
Diet	Omnivorous	Ingest both plant and animal material by feeding mainly on macrophytes, periphyton, epifauna and filamentous algae	↑	Pont et al. (2006)
	Piscivores	Feed mainly on fish but may include large nektonic invertebrates	↓	Pont et al. (2006)
	Planktivores	Feed predominantly on zooplankton and occasionally on phytoplankton in the water column, mainly by filter feeding.	↓	Hart (1992)
	Hyperbenthivores/ Zooplanktivores	Feed just over the bottom, predominantly either on smaller mobile invertebrates living over the bottom and zooplankton	↓	Pont et al. (2006)
	Hyperbenthivores/Pisciviores	Feed just over the bottom, predominantly either on larger mobile invertebrates living over the bottom and fish	↓	Pont et al. (2006)
	Detrivores	Feed on all the small organisms in or on the surface layer of the substratum and associated organic matter	↑	Khali and Kochzius (2002)
	Microbenthivores	Feed mainly on benthic, epibenthic and hyperbenthic fauna, with prey size < 1 cm	↓	Gabriels et al. (2010)
	Macrobenthivores	Feed mainly on benthic, epibenthic and hyperbenthic fauna, with prey size > 1 cm	↓	Gabriels et al. (2010)
	Herbivores	Graze predominantly on living macroalgal and macrophyte material	↓↑	Pont et al. (2006)
Reproductive mode	Lithophils	Spawning over gravel without preparing a nest or parental care	↓	Oberdorff and Hughes (1992).
	Phytophils	Fish needing plants to deposit adhesive eggs	↓↑	Pont et al. (2006)
	Ostracophils	Shell dwellers using empty shells for breeding	↓	Didier (1997)
	Phytolithophils	Adhesive eggs released over plants (nonobligatory)	↓↑	Growns (2004)
	Pelagophils	Free-floating eggs	↓	Perkin et al. (2015)
	Polyphils	Non-specialized spawners no preferred habitat	↑	Breine et al. (2008)
	Ariadnophilic	Adhesive eggs deposited within glued nest	↓	Kruk et al. (2017)
Migration guilds	Short range migrations within river "zones"	Freshwater migrants	↓	Pont et al. (2006)
	Intermediate range migrations between river "zones" but within river basin	Freshwater migrants	↓	Pont et al. (2006)
	Long range anadromous migrators	Migrate from the sea into the river to spawn	↓	Pont et al. (2006)
	Long range catadromous migrators	Migrate out to the sea to spawn	↓	Pont et al. (2006)

Scoring of the selected metrics

The individual fish metrics selected for the index development should be assigned to numbers without a unit, called scores, which are in turn integrated into index values. There are several approaches for scoring the selected metrics (Blocksom 2003). The reference sites have a key role in scoring; reference sites with minimum disturbance can be used to determine optimal scores (Quataert et al. 2011; Mostafavi et al. 2015). If the reference conditions are not available, preclassification systems can be used (e.g., Fame manual 2004; Quataert et al. 2011).

The scoring method may have a strong impact on the performance of the developed index. Discrete scoring was first used by Karr (1981) and it is one of the most frequently applied scoring methods for multimetric biotic indices. In discrete scoring methods, categorical scores (e.g., 1, 3, 5) are attributed to a certain range of metric values. On the contrary, in continuous scoring methods, metric values are scored on a continuous scale depending on the upper and lower thresholds of the observed values. The upper and lower thresholds can be 75th and 25th or 95th and 5th percentiles, respectively (Bozzetti and Schulz 2004; Dolph et al. 2010). The scores generally range between 0 to 100 or 0 to 10 depending on the scoring approach. In general, metric values increasing with better environmental quality get higher scores. Negative metrics get lower scores with increasing environmental quality. Scoring for these metrics is done in a similar way considering that their scores increase as the metric value decreases. It should be noted that under natural conditions a population should not consist of only one type of trophic guilds such as carnivorous fish. In such cases, a threshold for such metrics can be defined and higher values are assigned to lower scores not corresponding to the best situation (Breine et al. 2015).

The effect of natural variables (i.e., altitude, slope, etc.) on metric values should be assessed using appropriate statistical tools such as correlation tests. Calibrations are necessary when there is a significant relationship between natural variables and metric values (Karr 1981; Breine et al. 2004). The metrics to be scored can be grouped into different ranges of natural variables. For example; 3 different groups can be tabulated according to the altitude ranging from 0 to 250 m, 250 to 500 m and > 500 m. Thus, the upper and lower thresholds for metric values will be different in each group which enables an adjustment for scoring according to that of natural variable. However, this necessitates a very large dataset for each group to provide reliable estimates.

EQR calculation and determination of the thresholds for EQR

The sum of the scores of the selected metrics gives the total IBI for each sampling location. To be compatible with the WFD, the IBI values are then transformed to EQR values. EQR is a numerical value ranging between 0 (bad status) and 1 (high or pristine status). EQR is calculated with the following formulae:

$$\text{EQR} = (\text{IBI} - \text{lowest IBI possible})/(\text{maximum IBI possible} - \text{lowest IBI possible}).$$

The EQR value 1 corresponds to the upper limit of the metric value under reference conditions. On the other hand, when no reference sites are available the highest observed value for this particular metric can be scored as 1. If according to the preclassification no sites with a high or good quality are present, the value for score 1 can be obtained by extrapolation. EQR value 0 is defined by the lowest metric value for the sites preclassified as bad. Again extrapolation can be used if no bad sites are available.

Validation

The validity of the index can be tested by internal or external datasets (Oberdorff et al. 2002; Mostafavi et al. 2015). An independent dataset including both pristine and impacted sites from water bodies with features similar to those used to develop the index is required for the external validation. The preclassification system (based on independent physical and chemical assessments) can be used for internal validation (Mercado-Silva et al. 2002). A habitat status assignment for each sampling location based on pressure variables is needed for preclassification. To prevent circular reasoning, habitat status assignment and metric modeling should not be based on the same abiotic variables. In addition, abiotic parameters used for habitat status assignment should have a known effect on fish guilds. If available, a national water quality regulation scheme can be used directly or with slight modifications to preclassify the sampling locations.

A second approach to validate the index is the correlation test between habitat status assignment and IBI scores (or EQR values). The higher correlation coefficient gives the better fit for the developed index.

Graphical screening of the variation in the scores for selected metrics in different EQR classes and monitoring the responsiveness of the selected metrics with increasing environmental degradation is useful for validation. The box plots indicating metric score values should follow a logical trend in different integrity classes with improving ecological status. Metrics with a high degree of variation will create an overlap in different EQR classes and they are less effective for ecological status assessment of water bodies (van de Bund and Solimini 2007). According to the WFD, a fish-based index should clearly distinguish the high/good integrity classes from moderate-poor-bad integrity classes (EU WFD 2000). Hypothetical box plots showing the expected pattern of the score for a selected metric with improving environmental quality which presents a good separation and a metric which shows great variation (i.e., overlap) in different EQR classes are given in Fig. 2.

Discussion

The assessment of the ecological status of aquatic habitats is vital for developing management strategies and helping policy makers. Thus, environmental quality assessment tools are essential to concentrate rehabilitation and restoration efforts on locations where anthropogenic activities have adverse effects. Such tools are also crucial for countries where the financial sources are limited. National fish-based ecological evaluation tools for inland water bodies are not available yet in Turkey. It is a complicated task to develop a fish-based index for Turkish freshwaters due to the high variation in riverine and lacustrine habitat types, high endemism in fish fauna, lack of knowledge on the ecology and distribution of certain fish species (especially endemic fish) and the absence of long-term historical data sets. The major challenging issues related to develop a fish-based index for Turkish freshwaters are summarized below.

Abiotic data

The determination of physical and chemical characteristics of the aquatic environment and hydromorphological features of the catchment basin are crucial for ecological quality assessments. These parameters are

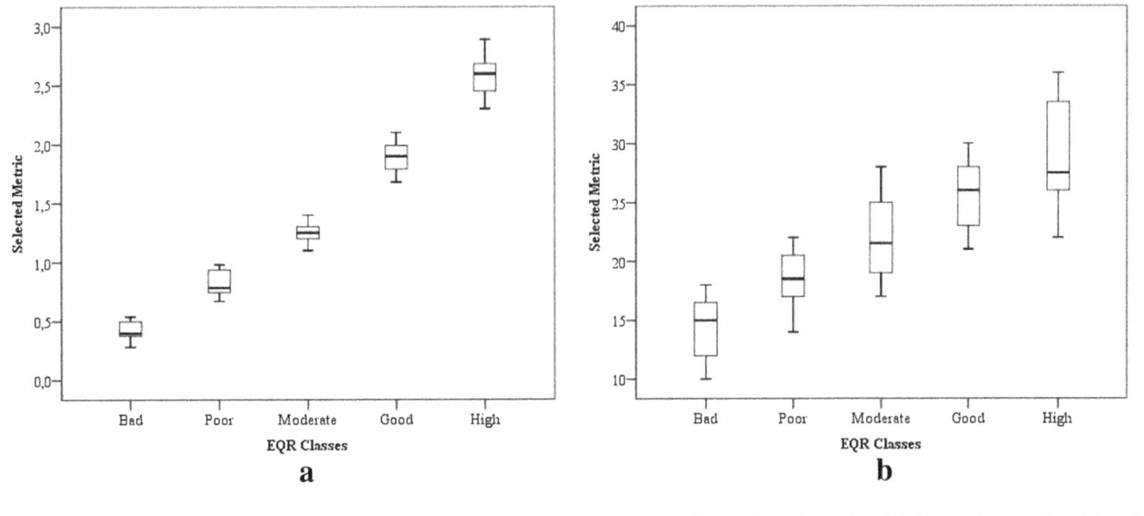

Fig. 2 Hypothetical box-plot diagrams showing a logical pattern of the score for a selected metric with improving quality (**a**) and a metric which shows great variation in different EQR classes (**b**)

highly variable in freshwater ecosystems and often have an impact on the community patterns in terms of species composition, abundance, biomass (van de Bund and Solimini 2007; Hawkins et al. 2010). The reliability and power of the assessments based on the analyses of physico-chemical parameters, depend on the quality of sampling. Instant measures of the physical and chemical variables (temperature, pH, dissolved oxygen, heavy metal concentrations, etc.) of rivers supply limited information on the immanent characteristics of those water bodies (Meybeck et al. 1996; Hunting et al. 2017). Several researchers have shown that the hydromorphological variables such as distance from the source, slope and stream order also have an effect on fish assemblages (Olaya-Marin et al. 2012; Civas et al. 2016; Wolter et al. 2016). Datasets for abiotic parameters are also useful to detect the behavior and fate of contaminants which may assist in developing management strategies. Sediment and biota are accumulative matrices for pollutants (e.g., hydrophobic and lipophilic substances); thus, they provide a less variable measurement of the pollutant load on a longer time scale (CIS 2009). The natural background concentrations have a crucial impact on the assessment of chemical status of the watersheds. The background concentrations of elements including nitrogen, phosphorus and heavy metals such as cobalt, zinc, and iron are affected by the annual runoff, atmospheric deposits, region-specific factors (Smith et al. 2003) and particularly sediment geology and geochemistry in catchment basins (Eker and Kılıç, 2018). The assignment of the thresholds for background concentrations needs datasets from pristine or minimally impacted sites in each basin or region.

Fish data and typology

Specific indices appraising different metrics may be needed for particular fish zones or (i.e., salmonid versus cyprinid rivers) or river segments (i.e., downstream/upstream sampling locations). Thus, species composition, relative abundance and biomass of particular fish species should be tested for significant differences. Currently, there is no regional information available from Turkey and Middle East for ecoregional delineations. Abell et al. (2008) defined 14 ecoregions in Turkey by generating species list for all drainage basins in the area and combining small catchments with similar faunistic elements. However, this ecoregion delination should be validated using geographic, climatic and biological characteristics (Bailey 2005). The freshwater fish fauna in Turkey is characterized by a high degree of endemism, and a great variation among regions. Thus, adopting available fish-based indices (such as EFI) without any modification is not possible. Several riverine (source waters, head waters, lowland and highland streams, streams flowing through forests and deep valleys, mountain rivers, closed or open river basins) and lacustrine (high mountain lakes, lowland lakes, lakes with a wide range of size and depth) habitats are observed in Turkey. Thus, there is a high diversity among the hydrologic, morphologic and climatic characteristics of those water bodies. Currently, a total of 56 riverine and 23 lacustrine habitats are recognized for freshwaters in Turkey. This typology is in compliance with the WFD (System A) and solely based on altitude, slope, precipitation, flow regime, drainage area and basin geology (RBPAP 2013). However, such an abiotic typology is incapable of reflecting the differences observed in fish community structure among different sampling locations. Melcher et al. (2007) also reported a similar finding for European freshwaters. Furthermore, high number of river and lake types limits the practical applicability of this approach. Thus, there is a need for a fish-based river and lake typology development. The first step of developing fish-based typology requires an abiotic classification (based on hydrology, hydromorphology, geology, climate, etc.) which reflects the differences in fish community structure. The second step includes the biotic validation using fish data (e.g., presence/absence of certain taxa, dominant taxa, co-occurrence of species, etc.) to assure the accuracy of these classifications and to avoid any misclassifications. The wetted width, depth, flow velocity, river bed material, source type, distance to the source, channel slope, altitude, climatic characteristics (mean annual air temperature, precipitation, etc.), lake depth, shore structure and conditions, trophic status, residence time may correspond to fish species composition and abundance in rivers and lakes (Owen et al. 2001, Fame Consortium 2004). The relevant abiotic parameters and their class boundaries reflecting distinct river types or segments and lake types with homogenous fish communities can be chosen with statistical tools (e.g., cluster analysis, twinspan analysis) and expert judgement (Noble et al. 2003). To develop a fish-based typology, fish sampling sites should include all representative freshwater habitats from unimpacted or less impacted sites (i.e., calibration sites). Sampling sites where local endemism (basin-specific endemics) is high can be grouped into a higher level of discrimination (e.g., sub-zones or subtypes) (Economou et al. 2004). A high degree of classification for a fish-based typology is possible with larger datasets; however, generic models are better to elucidate generalizations for datasets with gaps. There are several different approaches available for establishment of a fish-based river typology for several countries including Germany (Fieseler and Wolter 2006), France (Lasne et al. 2007) and Latvia (Birzaks 2012), for large rivers such as Danube River (Eros et al. 2017), for European rivers (Melcher et al. 2007), and for lakes in Ireland (Kelly et al. 2012). However, it is still possible to combine sampling locations with different characteristics in one single index, e.g., upstream index developed in Flanders (Breine et al. 2004).

Sampling

Each method of fish sampling has its own specific selectivity on a species level or size. It is always possible to miss a species unless considerable effort is exerted (Almeida et al. 2017). Personal experience has shown that small-bodied fish such as *Gambusia holbrooki* generally inhabit very shallow waters along the shoreline in lakes and avoid open waters. This lowers the probability of catching such fish. We also observed that multi-mesh gillnets are more likely to catch smaller fish than larger fish, while smaller fish are less vulnerable to electrofishing. Similar findings were also reported by Ruetz et al. (2007). Furthermore, we also observed that multi-mesh gillnets were usually unsuccessful to catch Northern pike (*Esox lucius*). Missing a top-carnivorous species (i.e. northern pike) which is very common in Turkish lakes and reservoirs will affect the estimates of the index developed. Thus, the limitations of the fish sampling methods should be well understood. Data collected by different sampling methods should not be combined or mixed to reduce the bias related to any single protocol (U.S. EPA 2002). Sampling locations should represent a gradient from highly impacted to less impacted sites reflecting all river and lake types. Since the flow velocity is very high in many rivers during spring due to snow melt and rainfall, fish sampling should be carried out during late summer which corresponds to low-flow periods in Turkish rivers. Since most of the natural lakes in Turkey are under heavy impact (RBPAP 2013, SAP 2016), it is a challenging factor to obtain a series of sampling sites for lakes from pristine locations. Paleolimnological methods including analysis of sedimentary e-DNA for fish are promising techniques (Turner et al. 2015; Domaizon et al. 2017) to elucidate unimpacted conditions in lakes. There are several river segments in Turkey without a major human disturbance (calibration sites); thus, monitoring data can be used for rivers to develop a fish-based typology. The number of multi-mesh nets offered by standard methods (CEN 2003b) is not feasible for large lakes. Thus, future research focuses on the assessment of optimal fishing method in such lakes. There is also a need for an optimal fishing method for large and non-wadable rivers.

Guild attribution

The ecological guild concept was originally defined by Root (1967) as "a group of species exploiting similar environmental resources in a similar way". The guild approach has been adopted by several researchers to assess the ecological integrity in freshwaters based on the fish data (Aarts and Nienhuis 2003; Welcomme et al. 2006; Noble et al. 2007; Schmutz et al. 2007). Any gaps or limited information related to the ecological requirements of certain fish will jeopardize the attribution of fish species to particular guilds and construction of candidate metric list. Thus, such cases should be handled carefully with the available literature and expert judgement; a consensus is needed on guild attribution of fish species. Since most of the previous ichthyological research in Turkey focused on faunistic records and species composition, there is a gap on the scientific literature mainly concerning feeding ecology and tolerance of fish (particularly endemic fish) species. A few individuals can be retained for stomach content analysis to elucidate the feeding biology of such fish. Overall tolerance can be used if there are partial data on the tolerance of fish to specific disturbances. It is better to keep such gaps blank, instead of processing with doubtful or wrong information. Metrics with patchy data are likely to be rejected by the models. However, we suggest a posterior evaluation whether to omit or to keep such metrics in candidate metric list. Note that metrics including fewer species may create a bias in scoring (Breine et al. 2004; Fame Consortium 2004). Thus, evaluation of the ecological quality of such water bodies with a naturally low species richness may be based on a limited number of metrics. A similar approach was used by Cheimonopoulou et al. (2011) for small rivers in Greece and Mostafavi et al. (2015) for cyprinid rivers in Iran.

Statistical analysis

The selection of responsive metrics can be done by several statistical methods including PCA, CCA and linear models. PCA uses variance to assess the importance of a particular dimension (pressure). High variance axes are, therefore, treated as principle components while low variance axes as noise. In a study comparing the abilities of generalized linear models (GLM) and CCA, it is stated that GLMs gave better predictions than CCA models (Guisan et al. 1999). The authors stated that GLMs provide specific predictors for each tested metric, while CCA uses same set of predictors for all tested metrics. They also noted that CCA provided more accurate results for rare or uncommon species among sampling sites. Linear mixed models are useful to demonstrate how independent variables (predictors) affect dependent variables (metrics) (Van Sickle et al. 2004). They are easy to develop and are considered reliable tools for ecological studies (Einheuser et al. 2013).

Eqr

The ecological status assessments based on EQR are useful for the assignment of a water body to a peculiar ecological status. EQR also enables comparability among different assessment methods through intercalibration and presents a standard ecological status assignment (van de Bund and Solimini 2007). This assignment has a crucial role in decision making for developing management strategies, efficient use of financial resources and diverting restoration efforts on precise spots.

The degree of deterioration of the sampling locations is important for EQR-based classifications. For a natural water body, the goal is to reach the "good ecological status (GES)". However, heavily modified water bodies (HMWB) which have substantial changes in their hydrogeomorphological characters resulting from human activities cannot reach this status. The best status an HMWB can obtain is the "maximum ecological potential" (MEP) or "Good Ecological Potential" (GEP). According to the WFD, the "MEP" is defined as the state where "the values of the relevant biological quality elements reflect, as far as possible, those associated with the closest comparable surface water body type, given the physical conditions, which result from the heavily modified characteristics of the body" (CIS 2003a, b). GEP is equivalent to achieving "Good Ecological Status" in unmodified water bodies once the impact of physical modifications necessary to support the use is taken into account. Thus, EQR for GEP or MEP is designated to 1, under which other integrity classes are defined. The thresholds for EQR classes are designated according to a particular degree of deviation from the reference conditions (van de Bund and Solimini 2007). Thus, type-specific reference conditions have a direct impact on the thresholds.

The main purpose of the use of EQR is to provide a comparable assessment among different methodologies. However, there is no agreement on the determination of class boundaries of EQR values yet (van de Bund and Solimini 2007). Intercalibration practices are crucial to ensure that the EQR class boundaries for high-good and good-moderate for different fish-based assessment methodologies correspond to comparable degree of habitat impairment. Thus, any misclassification due to a variety of errors, particularly related to assessment methods, should be eliminated or minimized by reliable statistical approaches.

Validation

The validation of the index is critical, however, often neglected part of the index development (Karr and Chu 1999, Mercado-Silva et al. 2002). The preclassification of the sampling sites (i.e., internal validation) is useful where the external independent datasets are not available. The preclassification of habitat status should be done on a 5 class system to be compatible with the WFD. The Turkish Surface Water Quality Directive (SWQD 2012) is based on 4 water quality classes. Thus, if the preclassification of the sampling locations will be done using this system, a slight adjustment is needed. Although, the preclassification is not an accurate system for habitat status assignment, it is accepted as a useful procedure to rank the sampling locations and develop an index (Quataert et al. 2011). The habitat assignment and EQR classification should match considering a one class divergence as acceptable errors (Type I or II error; see Breine et al. 2010).

Conclusion

Environmental quality assessments in aquatic habitats with reliable approaches are of central importance for effective use of financial funds in developing countries. Using inaccurate methods to evaluate ecological status in water bodies result in erroneous assignments which lead to waste of time and money. Although there are several challenging factors to develop a fish-based index for Turkish freshwaters, it is still possible to develop a reliable index using appropriate statistical methods. Metric selection, scoring and validation should be based on valid approaches including responsiveness tests and redundancy analyses.

Future studies should focus on the development of a fish-based typology for Turkish freshwaters, establishment of the guild attribution of freshwater fish (particularly endemic fish) in Turkey and optimal sampling method for non-wadable rivers and large lakes. Any gaps hindering the index development should be handled carefully considering available literature and expert judgement. It is better to leave such gaps blank where a consensus cannot be reached, instead of proceeding with questionable knowledge.

Acknowledgements A part of this study was conducted in INBO (Institute of Nature and Forest, Brussels). Mehmet Borga Ergönül was awarded by the scholarship of The Scientific and Technological Research Council of Turkey (2219- Research Fellowship Programme for Post-Doctoral Studies) during his stay in INBO in Belgium.

References

Aarts BGW, Nienhuis PH (2003) Fish zonations and guilds as the basis for assessment of ecological integrity of large rivers. Hydrobiologia 500:157–178

Aazami J, Esmaili-Sari A, Abdoli A, Van den Sohrabi H, Brink PJ (2015) Monitoring and assessment of water health quality in the Tajan River, Iran using physicochemical, fish and macroinvertebrates indices. J Environ Health Sci Eng. https://doi.org/10.1186/s40201-015-0186-y

Abell R, Thieme ML, Revenga C et al (2008) Freshwater ecoregions of the world: a new map of biogeographic units for freshwater biodiversity conservation. Bioscience 58:403–414

Almeida D, Alcaraz-Hernández JD, Merciai R, Benejam L, García-Berthou E (2017) Relationship of fish indices with sampling effort and land use change in a large Mediterranean river. Sci Tot Environ 605–606:1055–1063

Anonymous (2003) Rivers and Lakes—typology, reference conditions and classification systems. Common Implementation Strategy for the Water Framework Directive Guidance Document 10

Bailey RG (2005) Identifying Ecoregion Boundaries. Environ Manag 34(Suppl. 1):14–26

Bain M, Finn J, Booke H (1988) Streamflow regulation and fish community structure. Ecology 69(2):382–392

Bellan-Santini D (1980) Relationship between populations of amphipods and pollution. Mar Pollut Bull 11:224–227

Belpaire C, Smolders R, Vanden Auweele I, Ercken D, Breine J, Van Thuyne G, Ollevier F (2000) An index of biotic integrity characterizing fish populations and the ecological quality of Flandrian waterbodies. Hydrobiologia 434:17–33

Birzaks J (2012) Occurrence, abundance and biomass of fish in rivers of Latvia in accordance with river typology. Zool Ecol 22(1):9–19

Blocksom KA (2003) A performance comparison of metric scoring methods for a multimetric index for Mid-Atlantic Highland streams. Environ Manag 31(5):670–682

Boulton AJ (1999) An overview of river health assessment: philosophies, practice, problems and prognosis. Freshw Biol 41:469–479

Bozzetti M, Schulz UH (2004) An index of biotic integrity based on fish assemblages for subtropical streams in southern Brazil. Hydrobiologia 529:133–144

Breine J, Simoens I, Goethals P, Quataert P, Ercken D, Van Liefferinghe C, Belpaire C (2004) A fish-based index of biotic integrity for upstream brooks in Flanders (Belgium). Hydrobiologia 522:133–148

Breine J, Maes J, Stevens M, Simoens I, Elliott M, Hemingway K, Van den Bergh E (2008) Habitat needs to realise conservation goals for fish in estuaries: case study of the tidal Schelde. Harbasisns INBO. R 3:46

Breine J, Quataert P, Stevens M, Ollevier F, Volckaert FAM, Van den Bergh E, Maes J (2010) A zone-specific fish-based biotic index as a management tool for the Zeeschelde estuary (Belgium). Mar Pollut Bull 60:1099–1112

Breine J, Van Thuyne G, De Bruyn L (2015) Development of a fish-based index combining data from different types of fishing gear. A case study of reservoirs in Flanders (Belgium). Belgian J Zool 145:17–39

Camargo JA (1992) New diversity index for assessing structural alterations in aquatic communities. Bull Environ Contam Toxicol 48:428–434

CEN (European Committee for Standardization), EN 14011 (2003a) Water quality. Sampling of fish with electricity, EN 14011. CEN, Brussels

CEN (European Committee for Standardization), EN 14757 (2003b) Water quality. Guidance on the scope and selection of fish sampling methods, EN 14757. CEN, Brussels

Cheimonopoulou MT, Bobori DC, Theocharopoulos I, Lazaridou M (2011) Assessing ecological water quality with macroinvertebrates and fish: a case study from a small Mediterranean river. Environ Manag 47:279–290

CIS (Common Implementation Strategy) (2003a) Overall approach to the classification of ecological status and ecological potential. Water Framework Directive Common Implementation Strategy Working Group 2 A Ecological Status (ECOSTAT), 53, Rome

CIS (Common Implementation Strategy) (2003b) Identification and designation of heavily modified and artificial water bodies. Water Framework Directive Common Implementation Strategy Working Group, Guidance Document No. 4, Produced by Working Group 2.2—HMWB, European Communities, 118

CIS (Common Implementation Strategy) (2009) Guidance on Surface Water Chemical Monitoring under The Water Framework Directive. Water Framework Directive Common Implementation Strategy Working Group, Guidance Document No. 19. Produced by Working Group 2.2—HMWB, European Communities, 130

Civas L, Vytaustas K, Ruta B (2016) Impact of the riverbed morphology on the source of organic material and the trophic structure of fish community along the upper reaches of the Nevezis River. Lithuania. Acta Ichthyologica et Piscatoria; Szczecin 46(4):303–312

Clavel J, Poulet N, Porcher E, Blanchet S, Grenouillet G, Pavoine S, Biton A, Seon-Massin N, Argillier C, Daufresne M, Teillac Deschamps P, Julliard R (2013) A new freshwater biodiversity indicator based on fish community assemblages. PLoS ONE 8(11):e80968

Cunico AM, Allan JD, Agostinho AA (2011) Functional convergence of fish assemblages in urban streams of Brazil and the United States. Ecol Ind 11:1354–1359

Didier J (1997) Indice biotique d'integrity piscicole pour évaluer la qualité écologique des écosystèmes lotiques. Thèse de doctorat, Presses Universitaires de Namur, Namur

EU Water Framework Directive (2000) Directive of the European parliament and of the council 2000/60/EC establishing a framework for community action in the field of water policy. Official Journal of the European Communities

Dolph CL, Sheshukov AY, Chizinski CJ, Vondracek B, Wilson B (2010) The index of biological integrity and the bootstrap: can random sampling error affect stream impairment decisions? Ecol Ind 10:527–537

Domaizon I, Winegardner A, Capo E, Gauthier J, Gregory-Eaves I (2017) DNA-based methods in paleolimnology: new opportunities for investigating long-term dynamics of lacustrine biodiversity. J Paleolimnol. https://doi.org/10.1007/s10933-017-9958-y

Economou AN, Zogaris S, Giakoumi S, Barbieri R, Petridis D (2004) Developing a biotic river typology and defining reference conditions in the rivers of Greece: a spatially-based approach. Fame Consortium, 35

EFI + Consortium (2009) Manual for the application of the new European fish index-EFI + . A fish-based method to assess the ecological status of European running waters in support of the Water Framework Directive

Einheuser MD, Nejadhashemi AP, Wang L, Sowa SP, Woznicki SA (2013) Linking biological integrity and watershed models to assess the impacts of historical land use and climate changes on stream health. Environ Manage 51:1147–1163

Eker ÇS, Kılıç ED (2018) Geochemistry of Çoruh River Bed Sediments in NE Turkey: implications in Weathering-Sedimentary Cycle, Provenance, and Metal Pollution. Geochem Int 56(6):579–600

U.S. EPA (2002) Methods for Evaluating Wetland Condition: Developing Metrics and Indexes of Biological Integrity. Office of Water, U.S. Environmental Protection Agency, Washington, DC. EPA-822-R-02-016. 38

Eros T, Bammer V, Gyorgy AI, Pehlivanov L, Schabuss M, Zornig H, Weiperth A, Szaloky Z (2017) Typology of a great river using fish assemblages: implications for the bioassessment of the Danube river. River Res Appl 33(1):37–49

Fame Consortium (2004) Manual for the application of the European Fish index—EFI. A fish-based method to assess the ecological status of European rivers in support of the Water Framework Directive, 81

Fausch KD, Lyons J, Karr JR, Angermeier PL (1990) Fish communities as indicators of environmental degradation. Am Fish Soc Symp 8:123–144

Fieseler C, Wolter C (2006) A fish-based typology of small temperate rivers in the northeastern lowlands of Germany. Limnol—Ecol Manag Inland Waters 36(1):2–16

Freyhof J (2018) Photograph life or preserved fish in the field. http://www.fredie.eu/index.php/tutorials/55/55-photograph-life-or-preserved-fish-in-the-field. Accessed 20 May 2018

Gabriels W, Lock K, De Pauw N, Goethals PLM (2010) Multimetric Macroinvertebrate Index Flanders (MMIF) for biological assessment of rivers and lakes in Flanders (Belgium). Limnologica 40:199–207

Gallardo B, Clavero M, Sánchez MI, Vilà M (2016) Global ecological impacts of invasive species in aquatic ecosystems. Glob Change Biol 22:151–163

Gernes MC, Helgen JC (2002) Indices of biological integrity (IBI) for large depressional wetlands in Minnesota. Minnesota pollution control agency. Biological Monitoring Program, Environmental Outcomes Division, Minnesota

Growns I (2004) A numerical classification of reproductive guilds of the freshwater fishes of south-eastern Australia and their application to river management. Fish Manag Ecol 121:369–377

Guisan A, Weiss SB, Weiss AD (1999) GLM versus CCA spatial modeling of plant species distribution. Plant Ecol 143(1):107–122

Hart RC (1992) Experimental studies of food and suspended sediment effects on growth and reproduction of six planktonic cladocerans. J Plankton Res 14:1425–1448

Hawkins CP, Yong C, Roper B (2010) Method of predicting reference condition biota affects the performance and interpretation of ecological indices. Freshw Biol 55:1066–1085

Hering D, Feld CK, Moog O, Ofenböck T (2006) Cook book for the development of a multimetric index for biological condition of aquatic ecosystems: experiences from the European AQEM and STAR projects and related initiatives. Hydrobiologia 566:311–324

Hu JT, Wang HW, Lee HY (2007) Assessment of environmental conditions of Nan-Shih stream in Taiwan. Ecol Ind 7:430–441

Hughes JE, Deegan LA, Weaver MJ, Costa JE (2002) Regional application of an index of estuarine biotic integrity based on fish communities. Estuaries 25:250–263

Hugueny B, Camara S, Samoura B, Magassouba M (1996) Applying an index of biotic integrity based on fish assemblages in a West African river. Hydrobiologia 331:71–78

Hunting ER, de Jong S, Vijver MG (2017) Assessment of monitoring tools and strategies safeguarding aquatic ecosystems within the European water framework directive. CML-rapport 192. Leiden University, Institute of Environmental Sciences, Conservation Biology, Leiden

Irvine K (2004) Classifying ecological status under the European Water Framework Directive: the need for monitoring to account for natural variability. Aquatic Conserv: Mar Freshw Ecosyst 14:107–112

ISO 5667-1 (2010) Water quality- Sampling Part 1: Guidance on the design of sampling programmes and sampling techniques. International Organization for Standardization, Geneva, Switzerland

ISO 5667-3 (2012) Water quality-sampling. Part 3: Preservation and handling of water samples. International Organization for Standardization, Geneva, Switzerland

ISO 5667-6 (2014) Water quality-sampling. Part 6: Guidance on sampling of rivers and streams. International Organization for Standardization, Geneva, Switzerland

ISO 5667-4 (2016) Water quality-sampling. Part 4: Guidance on sampling from lakes, natural and man-made. International Organization for Standardization, Geneva, Switzerland

Jekel H (2005) Sustainable water management in Europe—the water framework directive. In: Vogtmann H, Dobretsov N (eds) Transboundary water resources: strategies for regional security and ecological stability. NATO Science Series (Series IV: Earth and Environmental Sciences), vol 46. Springer, Dordrecht

Joy MK, Death RG (2004) Application of the index of biotic integrity methodology to New Zealand freshwater fish communities. Environ Manag 34(3):415–428

Karr JR (1981) Assessment of biotic integrity using fish communities. Fisheries 6:21–27

Karr JR, Chu EW (1999) Restoring life in running waters: better biological monitoring. Island Press, Washington, D.C

Kelly FK, Harrison AJ, Allen M, Connor L, Rosell R (2012) Development and application of an ecological classification tool for fish in lakes in Ireland. Ecol Ind 18:608–619

Khali MA, Kochzius M (2002) Changes in trophic community structure of shore fishes at an industrial site of the Gulf of Aqaba, Red Sea. Mar Ecol Prog Ser 239:287–299

Kruk A, Ciepłucha M, Zięba G, Błońska D, Marszał L, Tybulczuk S, Tszydel M, Penczak T (2017) Disturbed fish fauna zonation as an indicator of large-scale human impact: a case study (2011–2012) of the large, lowland Warta River, Poland. J Appl Ichthyol 33:174–188

Kuru M, Yerli SV, Mangıt F, Ünlü E, Alp A (2014) Fish biodiversity in inland waters of Turkey. J Acad Doc Fish Aquacult 3:93–120

Lasne E, Bergerot B, Lek S, Laffaille P (2007) Fish zonation and indicator species for the evaluation of the ecological status of rivers: example of the Loire basin (France). River Res Appl 23(8):877–890

Lodge DM (1993) Biological invasions: lessons for ecology. Trends Ecol Evol 8:133–137

Lyons J (2012) Development and validation of two fish-based indices of biotic integrity for assessing perennial coolwater streams in Wisconsin, USA. Ecol Ind 23:402–412

Melcher A, Schmutz S, Haidvogl G (2007) Spatially based methods to assess the ecological status of European fish assemblage types. Fish Manag Ecol 14:453–463

Mercado-Silva N, Lyons JD, Maldonado GS, Medina M (2002) Validation of a fish-based index of biotic integrity for streams and rivers of central Mexico. Rev Fish Biol Fisheries 12:179–191

Meybeck M, Kuusisto E, Mäkelä A, Mälkki E (1996) Water Quality. In: Bartmen, J. and Ballance, R. (Eds) Water quality monitoring—a practical guide to the design and implementation of freshwater quality studies and monitoring programmes, Published on behalf of United Nations Environment Programme and the World Health Organization. UNEP/WHO

Moore JA, Lew AA, Cunningham JA, Kachuba M (1997) Fish communities as indicators of environmental quality in the West River Watershed. Bull-Yale For Environ Stud 100(2):178–196

Mostafavi H, Schinegger R, Melcher A, Moder K, Mielach C, Schmutz S (2015) A new fish-based multi-metric assessment index for cyprinid streams in the Iranian Caspian Sea Basin. Limnologica 51:37–52

Nakagawa S, Schielzeth H (2013) A general and simple method for obtaining R^2 from generalizes linear mixed-effects models. Methods Ecol Evol 4:133–142

Noble RAA, Cowx IG, Starkie A (2003) Fame Wp6/7 Uk/Ecoregion 18 Spatial approach—river and fish community typology, Metric selection and analysis, 64

Noble RAA, Cowx IG, Goffaux D, Kestemont P (2007) Assessing the health of European rivers using functional ecological guilds of fish communities: standardizing species classification and approaches to metric selection. Fish Manag Ecol 14:381–392

Oberdorff T, Hughes RM (1992) Modification of an index of biotic integrity based on fish assemblages to characterize rivers of the Seine Basin, France. Hydrobiologia 228(2):117–130

Oberdorff T, Pont D, Hugueny B, Porcher JP (2002) Development and validation of a fish-based index for the assessment of 'river health' in France. Freshwater Biol 47(9):1720–1734

Olaya-Marin EJ, Martínez-Capel F, Costa RMS, Alcaraz-Hernández JD (2012) Modelling native fish richness to evaluate the effects of hydromorphological changes and river restoration (Júcar River Basin, Spain). Sci Total Environ 440:95–105

Owen R, Duncan W, Pollard P (2001) Definition and establishment of reference conditions. REFCOND discussion paper for evaluation of techniques. Scottish Environment Protection Agency, Aberdeen, Scotland. 20

Perkin JS, Gido KB, Cooper AR, Turner TF, Osborne MJ, Johnson ER, Mayes KB (2015) Fragmentation and dewatering transform Great Plains stream fish communities. Ecol Monogr 85(1):73–92

Pont D, Hugueny B, Beier U, Goffaux D, Melcher A, Noble R, Rogers C, Roset N, Schmutz S (2006) Assessing river biotic condition at a continental scale: a European approach using functional metrics and fish assemblages. J Appl Ecol 43:70–80

Pont D, Hugueny B, Rogers C (2007) Development of a fish-based index for the assessment of river health in Europe: the European Fish Index. Fish Manag Ecol 14(6):427–439

Qadir A, Malik RN (2009) Assessment of an index of biological integrity (IBI) to quantify the quality of two tributaries of river Chenab, Sialkot, Pakistan. Hydrobiologia 621:127–153

Quataert P, Verschelde P, Breine J, Verbeke G, Goetghebeur E, Ollevier F (2011) A diagnostic modelling framework to construct indices of biotic integrity: a case study of fish in the Zeeschelde estuary (Belgium). Estuar Coast Shelf Sci 94:222–233

R Development Core Team (2012) R: a language and environment for statistical computing. Vienna, Austria. http://www.R-project.org, R: Foundation for Statistical Computing

RBPAP River Basin Protection Action Plans Project (2013) Republic of Turkey, Ministry of Forestry and Water Affairs. TUBITAK-MAM, Environment and Cleaner Production, Gebze, Kocaeli

Root RB (1967) The niche exploitation pattern of the bluegray gnatcatcher. Ecol Monogr 37:317–350

Ruetz CR, Uzarski DG, Krueger DM, Rutherford ES (2007) Sampling a littoral fish assemblage: comparison of small-mesh fyke netting and boat electrofishing. North Am J Fish Manag 3:825–831

SAP Sensitive Areas Project (2016) Determination of the sensitive areas and water quality objectives on the basis of watershed in Turkey. Republic of Turkey, Ministry of Forestry and Water Affairs. TUBITAK-MAM, Environment and Cleaner Production, Kocaeli (Turkey)

Saravanan TS, Mohamed MA, Chandrasekar R, Sundramoorthy M (2003) Freshwater fishes as indicators of Kaveri River pollution. J Environ Biol 24(4):381–389

Schinegger R, Trautwein C, Schmutz S (2013) Pressure-specific and multiple pressure response of fish assemblages in European running waters. Limnologia 43(5):348–361

Schmutz S, Cowx IG, Haidvogl G, Pont D (2007) Fish-based methods for assessing European running waters: a synthesis. Fish Manag Ecol 14:369–380

Schnier S, Cai X, Cao Y (2016) Importance of Natural and Anthropogenic environmental factors to fish communities of the Fox River in Illinois. Environ Manage 57:389–411

Schoolmaster DR Jr, Grace JB, Schweiger EW, Mitchell BR, Guntenspergen GR (2013) A causal examination of the effects of confounding factors on multimetric indices. Ecol Ind 29:411–419

Simpson EH (1949) Measurement of diversity. Nature 163:688–702

Smith RA, Alexander RB, Schwarz GE (2003) Natural background concentrations of nutrients in streams and rivers of the conterminous united states. Environ Sci Technol 37(14):3039–3047

Smolinski S, Calkiewicz J (2015) A fish-based index for assessing the ecological status of Polish transitional and coastal waters. Mar Pollut Bull 101(2):497–506

Stevens C, Council T (2008) A fish based index of biological integrity for assessing river condition in central Alberta. Technical Report, Produced by the Alberta Conservation Association, Sherwood Park and Lethbridge, Alberta, Canada, 29

Stirne J, Avcin AJ, Kerzan I, Marcotte BM, Meith-Avcin N, Vriser B, Vukovic S (1971) Selected biological methods for assessment of pollution. In: Pearson EAW, Defraja E (eds) Marine pollution and waste disposal. Pergamon Press, Oxford, pp 307–328

Sundermann A, Gerhardt M, Kappes H, Haase P (2013) Stressor prioritization in riverine ecosystems: which environmental factors shape benthic invertebrate assemblage metrics? Ecol Ind 27:83–96

Tableau A, Drouineau H, Delpech C, Pierrc M, Lobry J, Le Pape O, Breine J, Lepage M (2013) A fish-based index of estuarine ecological quality incorporating information from both scientific fish survey and experts knowledge. Ecol Ind 32:147–156

ter Braak CJF, Verdonschot PFM (1995) Canonical correspondence analysis and related multivariate methods in aquatic ecology. Aquat Sci 57(3):255–289

Turner CR, Uy KL, Everhart R (2015) Fish environmental DNA is more concentrated in aquatic sediments than surface water. Biol Cons 183:93–102

van de Bund W, Solimini AG (2007) Ecological Quality Ratios for Ecological Quality Assessment in Inland and Marine Waters. REBECCA Deliverable 10. European Commission, Directorate-General Joint Research Centre, Institute for Environment and Sustainability. Office for Official Publications of the European Communities, Luxembourg

Van Sickle J, Baker J, Herlihy A, Bayley P, Gregory S, Haggerty P, Ashkenas L, Li J (2004) Projecting the biological condition of streams under alternative scenarios of human land use. Ecol Appl 14:368–380

Wang L, Robertson DM, Garrison PJ (2007) Linkages between nutrients and assemblages of macroinvertebrates and fish in wadable streams: implication to nutrient criteria development. Environ Manag 39:194–212

Welcomme RL, Winemiller KO, Cowx IG (2006) Fish environmental guilds as a tool for assessment of ecological condition of rivers. Rivers Res Appl 22:377–396

Whitfield AK, Elliott M (2002) Fishes as indicators of environmental and ecological changes within estuaries: a review of progress and some suggestions for the future. J Fish Biol 61:229–250

Wolter C, Buijse AD, Parasiewicz P (2016) Temporal and spatial patterns of fish response to hydromorphological processes. River Res Appl 32:190–201

Wu W, Xu Z, Yin X, Zuo D (2014) Assessment of ecosystem health based on fish assemblages in the Wei River basin, China. Environ Monit Assess 186:3701–3716

Yerli SV, Kormaz M, Mangıt F (2016) Biological assessment by a fish-based index of biotic integrity for Turkish inland waters. In: Taylor WW, Bartley DM, Goddard CI., Leonard NJ, Welcomme R (Eds). Freshwater, fish and the future: Proceedings of the Global cross-sectoral conference. Food and Agriculture Organization of the United Nations, Rome; Michigan State University, East Lansing; and American Fisheries Society, Bethesda, Maryland, pp 91–97

2

Assessing functional diversity of macrobenthic assemblages in sewage-affected intertidal shores

G. V. Garaffo · M. L. Jaubet · E. N. Llanos · M. A. Saracho Bottero · R. Elías

Abstract Functional diversity is the component of diversity that influences ecosystem dynamics, stability, productivity, nutrient balance, and other aspects of ecosystem functioning. We evaluate the spatio-temporal variability of the functional traits (biological traits analysis), functional diversity indices (functional richness, functional evenness, functional divergence and functional dispersion) and AZTI marine biotic index in response to the distinct levels of sewage contamination at different sites on intertidal abrasion platforms. Variations in macrobenthic functional structure between reference and impacted sites were evident in the study area. The functional richness showed a pattern with higher values in the reference sites; Functional evenness showed a similar pattern, but only during spring. Functional divergence presented lower values in the impacted sites during autumn and spring. The functional diversity indices (functional richness and functional evenness) and the AZTI marine biotic index presented a negative relationship, indicating that functional diversity is related to benthic environmental health. Thus, biological traits analysis and functional diversity indices, especially FRic that reveals differences in macrobenthic functional structure between impacted and reference sites during all seasons, represent an informative and useful tool to describe the different aspects of sewage impacts in benthic assemblages on rocky shores. These analyses have a high potential for future monitoring programmes.

Keywords Functional traits · Functional diversity · Organic pollution · Benthic health · SW Atlantic

Introduction

Understanding the processes shaping biological communities under multiple disturbances is a core challenge in ecology and conservation science (Mouillot et al. 2013). Traditional approaches to assess anthropogenic disturbance have usually been focused on taxonomically based structural features, e.g., metrics based on species richness, density/biomass, diversity and environmental quality indices for coastal ecosystems (Borja et al. 2004; Muxika et al. 2007; Muniz et al. 2013). However, these metrics do not consider changes in functional structure, which is central to understand the effects of human disturbances on ecosystem

---

G. V. Garaffo (✉) · M. L. Jaubet · E. N. Llanos · M. A. S. Bottero · R. Elías
Departamento de Ciencias Marinas, Facultad de Ciencias Exactas y Naturales, Instituto de Investigaciones Marinas y Costeras (IIMyC), Universidad Nacional de Mar del Plata, Deán Funes 3350, 7602 AYL Mar del Plata, Argentina
e-mail: garaffo@mdp.edu.ar

G. V. Garaffo · M. L. Jaubet · E. N. Llanos · M. A. S. Bottero · R. Elías
Consejo Nacional de Investigaciones Científicas y Técnicas (CONICET), Av. Rivadavia 1917, C1033AAJ Buenos Aires, Argentina

functioning (Darr et al. 2014). The functional structure of a community is defined as the distribution of species and their abundances in the functional space (Mouillot et al. 2013).

Functional approach is not a new approach but has recently been popularized as being considered a more meaningful way of analyzing biodiversity compared to the traditional taxonomic analysis (Heino 2008). Functional diversity is the component of diversity that influences ecosystem dynamics, stability, productivity, nutrient balance, and other aspects of ecosystem functioning (Tilman 2001). Many methods and techniques have been proposed to assess the functional diversity of assemblages. The methods most widely used are analysis of biological traits analysis (BTA; Bremner 2008) and calculation of functional diversity indices (Laliberté and Legendre 2010; Villéger et al. 2008). Functional traits (biological characteristics of organisms) are those characteristics that define species in terms of their ecological roles, how they interact with the environment and with other species (Diaz and Cabido 2001). BTA is a multivariate approach that combines information on species distributions over space and time, with the multiple traits (life history, morphological, behavioral) they exhibit (Bremner 2008). On the other hand, the functional diversity indices have varying abilities to reflect the different aspects of assemblage functional trait structure. When the species in an assemblage differ greatly in their functional traits, higher values of these functional diversity indices are expected. Villéger et al. (2008) proposed a framework where functional diversity comprises three independent components: functional richness (FRic), functional evenness (FEve), and functional divergence (FDiv). The three independent components provide more detail in examining the mechanisms linking biodiversity to ecosystem functioning (Casanoves et al. 2008).

FRic represents the trait space filled by a given species assemblage by calculating the convex hull volume that comprises the entire trait space filled by all species of this assemblage (Villéger et al. 2008). The algorithm identifies the extreme species and then estimates the volume in the trait space. This index can thus be used as a proxy of the range of functional traits represented in an assemblage but does not take into account differences in species abundance (Schuldt et al. 2014). FRic is affected by the addition or removal of species with unique trait combinations. FEve index measures the regularity of the abundance distribution in trait space by summing the branch lengths of the minimum spanning tree that is required to connect all species in an assemblage weighted by the species abundances (Villéger et al. 2008). FDiv measures the distribution of the species abundances in relation to the gravity center of the functional trait space (Villéger et al. 2008). So functional divergence is related to how abundance is distributed within the volume of functional traits space (Casanoves et al. 2008). Laliberté and Legendre (2010) proposed the dispersion metric: FDis. It is a multi-dimensional index based on multitrait dispersion. FDis is the average distance of individual species to the centroid of all species in the community trait space taken into account the relative abundances of species by computing the weighted centroid.

Marine ecosystems are under anthropogenic pressures degrading both biodiversity and functional diversity. Coastal areas in particular are under the influence of multiple disturbances and stressors, naturally or anthropogenically driven, that impact their biodiversity and functioning (Micheli et al. 2016), thereby compromising their ability to sustain ecosystem services (Worm et al. 2006; Halpern et al. 2008). Despite studies with a functional approach have been scarce in marine environments. Francisco and de la Cueva (2017) consider that a more widespread use is urgently needed of this approach.

In aquatic environments, the largest amount of studies related to functional diversity has been developed in riparian macroinvertebrates (Heino 2005, 2008; Heino et al. 2007), as well as in coastal rocky substrate areas (Crowe and Russell 2009; Pacheco et al. 2011) and estuaries (Villéger et al. 2008, 2010; McLenaghan et al. 2011). In particular, in marine benthic environment, studies with functional approach have been proved useful to: (1) assess fishing effects on benthic fauna (e.g., Bremner et al. 2003; Tillin et al. 2006), (2) assess sewage discharges effects on macrobenthic assemblages (Gusmao et al. 2016; Vinagre et al. 2017), (3) investigate the effects of climate change (e.g., Neumann and Kröncke, 2010), (4) use for management and conservation purposes (e.g., Bremner 2008; Frid et al. 2008; Veríssimo et al. 2012), and (5) assess functional diversity in different species assemblages (e.g., Bremner et al. 2003; Hewitt et al. 2008; Van der Linden et al. 2016).

Mar del Plata city has a resident population of about 600,000 people, but receives almost 3,000,000 tourists during summer (December–March and it is placed at the Southwest Atlantic coast of Argentina (38°00′ S, 57°32′ W) (Bouvet et al. 2005). The city had the worst scenario for both its environmental status and health of people because raw sewage was directly discharged to intertidal, with a mean flow of 2.8 $m^3.s^{-1}$ (up to 3.5 $m^3.s^{-1}$ in summer) (Scagliola et al. 2006). This situation produced 15 km of beaches unfit for bathing because there was a risk for human health (Comino et al. 2010). To date, multiple studies have been conducted in the study area to evaluate the effect of sewage effluent on benthic communities (Elías et al. 2003, 2005, 2006, 2009; Elias et al. 2015; Vallarino and Elías 2006, Vallarino et al. 2014; Jaubet et al. 2011, 2013, 2015; Garaffo et al. 2012, 2016, 2017; Sánchez et al. 2013; Becherucci et al. 2016). However, all these studies had an ecological approach based on taxonomic diversity. The present study proposes to apply for the first time a functional approach and evaluate if this approach is concordant with the AZTI Marine Biotic Index (AMBI). The AMBI was the index that best reflected the environmental quality of the studio area (Garaffo et al. 2017).

From 2015, the sewage pre-treatment plant of Mar del Plata celebrated the setting up of a submarine outfall that moved away the discharge 3.5 km from coast. Therefore, to have information from both taxonomic and functional approaches of the study area prior to the operation of the submarine outfall will provide a useful background for future studies focusing on the recovery of the resident benthic communities in the area.

The aim of this study was to evaluate the spatial and temporal variability of the functional traits, FD indices and AMBI in response to the distinct levels of sewage contamination at different sites on intertidal abrasion platforms of Mar del Plata, Argentina. We predict that: (a) the functional traits' composition of macrofaunal assemblages varies significantly among impacted and no-impacted sites, (b) lower functional diversity values are expected in contaminated areas and (c) functional diversity indices will be negatively related to AMBI index (higher values of AMBI index indicate bad environmental quality).

Materials and methods

Study area

The coast of Buenos Aires Province is dominated by sandy beaches; however, around Mar del Plata city (38° S, 57°33′ W), there are quartzite outcrops and almost horizontal intertidal abrasion platforms (geological formation of consolidated loess, limestone, stony rocks or caliche). The tidal regime is semidiurnal with tidal amplitude ranging from approximately 0.8–1.6 m during exceptional tides. Sea surface temperature shows a great seasonal variation (9.3 °C in winter and 20 °C in summer) (Guerrero and Piola 1997), while sea pH remains between 7 and 8.5 (Isla et al. 1998). Until December 2014, the sewage outfall of Mar del Plata city was located 9 km towards the north of the city center (N°11 route, km 507). This intertidal urban effluent discharged 241,920 m^3 of untreated sewage daily during the winter (flow average rate of 2.8 $m^3\ s^{-1}$) and 302,400 m^3 daily during the summer (average of 3.5 $m^3\ s^{-1}$) into the coastal waters (Scagliola et al. 2006), when between two and three million people visit the city (Bouvet et al. 2005).

Sampling design and field and lab routines

Four sampling sites were delimited on intertidal abrasion platforms site 1 (R1), 9000 m to the north of the outfall; site 2 (R2), 8000 m to the north; site 3 (I1), 200 m south of the outfall; and site 4 (I2), 1000 m to the south (Fig. 1). The sites 3 and 4 are considered impacted by sewage inputs and the sites 1 and 2 are considered as reference locations (Elías et al. 2009; Jaubet et al. 2011, 2013; Sánchez et al. 2013; Garaffo et al. 2016). Sampling was conducted seasonally at each site (February 2008, April 2008, June 2008, November 2008, February 2009, April 2009, June 2009 and November 2009). In each sampling site, 12 sampling units (replicates) were randomly collected from independent rocks using a 10 cm diameter corer (78 cm^2), which was buried into the community matrix up to the basal rocky bottom (up to 20 cm). The samples were fixed in 7% neutralized formalin solution. In the laboratory, each sample was sieved through a 0.5-mm mesh and the retained organisms were identified, counted, and preserved in a 70% ethanol solution. Environmental variables (turbidity and temperature of the seawater) were measured in situ with a U10 Horiba equipment. The percentage of the total organic matter (TOM) in sediments was determined by the calcinations method (Byers et al. 1978).

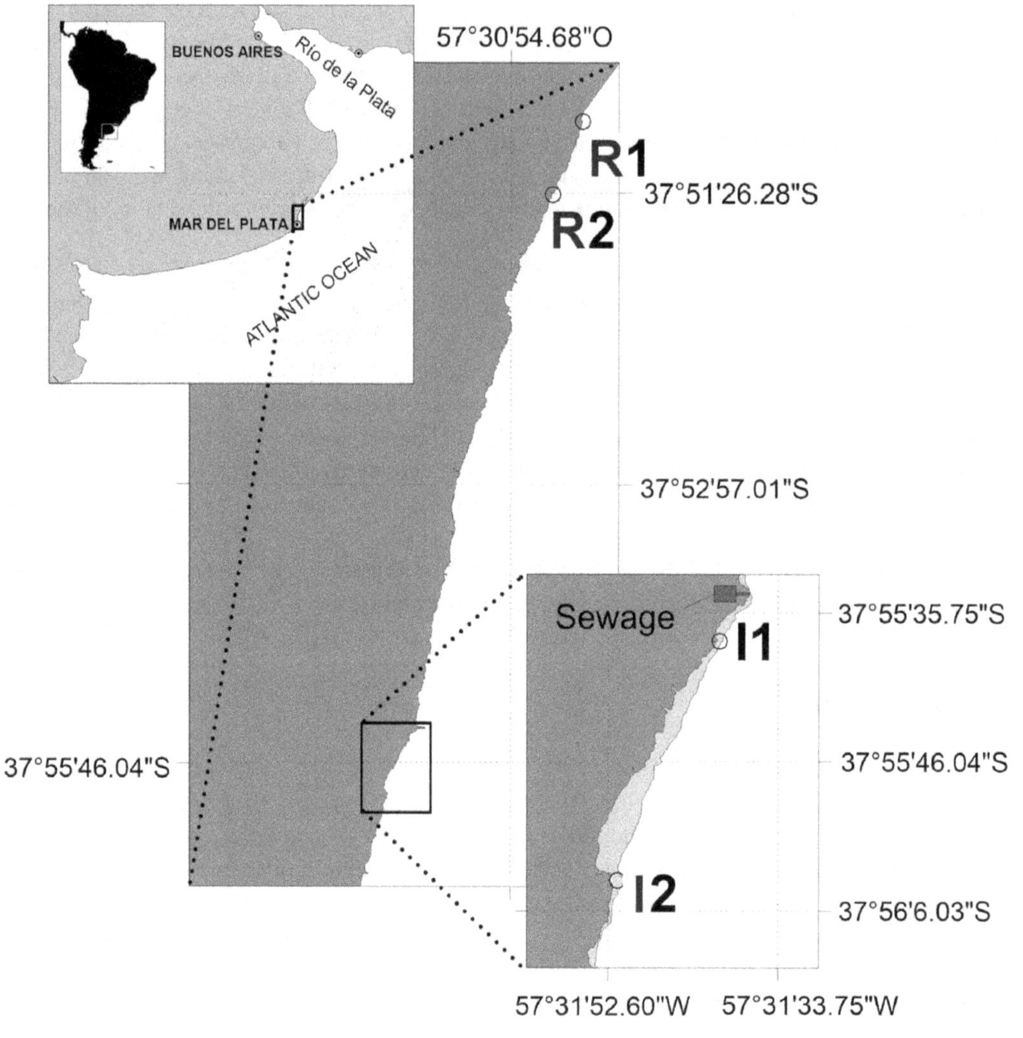

Fig. 1 Distribution of sampling sites (I1, I2, R1 and R2) and intertidal sewage outfall location in the study area. Site I1 was located at 200 m to the south of the outfall, site I2 at 1000 m south, R1 at 9000 m north and R2 8000 m to the north of the outfall

Biological traits analysis

To analyze how the macrofaunal functional trait composition varies among different sites and seasons, we used biological traits analysis (BTA). We chose to use seven functional traits, subdivided into several modalities to represent the range of variation for each functional trait, covering different aspects of the life history, morphology and behavior of each species: depth penetration, maximum size, development mode, living habit, feeding mode, relative adult mobility and tolerance to pollution (Table 1). Rationale for choosing traits was based on the potential of these traits to illustrate changes in ecosystem functioning and responses of benthos to organic pollution. Information on biological traits was gathered from various sources including identification guides (Boschi and Cousseau 2004; Vallarino and Elías 2009), scientific journals and ad hoc information from local specialists. When reliable information was missing, expert judgment and/or data from the nearest phylogenetic neighbor were considered. The "fuzzy coding" procedure (Chevenet et al. 1994), with a scoring ranging from 0 to 3, was used to classify each taxon according to its association with different modalities of functional traits. No affinity for a trait was coded as 0 and complete affinity as three. The scores for each trait (except for development mode) were assigned considering the species adult form on the shore. Trait scores of each taxon were multiplied by its abundance for every sample and subsequently summed to provide a matrix with the overall frequency of each trait modality per sample.

Table 1 List of biological traits and respective categories

Biological traits	Trait categories	Labels
Depth penetration	Surface	DS
	0–8 cm	D0-8
	> 8 cm	D > 8
Maximum size	Very small (< 1 cm)	SVS
	Small (1–3 cm)	SS
	Medium (> 3 cm)	SM
Development mode	Planktotrophic	DP
	Lecitotrophic	DL
	Direct	DD
Feeding mode	Deposit-feeder	FD
	Filter/suspension feeder	FF
	Opportunist/scavenger	FO
	Grazer	FG
	Predator	FP
Living habit	Burrow dweller	HB
	Attached	HA
	Tube dweller	HT
	Free living	HF
Relative adult mobility	None	MN
	Low	ML
	Medium	MM
	High	MH
Tolerance to pollution	1st order opportunistic	V
	2nd order opportunistic	IV
	Tolerant	III
	Indifferent	II
	Sensitive	I

AMBI and functional diversity metrics

For comparison with the FD indices, AMBI was calculated. The FD indices calculated to assess different components of the functional diversity were: FRic, FEve, FDiv and FDisp. The indices were calculated on the basis of matrix of functional traits with the FDiversity software (Di Rienzo et al. 2008). All taxonomic- and functional-based metrics were calculated per replicate.

Data analysis

The resemblance of the macroinvertebrates assemblages according to site and season was analyzed by PERMANOVA analysis on a Bray–Curtis similarity matrix after a fourth-root transformation [Clarke and Warwick 2001). The design included two fixed factors, "site" (four levels: R1, R2, I1 and I2)] and "season" (summer, autumn, winter and spring). Permutation of residuals under a reduced model (9999 permutations) was selected. The SIMPER (similarity percentage analysis) routine was used to determine the species accounting for the greatest contributions to the dissimilarity between assemblages. Taxa and sites were assessed by a direct gradient analysis of ordination technique (i.e., canonical correspondence analysis, CCA. To determine associations between the data and the main explanatory variables, a biplot from the CCA was obtained by overlaying a vector diagram, based on coefficients from the canonical functions describing each canonical axis, on the ordination graph. The CCA analysis was done using only species or taxa with constancy greater than 10%. Species or taxa with lower constancy are considered occasional.

Principal components analysis (PCA) was used to analyze the matrix with the overall frequency of each trait modality. This analysis was carried out to explore the relationships among sites and each trait modality. To favor the understanding of generated plot, the sampling units from the same sampling site were averaged.

The spatial and temporal variability for each functional index and AMBI was evaluated using ANOVA model. The design incorporated one spatial scale (sites) and one temporal scale (season). The factors of the linear model were: season (fixed) with four levels and sites (fixed) with four levels with twelve replicates each. Data for 2008 and 2009 were pooled together as replicates of seasons.

The data were combined since they did not show significant differences between years. Variance heterogeneity was analyzed using Cochran test and transformations were applied when necessary. Spearman correlations were used to test the relationship between each index of functional diversity and AMBI.

Results

Macroinvertebrate assemblages and environmental condition

PERMANOVA showed that the benthic community from impacted sites (I1 and I2) differs significantly from the reference sites (R1 and R2) in all seasons (Pseudo F site × season = 7.648, $p = 0.001$). SIMPER analyses showed that the species responsible, up to 90% of cumulative percentage, of the difference between assemblages of impacted sites and reference sites were *Boccardia proboscidea* (Polychaete), *Brachidontes rodriguezii* (Bivalve), *Syllis gracilis* (Polychaete), *Syllis prolixa* (Polychaete) and *Siphonaria lessonii* (Gastropoda). Numerically dominant species in the impacted sites was *B. proboscidea*, while the reference sites were numerically dominated by *B. rodriguezii* (Table 2) .

The CCA showed the main environmental variables associated with the occurrence of benthic species and sites. Species were sorted in relation to the first axis, showing a total inertia of 0.589 and explaining 99.3% of the total variance (eigenvalues for axis 1 was 0.493). Axis 1 separated *B. proboscidea* associated with impacted sites from *B. rodriguezii*, *S. gracilis*, *S. prolixa* and *S. lessoni*, which were associated with reference sites. Impacted sites showed the highest values of TOM, temperature and turbidity, thus indicating a higher degree of organic pollution than the remaining sites (Fig. 2).

Biological traits analysis

The PCA considering the functional traits displayed eigenvalues of 14.92 and 5.87 for axis 1 and 2, respectively (Fig. 3). The cumulative percentage of variance explained by the first two axes accounted for 86.58% (62.18% and 24.4% for the first and second axis, respectively). A clear separation was observed in assemblage functional trait composition between the impacted and reference sites (axis 1). Assemblage trait composition in the impacted sites was characterized by depth penetration in the sediment between 0 and 8 cm deep, lecithotrophic and direct development, deposit feeders, burrow dweller and tube dweller living habits, none to low relative mobility and opportunistic species. In the reference sites, an opposite pattern was founded. The separation between the reference sites (axis 2) was influenced mostly by relative adult mobility, tolerance to pollution, living habit and feeding mode. The R1 site (9000 m to the north of the outfall) was more related to organisms which show high relative mobility and are sensitive (EG I) to pollution. On the other hand, the R2 site (8000 m to the north of the outfall) was more related to organisms which show medium relative mobility, indifferent (EGII) regarding tolerance to pollution, grazer and predator feeding habit and free living habits.

AMBI and functional diversity metrics

The AMBI index values were significantly higher on the impacted sites than reference sites, although there was also variability across sites and season (Table 3, Fig. 4a). The FRic showed significantly higher values on the reference sites than impacted sites; these differences were observed in all seasons except in summer (Table 2, Fig. 4b). The site R1 presented higher FEve values than the rest of the sites during winter and spring. In addition, FEve values showed a greater difference between impacted and reference sites during spring

Table 2 Pairwise test and SIMPER result showing the taxa contribution to sampling sites dissimilarity according to sampling season (Summer, Autumn, Spring and Winter)

Pairwise test				SIMPER				
	Av. Dissim. (%)	*t*	*p*	Taxon	Av. Abund.	Av. Abund.	Contrib. (%)	Cum. (%)
Summer								
R1 ≠ I1	43.66	3.3206	0.001		R1	I1		
				Boccardia proboscidea	1.31	2.91	22.53	22.53
				Brachidontes rodriguezii	4.93	3.28	19.62	42.15
				Syllis prolixa	0.89	1.09	10.30	52.45
R1 ≠ I2	35.58	3.1085	0.001		R1	I2		
				Boccardia proboscidea	1.31	2.95	24.25	24.25
				Syllis prolixa	0.89	1.56	13.57	37.83
				Brachidontes rodriguezii	4.93	3.88	13.00	50.82
				Lineus bonaerensis	0.71	0.58	8.82	59.64
R1 = R2	26.42	1.2668	0.141					
I1 ≠ I2	39.95	1.8512	0.015		I1	I2		
				Boccardia proboscidea	2.91	2.95	23.55	23.55
				Syllis prolixa	1.09	1.56	12.86	36.41
				Brachidontes rodriguezii	3.28	3.88	10.63	47.04
				Siphonaria lessonii	0.83	1.05	8.38	55.42
I1 ≠ R2	41.87	3.0627	0.001		I1	R2		
				Boccardia proboscidea	2.91	1.81	22.44	22.44
				Brachidontes rodriguezii	3.28	4.63	16.82	39.26
				Syllis prolixa	0.81	1.62	10.98	50.24
I2 ≠ R2	33.21	2.5107	0.001		I2	R2		
				Boccardia proboscidea	2.95	1.81	24.60	24.60
				Syllis prolixa	1.56	0.82	14.64	39.24
				Brachidontes rodriguezii	3.88	4.63	11.02	50.26
				Siphonaria lessonii	1.05	1.48	9.53	59.80
Autumn								
					R1	I1		
R1 ≠ I1	49.62	5.8465	0.001	*Boccardia proboscidea*	0.10	3.52	46.50	46.50
R1 ≠ I2	53.27	7.6777	0.001		R1	I2		
				Boccardia proboscidea	0.10	4.60	49.21	49.21
R1 ≠ R2	44.98	7.9647	0.001		R1	R2		
				Boccardia proboscidea	0.10	2.19	23.96	23.96
				Syllis prolixa	0.51	2.21	20.13	44.09
				Syllis gracilis	0.74	2.06	15.12	59.21
I1 = I2	18.58	1.6027	0.108					
I1 ≠ R2	41.43	4.0849	0.001		I1	R2		
				Syllis prolixa	0	2.21	23.00	23.00
				Syllis gracilis	0	2.06	21.54	44.54
				Boccardia proboscidea	3.52	2.19	15.46	60.00

Table 2 continued

Pairwise test	Av. Dissim. (%)	t	p	SIMPER Taxon	Av. Abund.	Av. Abund.	Contrib. (%)	Cum. (%)
I2 ≠ R2	38.02	4.439	0.001		I2	R2		
				Boccardia proboscidea	4.60	2.19	26.06	26.06
				Syllis gracilis	0.46	2.06	17.14	43.20
				Brachidontes rodriguezii	3.31	4.80	15.38	58.58
R1 ≠ I1	41.02	4.2992	0.001		R1	I1		
				Boccardia proboscidea	1.93	4.71	26.14	26.14
				Alitta succinea	0	1.45	13.72	39.85
				Capitella capitata	0.18	1.39	11.86	51.71
R1 ≠ I2	47.04	2.9543	0.001		R1	I2		
				Boccardia proboscidea	1.93	4.14	27.52	27.52
				Brachidontes rodriguezii	3.85	2.47	20.67	48.20
R1 ≠ R2	30.78	2.227	0.001		R1	R2		
				Boccardia proboscidea	1.93	1.86	17.81	17.81
				Syllis prolixa	1.85	1.74	13.30	35.11
				Brachidontes rodriguezii	3.85	4.72	13.26	48.38
				Siphonaria lessonii	0.08	0.79	11.06	59.43
I1 ≠ I2	42.74	2.2766	0.003		I1	I2		
				Brachidontes rodriguezii	3.88	2.47	16.50	16.50
				Boccardia proboscidea	4.71	4.14	12.01	28.51
				Capitella capitata	1.39	0.17	11.50	40.01
				Alitta succinea	1.45	0.24	11.47	51.48
I1 ≠ R2	40.80	5.0779	0.001		I1	R2		
				Boccardia proboscidea	4.71	1.86	25.25	25.25
				Capitella capitata	1.39	0	11.76	37.00
				Alitta succinea	1.45	0.29	10.59	47.60
				Brachidontes rodriguezii	3.88	4.72	7.98	55.58
I2 ≠ R2	49.35	4.3547	0.001		I2	R2		
				Boccardia proboscidea	4.14	1.86	24.55	24.55
				Brachidontes rodriguezii	2.47	4.72	23.72	48.26
				Syllis prolixa	1.20	1.74	11.51	59.77
Spring								
R1 ≠ I1	63.35	6.4388	0.001		R1	I1		
				Boccardia proboscidea	1.72	6.05	36.28	36.28
				Brachidontes rodriguezii	4.29	1.88	23.72	60.00
R1 ≠ I2	58.36	6.4806	0.001		R1	I2		
				Boccardia proboscidea	1.72	5.67	35.26	35.26
				Brachidontes rodriguezii	4.29	2.06	23.45	58.71

Table 2 continued

Pairwise test				SIMPER				
	Av. Dissim. (%)	*t*	*p*	Taxon	Av. Abund.	Av. Abund.	Contrib. (%)	Cum. (%)
R1 ≠ R2	34.70	2.2146	0.002		R1	R2		
				Boccardia proboscidea	1.72	2.61	21.90	21.90
				Siphonaria lessonii	1.14	0.93	15.11	37.01
				Syllis gracilis	0.42	1.15	13.81	50.82
I1 = I2	37.04	0.8395	0.461					
I1 ≠ R2	56.87	5.3077	0.001		I1	R2		
				Boccardia proboscidea	6.05	2.61	29.97	29.97
				Brachidontes rodriguezii	1.88	4.32	25.06	55.03
I2 ≠ R2	51.80	5.1702	0.001		I2	R2		
				Boccardia proboscidea	5.67	2.61	28.83	28.83
				Brachidontes rodriguezii	2.06	4.32	25.08	53.91

Cumulative percentages that exceed 60% are not shown

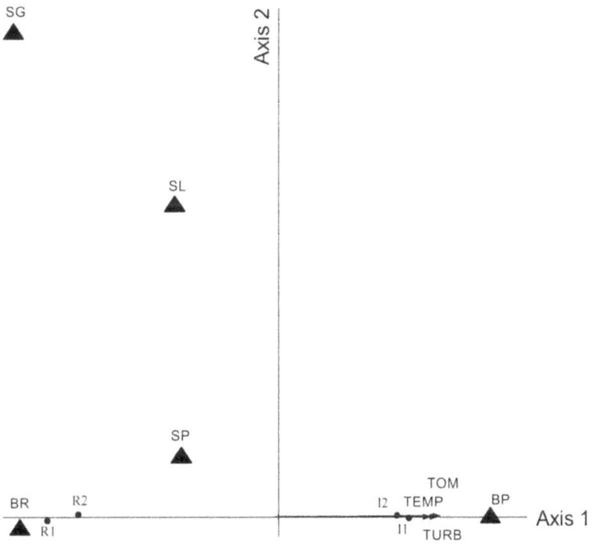

Fig. 2 Ordination diagram of species and sites in the two principal axes of the CCA. The relative lengths of the arrows indicate the importance of a variable. *BP Boccardia proboscidea*, *BR Brachidontes rodriguezii*, *SP Syllis prolixa*, *SG Syllis gracilis*, *SL Siphonaria lessoni*, *TOM* total organic matter, *TEMP* temperature and *TURB* turbity. The axis 1 explains 99.3% of the total variance

(Table 2, Fig. 4c). FDiv presented values with low variability for reference sites across seasons. In contrast, the impacted sites showed variability throughout the seasons. Impacted sites showed a marked decrease in FDiv values during autumn and spring (Table 2, Fig. 4d). The FDis showed significantly higher values on the impacted sites than reference sites during all seasons except for spring (Table 2, Fig. 4e). FRic and FEve were negatively related with the AMBI (EBR) ($p < 0.05$) ($r = -0.250$ and $r = -0.216$, respectively). FDis was positively related with the AMBI. Finally, FDiv showed no significant relationship with AMBI (EBR).

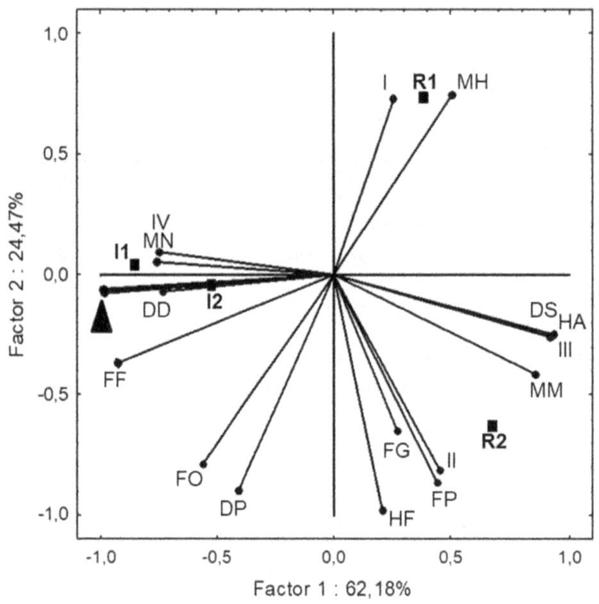

Fig. 3 Principal component analysis (PCA) depicting the trait categories for each site (I1, I2, R1 and R2). Triangle represents the following biological traits: D0-8, D > 8, DL, FD, HB, HT, ML and V. Traits codes: **a** depth penetration: *DS* surface, D0-8 = 0–8 cm, D > 8 = > 8 cm, **b** maximum size: *SVS* very small (< 1 cm), *SS* small (1–3 cm), *SM* medium (> 3 cm), **c** development mode: *DP* planktotrophic, *DL* lecitotrophic, *DD* direct, **d** feeding mode: *FD* deposit-feeder, *FF* filter/suspension feeder, *FO* opportunist/scavenger, *FG* grazer, *FP* predator, **e** living habit: *HB* burrow dweller, *HA* attached, *HT* tube dweller, *HF* free living, **f** relative adult mobility: *MN* none, *ML* low, *MM* medium, *MH* high and **g** tolerance to pollution: *V* 1st order opportunistic, *IV* 2nd order opportunistic, *III* tolerant, *II* indifferent and *I* sensitive

Table 3 Analysis of variance comparing the mean values AMBI index and the four functional diversity indices

Source	Df	AMBI	FRic	FEve	FDiv	FDis
Site (Si)	3	**< 0.001**	**< 0.001**	**0.0014**	**< 0.001**	**< 0.001**
Season (Se)	3	**< 0.001**	0.60	0.63	**< 0.001**	**0.011**
Si × Se	9	0.048	0.05	**< 0.001**	**< 0.001**	**< 0.001**

The numbers correspond to *p* values. Bold values indicate the significant results

Discussion

This is the first time that functional diversity of macrobenthic assemblages (BTA and functional diversity indices) was assessed in sewage-affected rocky shores of the SW Atlantic. To date, the use of functional-based approaches has been widely used in terrestrial and freshwater ecology (Grime 1974; Southwood 1977; Statzner et al. 1994; Charvet et al. 2000; Menezes et al. 2010) and more recently in soft-bottom benthos of transitional and coastal waters (Bremner et al. 2003; Bremner 2005, 2008; Veríssimo et al. 2012; Gusmao et al. 2016; Van der Linden et al. 2016). Only one work has used this approach on rocky shores in the western Portuguese coast (Vinagre et al. 2016).

Variations in functional trait composition between reference and impacted sites were evident in the study area. The index FRic showed a pattern with higher values in the reference sites; FEve showed a similar pattern, but only during spring. FDiv presented lower values in the impacted sites during fall and spring. In addition, the functional diversity indices (FRic and FEve) and AMBI presented a negative relationship, indicating that functional diversity is related to benthic environmental health. Thus, BTAs and functional diversity indices represent an informative and useful tool to describe the different aspects of sewage impacts in benthic assemblages on rocky shores, with high potential for the implementation of monitoring programmes. However, results for functional diversity indices showed more consistent patterns during spring, which indicated that this could be the season better reflecting the disturbance gradients and, therefore, could be the

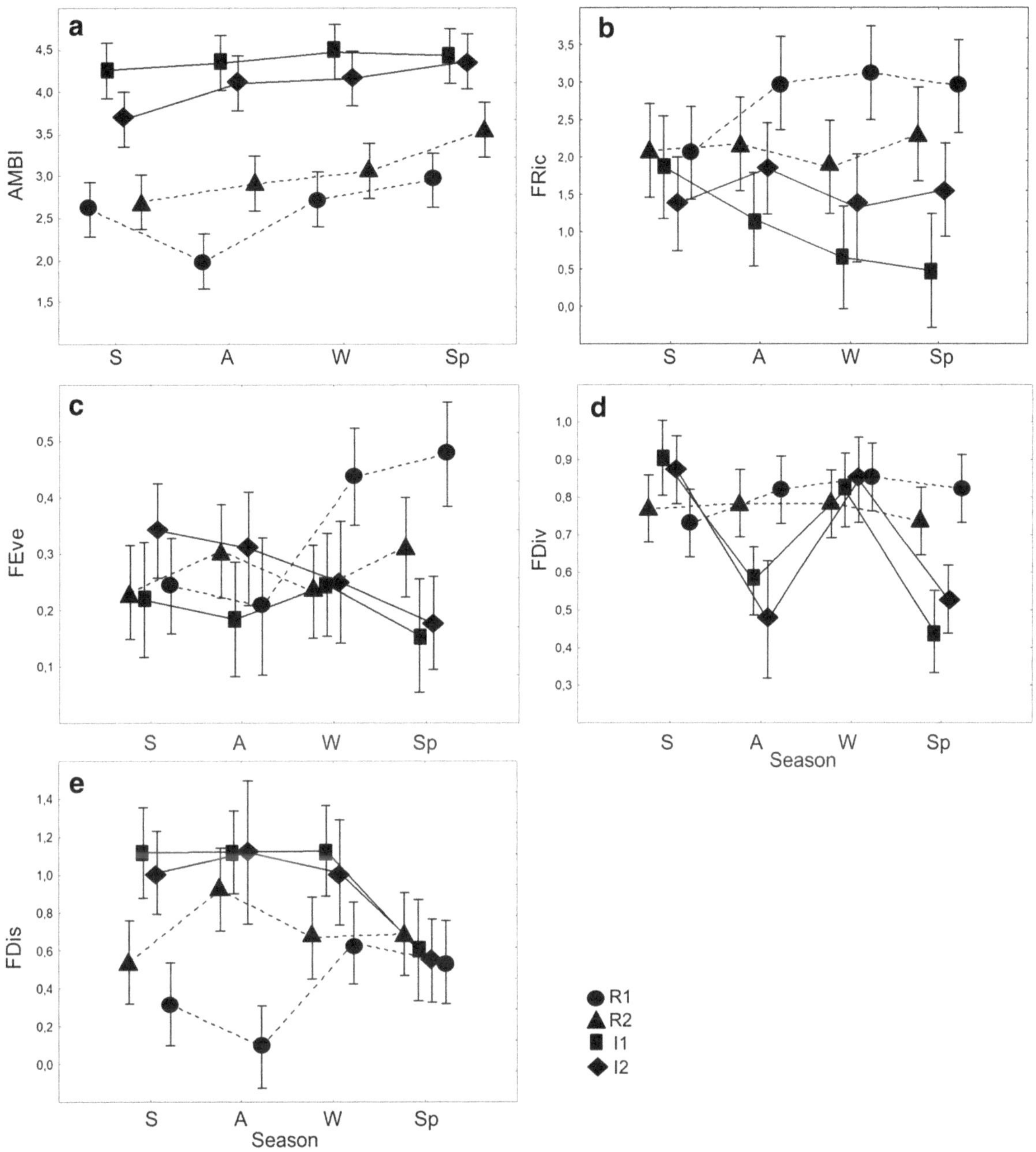

Fig. 4 Mean values (± SE) of AMBI index (**a**), functional richness (**b**), functional evenness (**c**), functional divergence (**d**) and functional dispersion (**e**) in each site along seasons

best period to undertake monitoring activities. Also, in other studies more consistent patterns were found in a particular season. Vinagre et al. (2017) concluded that summer could be the best season (compared to winter) to perform monitoring programmes.

BTA allowed the recognition of spatial changes in functioning on rocky shore macroinvertebrate communities within disturbance gradients; therefore, the first prediction was supported. Deposit feeders predominated in the impacted sites. Feeding habit describes the movement of energy and matter through the food web and determines abilities of species to utilize/tolerate different hydrodynamic conditions, e.g., with a switch from predominantly suspension feeders to surface deposit feeders indicating a potential reduction in water movement and siltation (Rosenberg 1995). The dominance of predators and grazer feeders in the reference sites could be related to increased sediment quality, which allows for the establishment of species with different feeding habits (Rosenberg 2001). Species with planktotrophic larval development predominated in the reference sites. Different types of recruitment confer different recovery potentials, e.g., communities recruiting mainly by pelagic dispersal are theoretically likely to recover from disturbance more quickly than those with predominantly direct benthic recruitment (Thrush and Whitlatch 2001). In this way, these sites would have a greater chance of recovering from a disturbance and, therefore, be characterized by a higher environmental quality.

Depth penetration in the sediment was between 0 and 8 cm deep in the impacted sites. This pattern was driven by the dominance of invasive polychaete *Boccardia proboscidea* in the impacted sites. In non-impacted conditions, the expected pattern is of not depth penetration since there is less amount of sediment to penetrate. The largest amount of sediment appears when *Boccardia proboscidea* forms the reefs. Finally, the tolerance to pollution (ecological groups: EG) reflects the response of communities to environmental stress (organic enrichment). In this study, the macrofaunal responses to pollution gradients agree with the pattern proposed by Borja et al. (2000).

The use of functional richness (FRic) and functional evenness (FEve) reveal differences in macrobenthic functional structure between impacted and reference sites. The indices FRic (all seasons) and FEve (during spring) showed patterns with higher values in the reference sites. Low functional richness indicates that some of the resources potentially available to the community are unused (Mason et al. 2005) and invasion resistance may also be lower since there will be gaps in niche space which an invader can exploit (Dukes 2001). The results of this study agree with this hypothesis, the impacted sites were dominated by invasive polychaete *B. proboscidea*, which is favored by organic contamination (Jaubet et al. 2011). In the same way, lower functional evenness indicates that some parts of niche space, whilst occupied, are under-utilized. This will tend to decrease productivity and reliability and increase opportunity for invaders (Mason et al. 2005).

With respect to functional divergence, high values of this index indicate a high degree of niche differentiation and thus low resource competition. Thus communities with high functional divergence may have increased ecosystem function as a result of more efficient resource use (Mason et al. 2005). The lowest values of this index were found in impacted sites (during fall and spring); it indicates assemblages with a lower relative abundance of species with extreme/unique categories of functional traits (Gerisch et al. 2012). For FDis index, an opposite pattern to the expected one was found. The lowest values of this index were found in the R1 site, which is furthest from the sewage discharge point. One possible explanation could be that the centroid for the calculation of this index is affected by highly dominant species (Laliberté and Legendre 2010). This site was highly dominated by the ecosystem engineer, the bivalve *Brachidontes rodriguezii*. According to the above, the second and third predictions were partially supported.

The submarine outfall that moved away the discharge 3.5 km far from coast was setting up in December-2015. This study contributed to a better understanding of the structure and functioning of intertidal rocky shore communities prior to the setting up of a submarine outfall. Future studies focusing on the recovery of the resident communities in the area will improve the knowledge on ecological process associated with marine pollution in the coast of Mar del Plata. The above provides an opportunity to assess whether the functional assemblage structure reflects the changes associated with the recovery of environmental quality. Finally, there seems to be potential to use BTA and the FD indices tested in the present study, together with more traditional methods, in future monitoring progammes to evaluate the evolution of the coast of Mar del Plata.

Conclusions

Based on the findings of the presented study, it can be concluded that BTAs and functional diversity indices, especially FRic that reveals differences in macrobenthic functional structure between impacted and reference sites during all seasons, represent an informative and useful tool to describe the different aspects of sewage impacts in benthic assemblages on rocky shores, with high potential for the implementation of monitoring programmes. There was a negative relationship between functional diversity indices (FRic and FEve) and the AMBI environmental quality index. This relationship indicates that functional diversity is related to benthic environmental health. Finally, we suggest that functional approach could be implemented in programmes to monitoring the evolution of the coast of Mar del Plata as well as in other coastal areas with similar problems.

Acknowledgements We are grateful to Dra. MA Sánchez who helped in sampling and sample processing, to Miss MS Rivero who helped in sampling and in the determination of TOM and to Dr. Julio Di Rienzo for helping with the installation FDiversity software. EN Llanos and MA Saracho Bottero were supported by a PhD fellowship from Consejo Nacional de Investigaciones Científicas y Técnicas (CONICET) of Argentina.

References

Becherucci ME, Llanos EN, Garaffo GV, Vallarino EA (2016) Succession in an intertidal benthic community affected by untreated sewage effluent: a case of study in the SW Atlantic shore. Mar Pollut Bull 109:95–103

Borja A, Franco J, Pérez V (2000) A marine biotic index to establish the ecological quality of soft-bottom benthos within European estuarine and coastal environments. Mar Pollut Bull 40:1100–1114

Borja A, Franco J, Valencia V, Bald J, Muxika I, Belzunce MJ, Solaun O (2004) Implementation of the European water framework directive from the Basque country (northern Spain): a methodological approach. Mar Pollut Bull 48:209–218

Boschi EE, Cousseau MB (2004) La vida entre mareas: vegetales y animales de las costas de Mar del Plata. Publicaciones especiales INIDEP, Mar del Plata, p 383

Bouvet Y, Desse RP, Morell P, Villar MC (2005) Mar del Plata (Argentina): la ciudad balnearia de los porteños en el Atlántico Suroccidental. Investigaciones Geográficas 36:61–80

Bremner J (2005) Assessing ecological functioning in marine benthic communities. University of Newcastle upon, Tyne **PhD dissertation**

Bremner J (2008) Species' traits and ecological functioning in marine conservation and management. J Exp Mar Biol Ecol 366(1–2):37–47

Bremner J, Rogers SI, Frid CLJ (2003) Assessing functional diversity in marine benthic ecosystems: a comparison of approaches. Mar Ecol Prog Ser 254:11–25

Byers S, Mills C, Stewart P (1978) Comparison of methods of determining organic carbon in marine sediments, with suggestions for a standard method. Hydrobiologia 58(1):43–47

Casanoves F, Di Rienzo JA, Pla L (2008) User manual fdiversity: statistical software for the analysis of functional diversity. First edition, Argentina Link: www.fdiversity.nucleodiversus.org. Accessed July 2017

Charvet S, Statzner B, Usseglio-Polatera P, Dumont B (2000) Traits of benthic macroinvertebrates in semi-natural French streams: an initial application to biomonitoring in Europe. Freshw Biol 43(2):277–296

Chevenet F, Dolédec S, Chessel D (1994) A fuzzy coding approach for the analysis of long-term ecological data. Freshw Biol 31(3):295–309

Clarke KR, Warwick RM (2001) Change in marine communities: an approach to statistical analysis and interpretation, 2nd edn. PRIMER-E, Plymouth

Comino AP, Scagliola M, Ge Z, Frick W (2010) Integrated coastal management strategy of Mar del Plata city and the use of virtual beach as a management tool. In: Proceedings of MWWD 2010, the 6th International Conference on Marine waste water disposal and marine environment (www.mwwd.org), Langkawi, Malaysia, 25–29 October, 2010, p 50–51. Accessed July 2017

Crowe TP, Russell R (2009) Functional and taxonomic perspectives of marine biodiversity: relevance to ecosystem processes. In: Wahl M (ed) Marine hard bottom communities: patterns, dynamics, diversity, change. Elsevier, Amsterdam, pp 375–390

Darr A, Gogina M, Zettler ML (2014) Functional changes in benthic communities along a salinity gradient—a western Baltic case study. J Sea Res 85:315–324

Di Rienzo JA, Casanoves F, Pla L (2008) FDiversity, version 2008. Cordoba, Argentina. www.fdiversity.nucleodiversus.org. Accessed July 2017

Diaz S, Cabido M (2001) Vive la difference: plant functional diversity matters to ecosystem processes. Trends Ecol Evol 16:646–655

Dukes JS (2001) Biodiversity and invasibility in grassland microcosms. Oecologia 126:563–568

Elias R, Jaubet ML, Llanos EN, Sánchez MA, Rivero MS, Garaffo GV, Sandrini-Neto L (2015) Effect of the invader *Boccardia proboscidea* (polychaeta: Spionidae) on richness, diversity and structure of SW Atlantic epilithic intertidal community. Mar Pollut Bull 91:530–536

Elías R, Rivero MS, Vallarino EA (2003) Sewage impact on the composition and distribution of polychaeta associated to intertidal mussel beds of the Mar del Plata rocky shore, Argentina. Iheringia, Serie Zoología 93(3):309–318

Elías R, Palacios JR, Rivero MS, Vallarino EA (2005) Short-term responses to sewage discharge and storms of subtidal sand-bottom macrozoobenthic assemblages off Mar del Plata City, Argentina (SW Atlantic). J Sea Res 53:231–242

Elías R, Rivero, Palacios JR, Vallarino EA (2006) Sewage-induced disturbance on polychaetes inhabiting intertidal mussel beds of *Brachidontes rodriguezii* off Mar del Plata (Southwestern Atlantic, Argentina). Sci Mar 70(3):187–196

Elías R, Sánchez MA, Jaubet ML, Rivero MS, Vallarino EA (2009) Do treatment sewage plants really work? The intertidal mussels' community of the southwestern Atlantic shore (38° S, 57° W) as a case study. Rev Biol Mar Oceanogr 44:357–368

Francisco V, de la Cueva H (2017) Nuevas perspectivas en la diversidad funcional de ambientes marinos. Lat Am J Aquat Res 45(2):261–275

Frid CLJ, Paramor OAL, Brockington S, Bremner J (2008) Incorporating ecological functioning into the designation and management of marine protected areas. Hydrobiologia 606:69–79

Garaffo GV, Jaubet ML, Sánchez MA, Rivero MS, Vallarino EA, Elías R (2012) Sewage induced polychaete reefs in a SW atlantic shore: rapid response to small scale disturbance. Mar Ecol 33:272–279

Garaffo GV, Jaubet ML, Sánchez MA, Llanos EN, Vallarino EA, Elías E (2016) Modelling the influence of environmental and weather factors on the density of the invasive polychaete *Boccardia proboscidea*. Mar Ecol 37(6):1256–1265

Garaffo GV, Jaubet ML, Becherucci ME, Elías R (2017) Assessing environmental health using ecological indices for soft-bottom in sewage-affected rocky shores: the case of the largest seaside resort of SW Atlantic. Mar Pollut Bull 115:233–239
Gerisch M, Agostinelli V, Henle K, Dziock F (2012) More species, but all do the same: contrasting effects of flood disturbance on ground beetle functional and species diversity. Oikos 121:508–515
Grime JP (1974) Vegetation classification by reference to strategies. Nature 250:26–31
Guerrero RA, Piola AR (1997) Masas de agua en la plataforma continental. In: Boschi EE (ed) El Mar Argentina y sus recursos pesqueros 1. Instituto nacional de investigación y desarrollo pesquero, Mar del Plata, pp 107–118
Gusmao JB, Brauko KM, Eriksson BK, Lana PC (2016) Functional diversity of macrobenthic assemblages decreases in response to sewage discharges. Ecol Indic 66:65–75
Halpern BS, Walbridge S, Selkoe KA, Kappel CV, Micheli F, D'Agrosa C, Bruno JF, Casey KS, Ebert C, Fox HE, Fujita R, Heinemann D, Lenihan HS, Madin EMP, Perry MT, Selig ER, Spalding M, Steneck R, Watson R (2008) A global map of human impact on marine ecosystems. Science 319(5865):948–952
Heino J (2005) Functional biodiversity of macroinvertebrate assemblages along major ecological gradients of boreal headwater streams. Freshw Biol 50:1578–1587
Heino J (2008) Patterns of functional biodiversity and function–environment relationships in lake littoral macroinvertebrates. Limnol Oceanogr 53(4):1446–1455
Heino J, Mykrä H, Kotanen J, Muotka T (2007) Ecological filters and variability in stream macroinvertebrate communities: do taxonomic and functional structure follow the same path? Ecography 30:217–230
Hewitt JE, Thrush SF, Dayton PD (2008) Habitat variation, species diversity and ecological functioning in a marine system. J Exp Mar Bio Ecol 366:116–122
Isla FI, Pérez Guzzi J, Folabella A (1998) Aportes de coliformes a la costa de Mar del Plata (Argentina) por vías naturales e inducidas. Thalassas 14:63–70
Jaubet ML, Sánchez MA, Rivero MS, Garaffo GV, Vallarino EA, Elías R (2011) Intertidal biogenic reefs build by invasive polychaete in sewage-impacted areas of Argentina, SW Atlantic. Mar Ecol 32:1–10
Jaubet ML, Garaffo GV, Sánchez MA, Elías R (2013) Reef-forming polychaetes outcompetes ecosystem engineering mussels. Mar Pollut Bull 71:216–221
Jaubet ML, Garaffo GV, Vallarino EA, Elías R (2015) Invasive polychaete *Boccardia proboscidea* Hartman, 1940 (polychaeta: spionidae) in sewage-impacted areas of the SW Atlantic coasts: morphological and reproductive patterns. Mar Ecol 36:611–622
Laliberté E, Legendre P (2010) A distance-based framework for measuring functional diversity from multiple traits. Ecology 91:299–305
Mason NWH, Mouillot D, Lee WG, Wilson JB (2005) Functional richness, functional evenness and functional divergence: the primary components of functional diversity. Oikos 111:112–118
McLenaghan NA, Tyler AC, Mahl UH, Howarth RW, Marino RM (2011) Benthic macroinvertebrate functional diversity regulates nutrient and algal dynamics in shallow estuary. Mar Ecol Prog Ser 426:171–184
Menezes S, Baird DJ, Soares AMVM (2010) Beyond taxonomy: a review of macroinvertebrate trait-based community descriptors as tools for freshwater biomonitoring. J Appl Ecol 47(4):711–719
Micheli F, Heiman KW, Kappel CV, Martone RG, Sethi SA, Osio GC, Fraschett S, Shelton AO, Tanner JM (2016) Combined impacts of natural and human disturbances on rocky shore communities. Ocean Coast Manag 126:42–50
Mouillot D, Graham NAJ, Villéger S, Mason NWH, Bellwood DR (2013) A functional approach reveals community responses to disturbances. Trends Ecol Evol 28:167–177
Muniz P, Lana P, Venturini N, Elias, Vallarino E, Bremec C, de Castro Martins C, Sandrini Neto L (2013) Un manual de protocolos para evaluar la contaminación marina por efluentes domésticos p 131 (ISBN: 978-9974-0-0899)
Muxika I, Borja A, Bald J (2007) Using historical data, expert judgement and multivariate analysis in assessing reference conditions and benthic ecological status, according to the European water framework directive. Mar Pollut Bull 55:16–29
Neumann H, Kröncke I (2010) The effect of temperature variability on ecological functioning of epifauna in the German bight. Mar Ecol 32(1):1–9
Pacheco AS, González MT, Bremner J, Oliva M, Heilmayer O, Laudien J, Riascos JM (2011) Functional diversity of marine macrobenthic communities from sub-littoral soft-sediment habitats off northern Chile. Helgol Mar Res 65:413–424
Rosenberg R (1995) Benthic marine fauna structured by hydrodynamic processes and food availability. Neth J Sea Res 34(4):303–317
Rosenberg R (2001) Marine benthic faunal successional stages and related sedimentary activity. Sci Mar 65:107–119
Sánchez MA, Jaubet ML, Garaffo GV, Elías R (2013) Spatial and long-term analyses of reference and sewage-impacted sites in the SW Atlantic (38° S, 57° W) for the assessment of sensitive and tolerant polychaetes. Mar Pollut Bull 74:325–333
Scagliola M, Furchi P, Von Haeften G, Comino AP, Moschione E, Gonzales R, Gayoso G, Caldararo A, Cerda G, Vergara S, Genga G, Elías R, Vallarino EA (2006) Sewage outfall project of Mar del Plata city (Argentina): an effective intervention to achieve quality objectives on the marine environment. In: Annals of the 4th international conference on marine waste water disposal and marine environment. Antalya, 6–10 November, p 22
Schuldt A, Bruelheide H, Durka W, Michalski SG, Purschke O, Assmann T (2014) Tree diversity promotes functional dissimilarity and maintains functional richness despite species loss in predator assemblages. Oecologia 174:533–543
Southwood TRE (1977) Habitat, the templet for ecological strategies? J Anim Ecol 46:337–365
Statzner B, Resh VH, Roux LA (1994) The synthesis of long-term ecological research in the context of concurrently developed ecological theory: design of a research strategy for the upper Rhone river and its floodplain. Freshw Biol 31(3):253–263
Thrush SF, Whitlatch RB (2001) Recovery dynamics in benthic communities: balancing detail with simplification. In: Reise K (ed) Ecological comparisons of sedimentary shores, ecological studies. Springer-Verlag, Berlin, pp 297–316

Tillin HM, Hiddink JG, Jennings S, Kaiser MJ (2006) Chronic bottom trawling alters the functional composition of benthic invertebrate communities on a sea-basin scale. Mar Ecol Prog Ser 318:31–45

Tilman D (2001) Functional diversity. In: Levin SA (ed) Encyclopaedia of biodiversity. Academic Press, San Diego, pp 109–120

Vallarino EA, Elías R (2006) High-diverse lowly variable sewage-impacted community, low diverse highly variable natural community: the paradox of the intertidal mussel beds of temperate areas of the SW Atlantic (38° S, 57° W). Curr Trends Ecol 1:77–91

Vallarino EA, Elías R (2009) La vida en las costas rocosas. Guía de campo. 1ed. Buenos Aires: Dunken 64:pp

Vallarino EA, Garaffo GV, Jaubet ML, Sánchez MA, Rivero MS, Elías R (2014) Are intertidal mussel population's indicators of sewage impact? The response of *Brachidontes rodriguezii* (Mytilidae) in the South western Atlantic rocky-stony shore. Thalassas 30(1):51–56

Van der Linden P, Marchini A, Dolbeth M, Patrício J, Veríssimo H, Marques JC (2016) The performance of trait-based indices in an estuarine environment. Ecol Indic 61(2):378–389

Veríssimo H, Bremner J, Garcia C, Patrício J, van der Linden P, Marques JC (2012) Assessment of the subtidal macrobenthic community functioning of a temperate estuary following environmental restoration. Ecol Indic 23:312–322

Villéger S, Mason NWH, Mouillot D (2008) New multidimensional functional diversity indices for a multifaceted framework in functional ecology. Ecology 89:2290–2301

Villéger S, Miranda JR, Hernández DF, Mouillot D (2010) Contrasting changes in taxonomic vs. functional diversity of tropical fish communities after habitat degradation. Ecol Appl 20:1512–1522

Vinagre PA, Pais-Costa AJ, Gaspar R, Borja A, Marques JC, Neto JM (2016) Response of macroalgae and macroinvertebrates to anthropogenic disturbance gradients in rocky shores. Ecol Indic 61(2):850–864

Vinagre PA, Veríssimo H, Pais-Costa AJ, Hawkins SJ, Borja A, Marques JC, Magalhães Neto J (2017) Do structural and functional attributes show concordant responses to disturbance? Evidence from rocky shore macroinvertebrate communities. Ecol Indic 75:57–72

Worm B, Barbier EB, Beaumont N, Duffy JE, Folke C, Halpern BS, Jackson JBC, Lotze HK, Micheli F, Palumbi SR, Sala E, Selkoe KA, Stachowicz JJ, Watson R (2006) Impacts of biodiversity loss on ocean ecosystem services. Science 314(5800):787–879

3

A new insight into the influence of habitat on the biochemical properties of three commercial sea cucumber species

Mercedes González-Wangüemert · Christina C. Roggatz · Maria João Rodrigues · Luísa Barreira · Manuela Moreira da Silva · Luísa Custódio

Abstract This work makes a comparative evaluation of the biochemical profile of three sea commercial cucumber species (*Holothuria mammata*, *H. polii* and *H. tubulosa*) caught from different locations of the Mediterranean Sea (SE Spain). All species had high levels of moisture (from 73.6% in *H. mammata* to 81.2% in *H. tubulosa*), crude ash (from 9.61% in *H. mammata* to 14.7% in *H. tubulosa*) and protein (3.01% in *H. tubulosa* to 11.1% in *H. mammata*). They also had a low fat content, from 0.21% in *H. tubulosa* to 0.55% in *H. mammata*. *Holothuria polii* had intermediate values between the other two species, for all considered variables. All species had adequate protein/lipid ratios (*H. mammata*, 20:1; *H. polii*, 23:1; *H. tubulosa*, 14:1) and low lipid levels, enriched in omega-3 polyunsaturated fatty acids, especially arachidonic acid. The fatty acid profile suggests that *H. polii* is feeding on sediments more influenced by terrestrial inputs than the remaining species. *Holothuria mammata* and *H. tubulosa* are feeding on marine food sources mainly, but also with some terrestrial influence. The most abundant amino acids detected were alanine, arginine, glutamic acid, and glycine. All species had similar contents of essential amino acids (EAA) and ratios of EAA/non-essential amino acids. *Holothuria tubulosa* had a high content of toxic metals including Cr, Pb and Ni. This work highlights differences in compositional characteristics between different species of the same genus (*Holothuria*) from different locations.

Keywords Bioaccumulation · Fatty acids · Habitat · PUFAs · Toxic minerals

Introduction

Sea cucumbers (Echinodermata: Holothuroidea) are an important fishery resource, mainly in the Indo-Pacific region. They are exported to different locations, such as Hong Kong and Mainland China, where they are consumed as a dried (*bêche-de-mer*) or fresh products (Kasai 2003; Purcell et al. 2012; Purcell 2014). In those

M. González-Wangüemert · C. C. Roggatz · M. J. Rodrigues · L. Barreira · L. Custódio (✉)
Center of Marine Sciences, Faculty of Sciences and Technology, University of Algarve, Ed. 7, Campus of Gambelas, 8005-139 Faro, Portugal
e-mail: lcustodio@ualg.pt

C. C. Roggatz
Department of Chemistry, University of Hull, Cottingham Road, Hull HU6 7RX, UK

M. M. da Silva
Center of Marine and Environmental Research (CIMA), University of Algarve, Campus de Gambelas, 8005-139 Faro, Portugal

Asian markets sea cucumbers are a high value product and can reach up to USD300–500/kg (Purcell 2010; Wen et al. 2010). The global sea cucumber trade aiming the food market is controlled largely by China, Hong Kong SAR, Singapore and Taiwan (Jaquemet and Conand 1999; FAO 2014; INFOPESCA and FAO 2013). However, sea cucumbers are being increasingly exported from Mauritania to the United States (Ann-Marie Holmes personal communication, from US Fish and Wildlife Service, US Department of Interior). Additionally, in Europe the areas and volume of sea cucumber capture are growing (Bordbar et al. 2011; FAO 2011; González-Wangüemert et al. 2016, 2018). The use of sea cucumbers has been mainly to fuel the increasing demand of Maiman based therapies also termed as zootherapies (Mahomoodally and Muthoorah 2014; Mootoosamy and Mahomoodally 2014).

It is estimated that 66 sea cucumber species are caught from more than 40 countries and exported to Asian markets (Conand and Byrne 1993; Choo 2008; Conand 2008; Kinch et al. 2008; Purcell 2010; Purcell et al. 2012). There has been a significant reduction in the sea cucumber stocks in many countries, or even a complete stock depletion in many tropical fisheries caused by the increasing market demand, unrestrained exploitation and/or inadequate fisheries (Purcell et al. 2012). Those situations boosted the catch and commercialization of species from alternative locations such as the Mediterranean Sea, being *Holothuria mammata*, *H. tubulosa* and *H. polii* some of the new target species (Aydin 2008; Aydin et al. 2011; González-Wangüemert et al. 2014; González-Wangüemert et al. 2016, 2018; González-Wangüemert and Domínguez-Godino 2016). Turkey is the main Mediterranean country actively harvesting and exporting sea cucumbers to Asian countries, including *H. mammata, H. polii, H. tubulosa* and *H. sanctori* (Çakli et al. 2004; Aydin 2008; González-Wangüemert et al. 2014; González-Wangüemert et al. 2015, 2016, 2018).

Sea cucumbers inhabit several sea floor habitats from intertidal zones to deep trenches, and from polar to tropical areas (Purcell et al. 2016). Those animals are mainly benthic and most of the exploited species are bottom detritus-feeders, consuming debris, bacteria and diatoms mixed with sediments on the seabed (Bruckner 2006; Purcell 2010; Purcell et al. 2016). Sea cucumbers have a low metabolism and its feeding source could influence their chemical composition, with strong implications for their nutritional properties (Aydin et al. 2011, Purcell et al. 2012, Yu et al. 2015). In this sense, this work makes a comparative evaluation of the chemical properties, including proximate composition, amino acids, fatty acids and essential and toxic metal contents of three commercial sea cucumber species (*H. mammata*, *H. polii* and *H. tubulosa*) caught from different locations of the Mediterranean Sea (SE Spain). Whenever possible results were compared with those obtained from the literature for the same species, caught from other Mediterranean areas such as Turkey and Italy. The amino acid levels of *H. mammata* is herein reported for the first time. Data was discussed through ecological and nutraceutical perspectives.

Materials and methods

Sample collection, processing and genetic identification

Holothuria mammata and *H. tubulosa* were collected in Los Cocedores (Murcia, SE Spain) while *H. polii* was caught in Los Urrutias (Mar Menor coastal lagoon, Murcia, SE Spain) (Fig. 1). All species were sampled during April of 2012 by scuba diving at 5 meters depth. Ten individuals of each target species were caught for analysis.

H. individuals had an average length of 19.32 ± 4.62 cm being located into rock crevice close to *Posidonia oceanica* meadows. *Holothuria tubulosa* specimens showing an average length of 24.71 ± 3.10 cm were caught on sandy and muddy bottom. *Holothuria polii* individuals with an average length of 14.02 ± 2.24 cm were sampled from Mar Menor coastal lagoon close to *Caulerpa prolifera* and *Cymodocea nodosa* meadows. This species is showing individuals with lower size than populations outside coastal lagoon (González-Wangüemert personal observation). Sea cucumbers were identified in situ considering its external morphology. Later this identification was confirmed using traditional taxonomic characters, such as ossicles, on the basis of the original descriptions of these species (Gmelin 1791; Grube 1840; Delle Chiaje 1824) and other relevant taxonomic contributions (Borrero-Pérez et al. 2009, 2010). This identification was further confirmed using genetic barcoding, by amplifying the cytochrome c oxidase I (COI) gene according to protocol from González-Wangüemert and Borrero-Pérez (2012). PCR fragments were sequenced and Basic Local Alignment

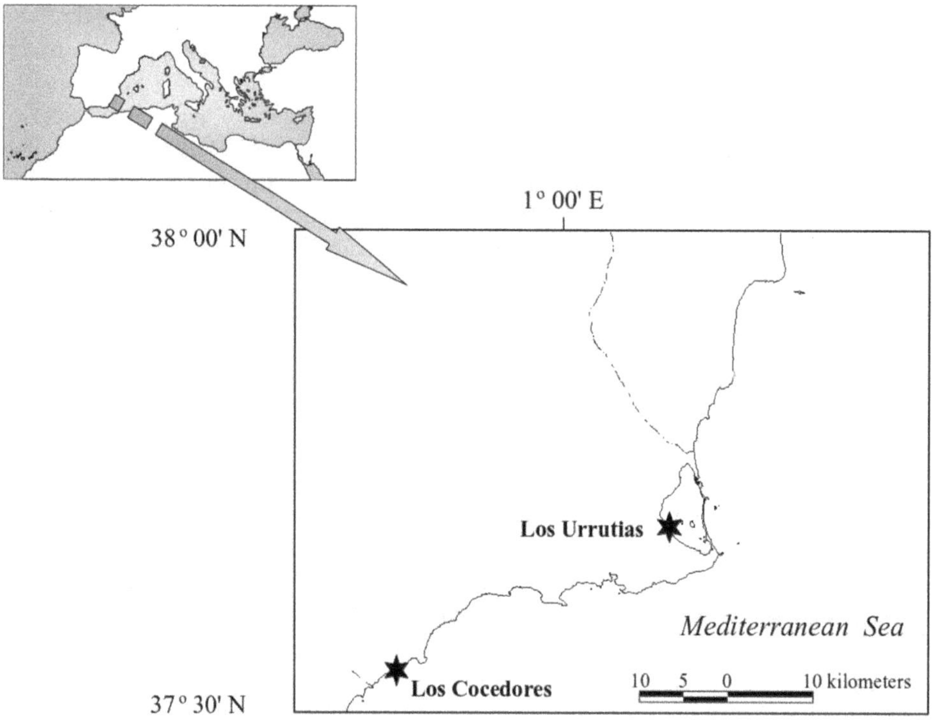

Fig. 1 Sampling sites of *H. mammata* and *H. tubulosa* (Los Cocedores/Almeria, Spain) and *H. polii* (Los Urrutias—Mar Menor coastal lagoon/Murcia, Spain)

Search Toll (BLAST) was performed against GenBank nucleotide database (www.ncbi.nlm.nih.gov/genbank). BLAST finds regions of local similarity between sequences and calculates statistical significance of matches. BLAST can be used to infer functional and evolutionary relationships between sequences as well as help identify members of gene families.

For the biochemical analysis fresh adult individuals were gutted, all inner organs were removed and the body walls were rinsed with fresh water. Samples were frozen at − 20 °C, freeze-dried, ground into a fine powder, pooled and stored at − 20 °C in tightly closed plastic bags until analyses. Each sample corresponded to ten adult individuals.

Biochemical analysis

Biochemical analysis included the determination of proximate composition, amino acids, fatty acid methyl esters (FAME) and minerals contents. These parameters were determined according to methods widely used to ascertain the biochemical profile of edible marine organisms in general (Njinkoue et al. 2016; Lah et al. 2017), and sea cucumber species in particular (Çakli et al. 2004; Zhong et al. 2007; Wen et al. 2010; Aydin et al. 2011; Roggatz et al. 2016; Gao et al. 2016).

Proximate composition

Moisture was determined by drying the samples at 52 °C (± 1 °C) until a constant weight was obtained (5 days). Ash was determined by incineration in a muffle furnace at 525 °C until the samples burned completely (5 h, AOAC 1990; Gressler et al. 2010) and the crude protein content ($N \times 6.25$) was estimated by the macro-Kjeldahl method. For this method approximately 1 g of dried samples were digested using a catalyst mixture (1 pellet of kjeldahl catalyst 0.3% in $CuSO_4$–$5H_2O$) and 25 mL of concentrated sulfuric acid in a digestion apparatus (BICASA Minerox MOD BE 97) under increasing temperature (200–400 °C) to convert all nitrogen forms present in the samples into ammonium sulfate. Sodium hydroxide (35% solution) was then added to liberate the ammonia. The samples were distilled (P Selecta PRO-NITRO II) into 250 mL-

Erlenmeyer flasks containing 10 mL of boric acid (4% solution) and a bromocresol green indicator. The amount of nitrogen (% *N*) in the sample (corresponding to the amount of fixed ammonia), was determined by titration with 0.5 N HCL according to the following formula:

$$\%N = \frac{V_{\mathrm{HCL}} \times f_{\mathrm{HCL}} \times 0.014 \times 100}{P_{\mathrm{sample}}}$$

where V_{HCL} is the volume of hydrogen chloride needed for the neutralization (mL), f_{HCL} the molarity of HCl used (mol L^{-1}) and P_{sample} is the exact weight of biomass (g) used for the analysis. The protein content was calculated by multiplying by 6.25, corresponding to approximately 16% of nitrogen content in proteins.

Crude fat was determined by a modified method of Bligh and Dyer (Pereira et al. 2013). In brief, a solution containing methanol, chloroform and water (2:2:1) was mixed with 100 mg of dried biomass and homogenised with an IKA Ultra-Turrax disperser. Then, samples were centrifuged and a known volume of the organic phase was moved into pre-weighed tubes. The solvent was then evaporated in a gentle nitrogen flow and the tubes were reweighed to calculate the lipid content. Special care was taken to obtain a full homogenization and performing repeated washings and extractions, having in mind that sea cucumbers generally have a low lipid content. Results are expressed as g per 100 g of wet weight biomass (WW) and dry weight biomass (DW).

Amino acids

Total amino acids were determined by gas chromatography/mass spectrometry (GC/MS) after sample derivatization with *N*-methyl-*N*-tert-butyldimethylsilyltrifluoroacetamide (BSTFA) and trimethylchlorosilane (TMCS) according to Deng et al. (2005). In brief, dried samples (100 mg) and the internal standard (L-norvaline, 20 mg) were mixed with 500 μL of acetonitrile and 500 μL of MBSTFA + TMCS (99:1, v/v) in screw vials under microwave irradiation (750 W, 60 s). After cooling to room temperature, samples (1 μL) were injected at 260 °C on an Agilent GC–MS, with a temperature profile of 50 °C (1 min) and 10 °C min^{-1} to 300 °C (6 min). To identify and quantify the amino acids, the total ion mode was used. A set of standards of four different concentrations containing all 22 proteinogenic amino acids was prepared and measured, and calibration curves were generated for each amino acid. Results are expressed as percentage of total amino acid content.

Fatty acid methyl esters (FAME) profile

The FAME content was determined by a modified protocol from Lepage and Roy (1984), as described previously (Pereira et al. 2013). In short, aliquots of dried samples (100 mg) were mixed with acetyl chloride and methanol (20:1, v/v) in reaction vessels and homogenized with an IKA Ultra-Turrax disperser for 2 min. Then, 1 mL of hexane was added to the mixture and heated (100 °C, 1 h). One mL of distilled water was added to the mixture, followed by centrifugation and removal of the organic phase, which was dried with anhydrous sodium sulphate. Samples were finally injected in a Varian 450-GC/240-MS (Varian 450-Gas Chromatograph/240-MS IT Mass Spectrometer, Varian Inc., Palo Alto, CA, USA), equipped with a BR-5MS capillary column (30 m × 0.25 mm internal diameter, 0.25 μM film thickness, Bruker). The injection temperature was 300 °C, and the trap, manifold and transfer line were established to 22, 50 and 250 °C, respectively. Helium was used as the carrier and the temperature program for the GC oven was 60 °C (1 min), 30 °C min^{-1} to 120 °C, 4 °C min^{-1} to 250 °C, 20 °C min^{-1} to 270 °C, and 2.5 °C min^{-1} to 300 °C. For identification and quantification of the FAME total ion mode was used. Because of differences in the response factors, for each FAME separate calibration curves were determined in triplicate, using the Supelco® 37 Component FAME Mix (Sigma-Aldrich, Sintra, Portugal) commercial standard. Values were expressed in amount (μg/g) of dry biomass.

Minerals

Minerals were analysed by atomic absorption spectrometry-AAS (GBC Avanta Sigma, Australia) provided with a deuterium background correction. Mg, Na, K, Ca, Fe, Mn and Zn were analysed by flame (F) AAS with an air-acetylene flame. Cd, Cr, Ni and Pb were analysed with electrothermal atomisation (ET) (GBC graphite furnace 3000) using an auto-sampler (PAL 3000) (Roggatz et al. 2016). The accuracy of the analytical procedure was assessed by the analysis of certified reference material, using Community Bureau of Reference BCR60 (Lagarosiphon major). Procedural blanks always accounted for less than 1% of the metal concentrations in samples. Values were expressed as g/kg dry biomass (Ca, Mg, Na and K) or mg/kg dry biomass (Fe, Mn, Zn, Cr, Pb, Ni and Cd).

Statistical analysis

Results were expressed as mean ± standard error of the mean (SEM). Significant differences were assessed by analysis of variance (ANOVA) or by Student–Newman–Keuls post hoc test ($p < 0.05$) (software "R", R Development Core Team 2013) with the packages GAD (Sandrini-Neto and Camargo 2011) and CAR (Fox and Weisberg 2011). For each species, three samples were analysed and the assays were carried out at least in triplicate.

Results

Morphological and genetic identification

Holothuria polii showed relative small buttons (ossicles) comparing with the other two species. *Holothuria polii* has buttons more rounded, with smooth surface and edged never wrinkled. *Holothuria tubulosa* shows buttons more slender with the surface an edges clearly wrinkled. *Holothuria mammata* was distinguished from *H. tubulosa* by its buttons, which are more elongated and slightly bigger.

COI gene amplification of the specimen caught in Los Urrutias (Mar Menor, SE Spain) allowed us to obtain a sequence of 484 bp in length. The BLAST in GENBANK, identified the individual from Los Urrutias as *H. polii* with a 100% maximum identity (100% query cover) with a specimen caught in Mar Menor (SE, Spain; EU750797) and 99% maximum identity (100% query cover) with a specimen sampled in Cabo de Palos (SE, Spain; GQ214758), but showing one mutational change in the 396 position. The haplotype for this specimen was recorded as EU750793. In the case of the individual found in Los Cocedores (Murcia, SE Spain) and identified as *H. tubulosa* using external morphology, the COI amplification (532 bp) and BLAST corroborated that identification, with a 99% maximum identity (100% query cover) with a specimen from Murcia (SE Spain; GQ214750) showing three mutational changes in 260, 452 and 521 positions; the haplotype from this specimen is GQ214748. Finally, the specimen identified as *H. mammata* (external morphology) also sampled in Los Cocedores, was assigned rightly to the *H. mammata* individual GQ214747 sampled in Murcia (SE Spain), showing 99% maximum identity (100% query cover) with only 3 mutational changes in the 12, 152 and 194 positions; this haplotype was recorded as GQ214746.

Proximate composition

Proximate composition of the three sea cucumbers species is shown in Table 1, and is expressed as % of total wet biomass (WW) and dry biomass (DW). Moisture ranged from 73.6% in *H. mammata* to 81.2% in *H. tubulosa*, and the crude ash content varied from 9.61% in *H. mammata* to 14.7% in *H. tubulosa*. In general all species had a low fat content, from 0.21% in *H. tubulosa* to 0.55% in *H. mammata*, and high protein levels, ranging from 3.01% in *H. tubulosa* to 11.1% in *H. mammata*. *Holothuria polii* showed intermediate values between the other two species, for all considered variables. All the species had adequate protein/lipid ratios (*H. mammata*, 20:1; *H. polii*, 23:1; *H. tubulosa*, 14:1).

Table 1 Proximate composition of the body wall of *H. mammata*, *H. polii* and *H. tubulosa*

	H. mammata		*H. polii*		*H. tubulosa*	
	WW	DW	WW	DW	WW	DW
Moisture[1]	73.6 ± 1.73[b]	–	76.3 ± 1.52[b]	–	81.2 ± 1.55[a]	–
Ash[2]	9.61 ± 0.32[b]	34.4 ± 1.21[c]	13.4 ± 0.43[a]	56.54 ± 1.81[b]	14.7 ± 0.34[a]	78.1 ± 1.80[a]
Protein[2]	11.1 ± 0.27[a]	42.0 ± 1.02[a]	7.37 ± 0.24[b]	31.09 ± 1.01[b]	3.01 ± 0.23[c]	16.0 ± 1.11[c]
Fat[2]	0.55 ± 0.05[a]	2.08 ± 0.18[a]	0.32 ± 0.03[b]	1.35 ± 0.12[b]	0.21 ± 0.04[b]	1.11 ± 0.21[b]

Data represent the mean ± SD (n = 3)

In the same row values followed by different letters are significantly different at $p < 0.05$ (one-way ANOVA with Student–Newman–Keuls post hoc test)

[1]Percentage (%)

[2]g/100 g of wet weight (WW) and dry weight (DW)

Amino acids profile

The composition of total amino acids of dried sea cucumber is shown in Table 2, and results are given in % of total amino acid content. Nineteen amino acids were detected, including 8 essential (EAA) and 11 non-

Table 2 Amino acid profile of *H. mammata*, *H. polii* and *H. tubulosa* in % of total amino acid content

	H. mammata	*H. polii*	*H. tubulosa*
ALA	12.2 ± 0.3[a]	15.1 ± 2.6[a]	14.5 ± 2.7[a]
ARG	6.8 ± 0.2[d]	13.4 ± 2.4[a]	12.9 ± 1.7[b]
ASN	1.0 ± 0.2[a]	nd	nd
ASP	5.3 ± 0.8[a]	4.5 ± 0.4[b]	4.8 ± 0.7[b]
CYS	8.2 ± 2.9[a]	2.4 ± 0.8[b]	2.4 ± 0.9[b]
GLU	12.1 ± 1.9[ab]	8.4 ± 1.4[b]	8.6 ± 2.1[b]
GLN	2.3 ± 0.7[b]	1.0 ± 0.2[b]	1.3 ± 0.8[b]
GLY	11.8 ± 0.6[a]	10.6 ± 1.2[a]	10.6 ± 1.1[a]
PRO	8.2 ± 0.4[a]	10.1 ± 2.4[a]	9.5 ± 2.0[a]
SER	3.2 ± 0.3[b]	3.0 ± 0.5[b]	3.5 ± 0.7[b]
TYR	7.1 ± 1.3[a]	3.8 ± 1.4[b]	3.3 ± 1.4[b]
Non-essential (NEAA)	78.5 ± 9.7[a]	72.6 ± 13.0[a]	71.9 ± 14.1[a]
HIS	2.0 ± 0.6[b]	nd	nd
ILE	2.0 ± 0.1[b]	2.9 ± 0.1[a]	3.3 ± 0.3[a]
LEU	3.7 ± 0.1[c]	5.4 ± 0.4[b]	5.8 ± 0.6[b]
LYS	2.2 ± 0.7[b]	1.1 ± 0.2[b]	1.6 ± 0.7[b]
MET	1.0 ± 0.0[b]	0.7 ± 0.3[b]	0.1 ± 0.1[d]
PHE	3.3 ± 2.0[b]	8.1 ± 1.5[a]	7.9 ± 2.3[a]
THR	3.3 ± 0.1[a]	3.4 ± 0.6[a]	3.7 ± 0.7[a]
VAL	3.7 ± 0.1[b]	5.4 ± 0.6[a]	5.5 ± 0.7[a]
Essential (EAA)	21.4 ± 3.6[a]	27.3 ± 3.6[a]	28.0 ± 5.5[a]
EAA/NEAA	0.27	0.38	0.39
LYS/ARG	0.33	0.09	0.12

Data represent the mean ± SD (n = 2)

nd not detected, *ALA* alanine, *ARG* arginine, *ASN* asparagine, *ASP* aspartic acid, *CYS* cysteine, *GLN* glutamine, *GLU* glutamic acid, *GLY* glycine, *HIS* histidine, *ILE* isoleucine, *LEU* leucine, *LYS* lysine, *MET* methionine, *PHE* phenylalanine, *PRO* proline, *SER* serine, *THR* threonine, *TRP* tryptophan, *TYR* tyrosine, *VAL* valine, *TAA* total amino acids, *EAA* essential amino acids, *NEAA* non-essential amino acids, *nd* not detected

In the same row, values followed by different letters are significantly different at $p < 0.05$ (one-way ANOVA with Student–Newman–Keuls post hoc test)

essential amino acids (NEAA). The most abundant amino acids were alanine, arginine, glutamic acid, and glycine. All species had similar contents of EAA (21.4–28%), and as a consequence, similar ratios of EAA/ NEAA. The lysine/arginine ratios ranged from 0.09 in *H. polii* to 0.33 in *H. mammata*.

FA content and profile

The FA contents of the three sea cucumber species included in this work are represented in Table 3. They had high amounts of polyunsaturated FA (PUFA) ranging from 402 μg/g in *H. polii* to 1028 μg/g in *H. mammata*. PUFA accounted for 32–42% of total FA content. The least concentrated FA in these species was the SFA (321–526 μg/g) which accounted for 21–27% of total FA. Arachidonic acid (AA, C20:4*n*-6) was the most abundant FA in *H. mammata* (643 μg/g) and *H. tubulosa* (365 μg/g), while oleic acid (C18:1*n*-9*c*) was predominant in *H. polii* (246 μg/g). All species had similar amounts of EPA (115–221 μg/g). In contrast, regarding docosahexaenoic acid (DHA, C22:6*n*-3) *H. mammata* (89.7 μg/g) and *H. tubulosa* (63.7 μg/g) had the highest contents. Besides palmitic acid (C16:0), and the two monounsaturated FA (MUFA) eicosenoic (C20:1*n*-9) and nervonic acids (C24:1*n*-9), the other FA contributed to less than 5% of the total amount of FA.

Table 3 FAME profile (μg/g DW) of *H. mammata*, *H. polii* and *H. tubulosa*

	Common name	*H. mammata*	*H. polii*	*H. tubulosa*
C14:0	Myristic acid	45.5 ± 1.7^{a}	31.3 ± 6.5^{b}	47.5 ± 1.5^{a}
C15:0	Pentadecanoic acid	21.3 ± 0.7^{a}	10.0 ± 1.1^{c}	19.7 ± 1.7^{b}
C16:0	Palmitic acid	107 ± 4^{a}	40.9 ± 9.2^{c}	91.9 ± 11.3^{b}
C17:0	Margaric acid	25.3 ± 2.0^{a}	14.6 ± 2.9^{b}	24.8 ± 3.4^{a}
C18:0	Stearic acid	88.7 ± 9.6^{a}	39.9 ± 8.5^{c}	73.1 ± 6.5^{b}
C19:0	Nonadecanoic acid	45.9 ± 3.8^{a}	27.7 ± 4.2^{b}	39.2 ± 3.2^{b}
C20:0	Arachidic acid	67.9 ± 4.3^{a}	42.0 ± 6.9^{b}	51.2 ± 1.8^{a}
C21:0	Heneicosanoic acid	61.9 ± 5.4^{a}	35.9 ± 7.8^{b}	41.5 ± 4.0^{c}
C22:0	Behenic acid	nd	37.4 ± 7.0^{b}	42.8 ± 3.3^{a}
C23:0	Tricosanoicc acid	27.5 ± 2.7^{b}	16.9 ± 5.0^{a}	18.9 ± 2.6^{b}
C24:0	Lignoceric acid	34.9 ± 7.9^{b}	24.8 ± 1.7^{a}	27.7 ± 0.9^{b}
Σ SFA		526 ± 35^{a}	321 ± 17^{c}	478 ± 29^{b}
C16:1	Palmitoleic acid	76.0 ± 3.1^{b}	23.5 ± 7.3^{c}	60.1 ± 6.8^{ab}
C18:1n-9 c	Oleic acid	350 ± 23^{c}	246 ± 32^{a}	288 ± 18^{b}
C18:1n-9 t	Elaidic acid	58.9 ± 5.9^{b}	24.1 ± 5.2^{c}	49.1 ± 6.6^{a}
C20:1n-9	Eicosenoic acid	137 ± 12^{b}	90.1 ± 18.3^{a}	91.2 ± 14.6^{c}
C22:1n-9	Docosenoic acid	67.0 ± 6.3^{b}	44.1 ± 7.0^{a}	43.0 ± 4.8^{c}
C24:1n-9	Nervonic acid	182 ± 19^{a}	85.6 ± 17.4^{b}	99.1 ± 10.9^{c}
Σ MUFA		870 ± 113^{b}	514 ± 79^{a}	631 ± 90^{b}
C18:2n-6	Linoleic acid	35.8 ± 3.4^{a}	16.1 ± 1.8^{b}	27.3 ± 1.5^{a}
C20:4n-6	Arachidonic acid (AA)	643 ± 85^{a}	228 ± 54^{c}	365 ± 46^{b}
C20:5n-3	Eicosapentaenoic acid (EPA)	221 ± 31^{a}	115 ± 27^{a}	157 ± 19^{a}
C20:2n-6	Eicosadienoic acid	38.2 ± 3.3^{a}	16.3 ± 1.8^{b}	30.4 ± 3.6^{a}
C22:6n-3	Docosahexaenoic acid (DHA)	89.7 ± 9.5^{a}	26.1 ± 4.3^{b}	63.7 ± 6.4^{a}
Σ PUFA		1028 ± 202^{a}	402 ± 75^{b}	643 ± 116^{ab}
Σ *n*-3		311 ± 32^{a}	141 ± 27^{b}	220 ± 20^{a}
Σ *n-6*		717 ± 85^{a}	260 ± 54^{c}	422 ± 46^{b}
Σ *n-6*/*n*-3		2.3	1.8	1.9

Data represent the mean ± SD ($n = 3$)

ΣSFA total saturated fatty acids (FAs), *ΣMUFA* total monounsaturated FAs, *ΣPUFA* total polyunsaturated FAs, *Σn-3* total omega-3 PUFAs; *Σn-6* total omega-6 PUFAs; *Σn-6/n-3* ratio of omega-6 to omega-3 fatty acids, *nd* not detected

In the same row, values followed by different letters are significantly different at $p < 0.05$ (one-way ANOVA with Student–Newman–Keuls post hoc test)

The amount of Σ *n*-6 FAME (260–717 μg/g) was always higher than the percentage of Σ *n*-3 (141–311 μg/g) in all species. The ratio between the Σ *n*-6 and Σ *n*-3 (1.8–2.3) was similar for the three species.

Minerals

As shown in Table 4, the sea cucumbers under study contain many dietary essential minerals, such as sodium (56.9–75.2 g/kg), potassium (3.37–5.58 g/kg), calcium (41.1–145 g/kg), magnesium (12.7–21.4 g/kg), iron (33.7–4400 g/kg), zinc (8.9–227.7 mg/kg), and manganese (4.53–86.6 mg/kg). *Holothuria polii* had the highest levels of calcium (145 g/kg) and magnesium (21.4 g/kg), while *H. tubulosa* had the highest content of residual metals; especially iron (4400 mg/kg), which was 100 times higher than the amounts found in the remaining species. A similar trend was observed with the other residual minerals such as manganese and zinc and also with the toxic metals chromium, lead and nickel (Table 4).

Discussion

The proximate composition of fresh sea cucumbers differs between species, catching season and feeding patterns (Bordbar et al. 2011). Holothurians are on the bottom of the food chain and help to recycle detritus, since most of them are deposit feeders that consume bacteria and diatoms mixed with sediments on the seabed (Purcell 2010; Purcell et al. 2016). Bacterial densities in sediments can differ greatly from site to site being generally associated with the type of organic compounds found in those sediments. For example, significant differences in chlorophyll *a* concentration and bacterial cell concentrations were observed between the Eastern and Western Mediterranean sediments and water, associated with a shift from carbohydrates to amino acids as the dominant biochemical components (Danovaro et al. 1999). Such diversity in feeding source could explain the variation in the chemical composition of sea cucumbers (Aydin et al. 2011).

The high-moisture levels registered in the sea cucumbers studied on this work are in accordance with the values found in the literature for the same species sampled in the Eastern Mediterranean regions (Aydin et al. 2011), and in the Southern Adriatic Sea (Sicuro et al. 2012), and also for other sea cucumber species belonging to the same genus and inhabiting the SW Mediterranean region, as for example, *H. arguinensis* (Roggatz et al. 2016). The crude ash contents was higher than the values reported by other authors in the same species from Turkey (Çakli et al. 2004; Aydin et al. 2011) and from the Southern Adriatic Sea (Sicuro et al. 2012), and also than those reported for important commercial sea cucumbers, such as *Stichopus japonicus* L. (Gao et al. 2016).

Table 4 Mineral content of the body wall of *H. mammata, H. polii* and *H. tubulosa* in g/kg dry biomass (Ca, Mg, Na and K) or mg/kg dry biomass (Fe, Mn, Zn, Cr, Pb, Ni and Cd)

	Symbol	*H. mammata*	*H. polii*	*H. tubulosa*
Calcium	Ca	41.1 ± 0.05^{c}	145 ± 0.13a	95.8 ± 0.86^{b}
Magnesium	Mg	12.7 ± 0.04^{c}	21.4 ± 0.10^{a}	16.6 ± 0.09^{b}
Sodium	Na	66.5 ± 0.45^{b}	56.9 ± 0.38^{c}	75.2 ± 1.01^{a}
Potassium	K	3.86 ± 0.01^{b}	3.27 ± 0.07^{b}	5.58 ± 0.15^{a}
Iron	Fe	33.7 ± 2.28^{c}	40.6 ± 0.30^{b}	4400 ± 77.94^{a}
Manganese	Mn	4.53 ± 0.14^{c}	46.2 ± 0.83^{b}	86.6 ± 3.47^{a}
Zinc	Zn	10.5 ± 0.08^{b}	8.90 ± 0.04^{c}	227.7 ± 0.48^{a}
Chromium	Cr	0.87 ± 0.02^{b}	0.77 ± 0.02^{b}	15.2 ± 0.42^{a}
Lead	Pb	nd	3.07 ± 0.02^{b}	6.5 ± 0.07^{a}
Nikel	Ni	0.46 ± 0.03^{b}	nd	2.5 ± 0.07^{a}
Cadmium	Cd	nd	0.09 ± 0.01^{a}	nd

Data represent the mean ± SD (n = 3)

nd not detected

In the same row, values followed by different letters are significantly different at $p < 0.05$ (one-way ANOVA with Student–Newman–Keuls post hoc test)

These results could be related to higher levels of minerals in the tissues, which could be linked with the higher availability of diatoms and microalgae in this geographical area (SW Mediterranean Sea; Rohling et al. 2015), which are consumed by cucumbers.

Sea cucumbers are generally characterized by low amounts of fat and high protein contents (Bordbar et al. 2011). The protein levels of the sea cucumbers were high, although slightly lower than those reported in the literature for different sea cucumber species, as for example, *Actinopyga mauritiana* and *Bohadschia argus* (Wen et al. 2010). *Holothuria tubulosa* had a lower protein content than the same species caught in Turkey (Aydin et al. 2011) and in the Southern Adriatic Sea (Sicuro et al. 2012). *Holothuria polii* had a similar crude protein content that the same species caught in the later area (Sicuro et al. 2012). The sampling location had little effect on the protein levels of *H. mammata*. The species under study had a low crude fat content, but higher than the values reported for the same species caught in the Southern Adriatic Sea (Sicuro et al. 2012), and also than the reference amount reported for other commercial sea cucumber species (Bordbar et al. 2011). The fat contents was, however, lower than the values reported for other seafood products, namely skipjack tuna (*Katsuwonus pelamis*; Khodabux et al. 2007). All species had an appropriate protein/lipid ratio which is nutritionally relevant in view of the important role of proteins for the human body.

The amino acid profile of the studied species agrees with the reported in literature for the same species (Sicuro et al. 2012) and also for different sea cucumbers genera, such as *Cucumaria* and *Actinopygia* (Zhong et al. 2007; Wen et al. 2010). The richness of glycine is highly relevant, since the intake of food rich in that amino acid is associated with the reduction in total cholesterol levels in serum (Aljawad et al. 1991; Ikeda et al. 1993). The lysine/arginine ratios are usually linked to hypocholesterolemic effects (Bordbar et al. 2011) and were lower than those usually described in different sea cucumbers species (Bordbar et al. 2011).

All species showed a higher proportion of PUFA relatively to other FA, while SFA were present in the lowest relative amount. PUFA are significant components in tissue repair and wound healing, inflammation prevention, reduction of the incidence of coronary heart disease and cancer (Fredalina et al. 1999; Gill and Valivety 1997; Roynette et al. 2004; Harper and Jacobson 2005). AA was the most representative FA, as observed for most species of sea cucumbers (Bordbar et al. 2011). This FA is a major component of the cell membrane phospholipids and has a potential role in growth and in blood clotting process leading to wound healing (Gill and Valivety 1997; Roynette et al. 2004; Harper and Jacobson 2005; Harris et al. 2008). The levels of EPA and DHA together ranged from 141 μg/g in *H. polii* to 311 μg/g in *H. mammata*. Both PUFAs are associated with several health benefits, from the prevention of cardiovascular diseases to the improvement of brain and eye function (Harris et al. 2008). The levels of such PUFA in these species are, however, very low considering the daily intake requirements advised by the WHO (around 500 mg/d of EPA + DHA). Nonetheless, the FA profile of these species, with n-6 to n-3 FA ratios (1.8–2.3) considerably lower than 10 is highly beneficial to human health (Sánchez-Machado et al. 2004; Schmitz and Ecker 2008). Hence, the consumption of these species is not likely to have an adverse effect on cardiovascular diseases.

When compared to previously reported studies, some significant differences can be deduced from the profiles obtained herein, probably due to the distinct geographical origins of the species used. The FA profile of *H. polii*, *H. tubulosa* and *H. mammata* reported by Aydın et al. (2011) for example, shows a similar FA pattern to our results in which AA and EPA are among the most abundant FA. However, contrary to our results, in that study, oleic acid was not a dominant FA and α-linolenic (C18:3n-3) and eicosadienoic (C20:2) acids were also detected (Aydın et al. 2011). Our results were not comparable with those reported for *H. tubulosa* and *H. polii* caught from Italian Southern Adriatic Sea due to the different processing method used, which consisted in drying samples at 70 °C for 24 h prior to chemical analysis (Sicuro et al. 2012), since this drying method can modify the fatty acids contents of the samples (Bordbar et al. 2011).

Fatty acids have been increasingly used as biomarkers by ecologists to trace the origin and trajectory of organic matter in ecosystems (Parrish et al. 2000). According to Alfaro et al. (2006) long chain FA (with 24 or more carbons) are characteristic of terrestrial plants and thus their presence indicates organic material of terrestrial origin. On the other hand, the presence of EPA and DHA are indicative of diatoms and dinoflagellates, respectively, in the sea cucumbers food sources. As sea cucumbers are bottom feeders with low metabolism, it is probable that the FA composition of their body walls reflects the FA composition of their food source (Ramon et al. 2010; Aydin et al. 2011). In this context, species enriched in C24:0 FA would feed mainly on food sources of terrestrial origin, while those with prevalence of EPA and DHA would feed mainly on marine food sources. *Holothuria polii* was collected in a location more influenced by terrestrial inputs than the remaining species. In fact, *H. polii* was collected from Los Urrutias (SW Mar Menor coastal lagoon, SE Spain), a location close to several temporary and stable watercourses harbouring waters and sediments from agriculture and urban areas (González-Wangüemert and Vergara-Chen 2014). *Holothuria mammata* and *H. tubulosa* sampled from Los Cocedores (Murcia, Mediterranean Sea, SE Spain) presented a mixed signature, because of the terrestrial inputs in this marine region are limited and only produced during the rain season. In general, bacteria present a distinct lipid profile, usually enriched in SFA and MUFA, contrary to microalgae, which are showing high content in PUFA (Lebreton et al. 2011).

Sea cucumbers usually contain a good mineral profile, with high amounts of Ca, Mg, Fe and Zn (Bordbar et al. 2011). In this work, *H. polii* had the highest content of Ca, Mg and K, but the highest levels of Fe and Zn were observed in *H. tubulosa* (Table 4). The human consumption of ingredients rich in nutrients such as K, Ca, Mg, Fe, Zn, and Mn can have positive health implications, since they are an essential part of many important enzymes and they play different roles as catalysts and antioxidants. *Holothuria tubulosa* had a very high content of these essential metals, but also the highest levels of toxic metals such as Cr, Pb and Ni. *Holothuria tubulosa* had higher amounts of Cr, Fe, Pb and Zn than the same species from Southern Adriatic Sea (Sicuro et al. 2012); while *H. polii* had lower levels of Fe and Zn, and a higher content of Pb (Sicuro et al. 2012). Working with the same species caught in the Northern Mediterranean Sea, Tunca et al. (2016) observed that *H. mammata* significantly accumulated some metals, such as Cd and Mn in its body wall, while *H. polii* accumulated Cu, Cd, Cr and Mn at the lowest amount. The levels of Pb found in our study for *H. tubulosa* are lower or similar than those reported for the same species caught in the NW Mediterranean (Warnau et al. 2006), while the Fe and Zn contents were much higher. The accumulation of minerals in living organisms is influenced by several factors, as for example, the environment, concentration and type of the mineral in the bottom, exposure time, presence of other elements in the environment, condition, metabolic rate and/or sex of the organism, physico-chemical characteristics of the medium and feeding habits (Warnau et al. 2006). The differences in the levels of particular minerals observed in our study can be ascribed, for example, to possible differences in feeding habits between species (Aydın 2008; Aydın and Erkan 2015). Previous works showed that the decomposition of plant tissues, including seagrass, may be a source of metals to the marine environment, which are released through leaching and mineralization, and sink after adsorption to litter (Weis and Weis 2004). The high metal contents in *H. tubulosa* tissues could, therefore, be partially explained by its feeding behaviour associated to sediment from seagrass meadows which are showing a potential high metal concentration (Marín-Guirao et al. 2005; Warnau et al. 2006). In fact, it has been suggested that due to the peculiar ecological characteristics of *H. tubulosa*, this species could be used to complement the small set of bioindicators available so far for surveying metal contamination in the *Posidonia oceanica* ecosystem from Mediterranean Sea (Warnau et al. 2006). *Holothuria tubulosa* was caught from Los Cocedores, a location with vast areas of *P. oceanica* and *C. nodosa* meadows; some authors have described the bio-accumulation of heavy metals on these marine phanerogam species (de Leon et al. 1982; Marín-Guirao et al. 2005). *Holothuria tubulosa* is usually associated with those meadows with 24 h feeding activity (Costa et al. 2014), while *H. mammata* is found during the day under the rocks, and only gets out at night for feeding in areas not close to *P. oceanica* (González-Wangüemert et al. 2016). Moreover, *H. tubulosa* could be a more efficient metal bio-accumulator than *H. mammata* (Warnau et al. 2006). Although 50% lower than in *H. tubulosa*, the Pb concentration in *H. polii* was significantly lower than those quantified on the body wall tissues of sea cucumbers from the Southern Adriatic Sea (Storelli et al. 2001). Moreover, *H. polii* was the only species in which Cd was found in concentrations similar to those reported by other authors (Storelli et al. 2001). *Holothuria polii* was collected in the south basin of the Mar Menor lagoon, which is one of the largest coastal lagoons in Europe. Mar Menor has fresh-water inputs and its southern part is enclosed by mountains, which were subjected to intense mining activity in the last two centuries (Marín-Guirao et al. 2005). Although mining activity stopped in 1991, during flood seasons the metals of mine tailings, including Pb and Cd, are released to Mar Menor and accumulate in the sediments and in the seagrass *C. nodosa* (de Leon et al. 1982; Marín-Guirao et al. 2005) which could explain the high levels of those elements detected on *H. polii*.

Conclusions

All species had high moisture, ash and protein levels, low lipid content, and adequate amounts of most essential amino acids, coupled with low lysine to arginine ratios. The FA profiles of all species were characterized by high levels of omega-3 PUFA and considerable amounts of DHA and EPA. *Holothuria polii* had a FA profile enriched in C24:0 FA, while *H. mammata* and *H. tubulosa* had a prevalence of EPA and DHA. *Holothuria tubulosa* had a high content of toxic metals such as Cr, Pb and Ni. This work pinpoints the influence of some physical parameters of the environment, namely mineral loads and terrestrial inputs, on the nutritional properties of sea cucumbers.

Acknowledgements This work was supported by the CCMAR/Multi/04326/2013, SEABIOMED (PTDC/MAR/103957/2008), CUMFISH (PTDC/MAR/119363/2010) and CUMARSUR (PTDC/MAR-BIO/5948/2014) projects funded by the Foundation for Science and Technology (FCT) and the Portuguese National Budget. Luísa Custódio and Mercedes González-Wangüemert were supported by FCT Investigator Programme (IF/00049/2012 and IF/00998/2014, respectively). The authors acknowledge Dr. Zélia Velez (CCMAR) and Dr. Maribela Pestana (FCT) for their assistance with the analysis of amino acids and total proteins, respectively.

References

Alfaro AC, Thomas F, Sergent L, Duxbury M (2006) Identification of trophic interactions within an estuarine food web (northern New Zealand) using fatty acid biomarkers and stable isotopes. Estuar Coast Shelf Sci 70:271–286

Aljawad NS, Beth FE, Fryer HC (1991) Effects of casein, soy, and whey proteins and amino acid supplementation on cholesterol metabolism in rats. J Nutr Biochem 2:150–155

AOAC (1990) Official methods of analyses of association of analytical chemists, 15th edn. AOAC, Washington

Aydin M (2008) The commercial sea cucumber fishery in Turkey. SPC Bêche de Mer Bull 28:40–41

Aydın M, Erkan S (2015) Identification and some biological characteristics of commercial sea cucumber in the Turkey coast waters. Int J Fish Aquat Stud 3:260–265

Aydin M, Sevgili H, Tufan B, Emre Y, Köse S (2011) Proximate composition and fatty acid profile of three different fresh and dried commercial sea cucumbers from Turkey. Int J Food Sci Technol 46:500–508

Bordbar S, Anwar F, Saari N (2011) High-value components and bioactives from sea cucumbers for functional foods—a review. Mar Drugs 9:1761–1805

Borrero-Pérez G, Pérez-Ruzafa Á, Marcos C, González-Wangüemert M (2009) The taxonomic status of some Atlanto-Mediterranean species in the subgenus Holothuria (Echinodermata: *Holothuroidea*: *Holothuriidae*) based on molecular evidence. Zool J Linn Soc 157:51–69

Borrero-Pérez G, Gómez-Zurita J, González-Wangüemert M, Marcos C, Pérez-Ruzafa A (2010) Molecular systematic of the genus Holothuria in the Mediterranean and Northeastern Atlantic and a molecular clock for the diversification of the Holothuriidae (Echinodermata: *Holothuroidea*). Mol Phylogenet Evol 57:899–906

Bruckner AW (ed) (2006) Proceedings of the CITES workshop on the conservation of sea cucumbers in the families Holothuriidae and Stichopodidae. NOAA technical memorandum NMFS-OPR 34, Silver Spring

Çakli Ş, Cadun A, Kişla D, Dinçer T (2004) Determination of quality characteristics of *Holothuria tubulosa*, (Gmelin, 1788) in Turkish Sea (Aegean Region) depending on sun drying process step used in Turkey. J Aquat Food Prod Technol 13:69–78

Choo PS (2008) Population status, fisheries and trade of sea cucumbers in Asia. In: Toral-Granda V, Lovatelli A, Vasconcellos M (eds) Sea cucumbers: a global review of fisheries and trade. FAO fisheries and aquaculture technical paper no. 516. FAO, Rome, pp 81–118

Conand C, Byrne M (1993) A review of recent developments in the world sea cucumber fisheries. Mar Fish Rev 55:1–13

Conand C (2008) Population status, fisheries and trade of sea cucumbers in Africa and the Indian Ocean. In: Toral-Granda V, Lovatelli A, Vasconcellos M (eds) Sea cucumbers: a global review of fisheries and trade, vol 516. FAO fisheries and aquaculture technical paper no. 516. FAO, Rome, pp 143–193

Costa V, Mazzola A, Vizzini S (2014) *Holothuria tubulosa* Gmelin 1791 (Holothuroidea, Echinodermata) enhances organic matter recycling in *Posidonia oceanica* meadows. J Exp Mar Biol Ecol 461:226–232

Danovaro R, Dinet A, Duineveld G, Tselepides A (1999) Benthic response to particulate fluxes in different trophic environments: a comparison between the Gulf of Lions–Catalan Sea (western-Mediterranean) and the Cretan Sea (eastern-Mediterranean). Prog Oceanogr 44:287–312

de Leon AR, Guerrero J, Faraco F (1982) Evolution of the pollution of the coastal lagoon of Mar Menor. VI Journées Étud. Pollutions. CIESM, Cannes

Deng C, Yin X, Zhang L, Zhang X (2005) Development of microwave-assisted derivatization followed by gas chromatography/mass spectrometry for fast determination of amino acids in neonatal blood samples. Rapid Commun Mass Spectrom 19:2227–2234

FAO (2011) Fisheries statistics. http://www.fao.org/fishery/statistics/en. Accessed 20 Dec 2017

FAO (2014) The state of world fisheries and aquaculture, Rome

Fox J, Weisberg S (2011) An {R} companion to applied regression, 2nd edn. Sage, Thousand Oaks. http://socserv.socsci.mcmaster.ca/jfox/Books/Companion. Retrieved 01 May 2012

Fredalina B, Ridzwan B, Abidin A, Kaswandi M, Zaiton H, Zali I, Kittakoop P, Jais A (1999) Fatty acid compositions in local sea cucumber: *Stichopus chloronotus*, for wound healing. Gen Pharmacol 33:337–340

Gao Y, Li Z, Qi Y, Guo Z, Lin Y, Li W, Hu Y, Qiancheng Zhao Q (2016) Proximate composition and nutritional quality of deep sea growth sea cucumbers (*Stichopus japonicus*) from different origins. J Sci Food Agric 96:2378–2383

Gill I, Valivety R (1997) Polyunsaturated fatty acids, part 1: occurrence, biological activities and applications. Trends Biotechnol 15:401–409

González-Wangüemert M, Borrero-Pérez G (2012) A new record of *Holothuria arguinensis* colonizing the Mediterranean Sea. Mar Biodivers Rec 5:e105

González-Wangüemert M, Domínguez-Godino JA (2016) Sea cucumbers as new marine resource in Europe. Front Mar Sci. https://doi.org/10.3389/conf.fmars.2016.04.00112

González-Wangüemert M, Vergara-Chen C (2014) Environmental variables, habitat discontinuity and life history shaping the genetic structure of *Pomatoschistus marmoratus*. Helgol Mar Res 68:357–371

González-Wangüemert M, Aydin M, Chantal C (2014) Assessment of target sea cucumber populations from Aegean Sea (Turkey): first insights for a right management of their fisheries. Ocean Coast Manag 92:87–94

González-Wangüemert M, Valente S, Aydin M (2015) Effects of fishery protection on biometry and genetic structure of two target sea cucumber species from the Mediterranean Sea. Hydrobiologia 743:65–74

González-Wangüemert M, Valente S, Henriques F, Domínguez-Godino J, Serrão E (2016) Setting preliminary biometric baselines for new target sea cucumbers species of the NE Atlantic and Mediterranean fisheries. Fish Res 179:57–66

González-Wangüemert M, Domínguez-Godino J, Cánovas F (2018) The fast development of sea cucumber fisheries in the Mediterranean and NE Atlantic waters: from a new marine resource to its over-exploitation. Ocean Coast Man J 151:165–177

Gressler V, Yokoya NS, Fujii MT, Colepicolo P, Torres RP, Pinto E (2010) Lipid, fatty acid, protein, amino acid and ash contents in four Brazilian red algae species. Food Chem 120:585–590

Harper CR, Jacobson TA (2005) Usefulness of omega-3 fatty acids and the prevention of coronary heart disease. Am J Cardiol 96:1521–1529

Harris WS, Miller M, Tighe AP, Davidson MH, Schaefer EJ (2008) Omega-3 fatty acids and coronary heart disease risk: clinical and mechanistic perspectives. Atherosclerosis 197:12–24

Ikeda A, Imaizumi K, Sugano M (1993) Interaction of dietary protein and fat on plasma cholesterol and amino acid levels, fatty acid desaturation, and prostacyclin production in exogenous hypercholesterolemic rats. Biosci Biotechnol Biochem 57:1867–1872

INFOPESCA and FAO (2013) http://www.infopesca.org/. Accessed 20 Dec 2017

Jaquemet S, Conand C (1999) The *bêche-de-mer* trade in 1995–96 and an assessment of exchanges between the main world markets. SPC Bêche-de-Mer Inf Bull 12:11–14

Kasai T (2003) Lipid contents and fatty acid composition of total lipid of sea cucumber *Stichopus japonicus* and *konowata* (salted sea cucumber entrails). Food Sci Technol Res 9:45–48

Khodabux K, L'Omelette MSS, Jhaumeer-Laulloo S, Ramasami P, Rondeau P (2007) Chemical and near-infrared determination of moisture, fat and protein in tuna fishes. Food Chem 102:669–675

Kinch J, Purcell S, Uthicke S, Friedman K (2008) Population status, fisheries and trade of sea cucumbers in the Western Central Pacific. In: Toral-Granda V, Lovatelli A, Vasconcellos M (eds) Sea cucumbers: a global review of fisheries and trade. FAO fisheries and aquaculture technical paper no. 516. FAO, Rome, pp 7–55

Lah RA, Smith J, Savins D, Dowell A, Bucher D, Benkendorff K (2017) Investigation of nutritional properties of three species of marine turban snails for human consumption. Food Sci Nutr 5:14–30

Lebreton B, Richard P, Galois R, Radenac G, Pfléger C, Guillou G, Mornet F, Blanchard GF (2011) Trophic importance of diatoms in an intertidal *Zostera noltii* seagrass bed: evidence from stable isotope and fatty acid analyses. Estuar Coast Shelf Sci 92:140–153

Lepage G, Roy CC (1984) Improved recovery of fatty acid through direct transesterification without prior extraction or purification. J Lipid Res 25:1391–1396

Mahomoodally MF, Muthoorah LD (2014) An ethnopharmacological survey of natural remedies used by the Chinese community in Mauritius. Asian Pac J Trop Biomed 4:S387–S399

Marín-Guirao L, Cesar A, Marín A, Lloret J, Vita R (2005) Establishing the ecological quality status of soft-bottom mining-impacted coastal water bodies in the scope of the Water Framework Directive. Mar Pollut Bull 50:374–387

Mootoosamy A, Mahomoodally MF (2014) A quantitative ethnozoological assessment of traditionally used animal-based therapies in the tropical island of Mauritius. J Ethnopharmacol 154:847–857

Njinkoue JM, Gouado I, Tchoumbougnang F, Yanga Ngueguim JH, Ndinteh DT, Fomogne-Fodjo CY, Schweigert FJ (2016) Proximate composition, mineral content and fatty acid profile of two marine fishes from Cameroonian coast: *Pseudotolithus typus* (Bleeker, 1863) and *Pseudotolithus elongatus* (Bowdich, 1825). NFS J 4:27–31

Parrish C, Abrajano T, Budge S, Helleur R, Hudson E, Pulchan K, Ramos C (2000) Lipid and phenolic biomarkers in marine ecosystems: analysis and applications. Mar Chem 5:193–223

Pereira H, Barreira L, Custódio L, Alrokayan S, Mouffouk F, Varela J, Abu-Salah KM, Ben-Hamadou R (2013) Isolation and fatty acid profile of selected microalgae strains from the Red Sea for biofuel production. Energies 6:2773–2783

Purcell S (2010) Managing sea cucumber fisheries with an ecosystem approach. FAO fisheries and aquaculture technical paper, 520

Purcell SW (2014) Value, Market preferences and trade of beche-de-mer from Pacific Island sea cucumbers. PLoS ONE 9:e95075. https://doi.org/10.1371/journal.pone.0095075

Purcell SW, Samyn Y, Conand C (2012) Commercially important sea cucumbers of the world, vol 6. FAO species catalogue for fishery purposes. FAO, Rome

Purcell SW, Conand C, Uthicke S, Byrne M (2016) Ecological roles of exploited sea cucumbers. Oceanogr Mar Biol Annu Rev 54:367–386

Ramón M, Lleonart J, Massutí E (2010) Royal cucumber (*Stichopus regalis*) in the northwestern Mediterranean: distribution pattern and fishery. Fish Res 105:21–27

Roggatz C, González-Wangüemert M, Pereira H, Barreira L, Rodrigues MJ, Moreira da Silva M, Varela J, Custódio L (2016) First report of the nutritional profile and antioxidant potential of *Holothuria arguinensis*, a new resource for aquaculture in Europe. Nat Prod Res 9:1–7

Rohling EJ, Marino G, Grant KM (2015) Mediterranean climate and oceanography, and the periodic development of anoxic events (sapropels). Earth Sci Rev 143:62–97

Roynette CE, Calder PC, Dupertuis YM, Pichard C (2004) n-3 Polyunsaturated fatty acids and colon cancer prevention. Clin Nutr 23:139–151

Sánchez-Machado DI, López-Cervantes J, López-Hernández J, Paseiro-Losada P (2004) Fatty acids, total lipid, protein and ash contents of processed edible seaweeds. Food Chem 85:439–444

Sandrini-Neto L, Camargo MG (2011) GAD: an R package for ANOVA designs from general principles. Available on CRAN. http://ftp5.gwdg.de/pub/misc/cran/. Retrieved 01 May 2012

Schmitz G, Ecker J (2008) The opposing effects of n-3 and n-6 fatty acids. Prog Lipid Res 47:147–155

Sicuro B, Piccino M, Gai F, Abete MC, Danieli A, Dapra F, Miolettiu S, Vilella S (2012) Food quality and safety of Mediterranean sea cucumbers *Holothuria tubulosa* and *Holothuria polii* in Southern Adriatic Sea. Asian J Anim Vet Adv 7:851–859

Storelli MM, Storelli A, Marcotrigiano GO (2001) Heavy metals in the aquatic environment of the Southern Adriatic Sea, Italy: macroalgae, sediments and benthic species. Environ Int 26:505–509

Tunca E, Aydın M, Şahin Ü (2016) Interactions and accumulation differences of metal(loid)s in three seacucumber species collected from the Northern Mediterranean Sea. Environ Sci Pollut Res Int 23(20):21020–21031

Warnau M, Dutrieux S, Ledent G, Rodriguez y Baena AM, Dúbois P (2006) Heavy metals in the sea cucumber *Holothuria tubulosa* (Echinodermata) from the Mediterranean *Posidonia oceanica* ecosystem: body compartment, seasonal, geographical and batymetric variations. Environ Bioindic 1:268–285

Weis J, Weis P (2004) Metal uptake, transport and release by wetland plants: implications for phytoremediation and restoration. Environ Int 30:685–700

Wen J, Hu C, Fan S (2010) Chemical composition and nutritional quality of sea cucumbers. J Sci Food Agric 90:2469–2474

Yu HB, Gao QF, Dong SL, Wen B (2015) Changes in fatty acid profiles of sea cucumber *Apostichopus japonicus* (Selenka) induced by terrestrial plants in diets. Aquaculture 442:119–124

Zhong Y, Khan M, Shaidi F (2007) Compositional characteristics and antioxidant properties of fresh and processed sea cucumber (*Cucumaria frondosa*). J Agric Food Chem 55:1188–1192

4

Seasonal species diversity and abundance of phytoplankton from the southwestern Caspian Sea

N. Heydari · S. M. R. Fatemi · A. Mashinchian · R. Musavi Nadushan · B. Raeisi

Abstract In the present investigation, we aimed to study diversity and cell abundance of phytoplankton in the southwest of the Caspian Sea. This survey included 4 transects and 12 stations. A total of 48 samples were collected during spring 2012 and spring 2013. Finally, 72 species of phytoplankton were identified including phyla Diatoms (29 species), Chlorophyta (17 species), Pyrrophyta (16 species), Cyanophyta (6 species), and Euglenophyta (4 species). The total average of cell abundance was approximately 43.55 ± 5.10 × 104 cells/m^3. The number of recorded species in spring, summer, autumn, and winter was 46, 50, 41 and 39, respectively. Diatoms had the highest species diversity and cell abundance in winter related to *Thalassionema nitzschiodes* at a depth of 10 m. Seasonal changes in diversity of phytoplankton significantly differed, showing maximum in autumn with high diversity index (2.532) and minimum in spring (2.201) on Shannon Diversity Index. Finally, the quality of water is classified according to this index in the middle level.

Keywords Phytoplankton diversity · Abundance · Seasonality · The Caspian Sea

Introduction

The Caspian Sea is regarded as the largest close lake with brakishwater on Earth (Grigorovich et al. 2003). Ecological and morphological traits of the open seas are not seen in this ecosystem; this is regarded as a sea because of its vast area. Based on physical, geographical, and hydrological characteristics, the Caspian Sea is classified into three parts: Northern, Central and Southern Caspian. Rivers like Sefidrud, Gorganrud, Tajan, Haraz, Shiroud, Sardabrud, Talar, Babolrud, etc. are discharged into the Caspian Sea. One of the unique features of the Caspian Sea habitat is the diversity of its origins. The most abundant species are endemic (about 75% of the total number) and a few from Mediterranean (6%) and the Arctic (3%), and also the immigrants from freshwater. About 15 phytoplankton species have been introduced to the Caspian Sea, as the

N. Heydari · S. M. R. Fatemi (✉) · A. Mashinchian
Department of Natural Resources and Environment, Science and Research Branch, Islamic Azad University, Tehran, Iran
e-mail: reza_fatemi@hotmail.com

R. M. Nadushan
Department of Fishery, Marine Science and Technology Campus, Islamic Azad University, North Branch of Tehran, Zaaferaniyeh, Tehran, Iran

B. Raeisi
Department of Mathematics, Basic Sciences School, Shahed University, Tehran, Iran

result of the artificial connecting of the Caspian Sea with the Azov-Black Sea basin by the Volga-Don Canal since 1952 (Kosarev and Yablonskaya 1994).

Bandar-e Anzali is the largest port in northern Iran and is one of the most active ports of the Caspian Sea. Anzali's main importance is its harbor location, which plays an important role in the transportation of water of the Caspian Sea, fisheries and aquatic environments on the southern shores of the Caspian Sea (Dumont 1998). Phytoplankton is considered as an important and main element in aquatic environments, since it composes the first ring of the food chain in the ecosystem (Boni 1925). Phytoplanktons are useful bioindicators of changing environmental conditions and the advantage of using biological indicators over chemical or physical ones (Griffith et al. 2002). On the other hand, a number of introduced species have ecological negative effects on ecosystems. Therefore, the identification and qualitative studies of phytoplankton populations not only have an important role in a collection of historical data, but are also necessary to understand the polluted and under stress environment of the Caspian Sea (Nasrollahzadeh Saravi et al. 2014).

Phytoplankton is beneficial to evaluate the lakes because of suitable characters such as short life cycle, high cell abundance and species diversity, sensitivity to different kinds of organic and inorganic pollution, various zone of tolerance against contaminates and environmental stressors (Castro and Huber 1991).

Primary production in the Caspian Sea is dependent on nutrients content and type of compounds in sea water. Most nutrient compounds are entered through rivers and drainages, as annual discharge of silica, phosphorous and nitrogenous into the Caspian Sea by the rivers is 95, 90, and 80%, respectively. Vertical distribution of the nutrients in deep waters of the Central and Southern Caspian Sea is influenced by the intense of water mixing resulted from convection flow and increased with increased depth (Ghasemov and Bagherov 1983). In the Caspian Sea, 449 phytoplankton species consist of 163 diatoms, 139 Chlorophytes, 102 Cyanophytes, 39 Dinofagellates, 5 Euglenophytes and 1 Chrysophyta during 1962–1974. In addition, different kinds of algae have been identified as 414 species in the north, 225 species in the center and 71 species in the south.

Therefore, the number of phytoplankton is decreased from north to south due to a decline in the number of freshwater phytoplankton species (Proshkina-Lavrenko and Makarova 1968; Kosarev and Yablonskaya 1994). Of the Caspian phytoplankton compounds, diatoms are dominant in species number (163 species), but their population becomes less than green algae at the northern Caspian in summer for some years (Ghasemov and Bagherov 1983; Salmanov 1987).

Phytoplankton distribution in this ecosystem is not uniform during different seasons of the year (Salmanov 1987). At present, there are several problems in this sensitive and vulnerable ecosystem which are considered as major environmental concerns.

Population growth of cities in the Caspian basin, increased discharge of different wastewaters (industrial, agricultural and civil effluents), oil extraction development and the connection of this sea with the open seas through Volga-den canal have caused many problems (Bagheri et al. 2010) like as increased water temperature, comb jellyfish invasion, and eutrophication at the end of twentieth century (Izaguirre et al. 2001). Phytoplankton is the main group that plays a dynamic and active role in the marine food web, in exploiting primary products at higher levels of nutrients and elements in the global cycle of crops. By itself, it contains 1% of organic carbon and 45% of annual primary production on Earth. Phytoplankton include proteins, carbohydrates, lipids, vitamins and mineral salts, that they are directly and indirectly considered as food sources for all marine organisms. In addition, the oxygen produced by the phytoplankton is a result of photosynthesis, a vital component of the system on the Earth (Falkowski and Raven 1997).

Hence, the goal of this investigation was to identify species diversity and abundance of the main phytoplankton groups in south west of the Caspian. Finally, this present study shows data of phytoplankton in recent years.

Materials and methods

The present survey was conducted in Bandar-e Anzali in the west southern Caspian Sea (37°0.27′59″N 49°0.27′59″E) and has a coastline of 40 km. Sampling was performed at 12 stations located at four transects (Hasanrod, Taleb abad, Bandar coast, Sangachin) (a half line perpendicular to the coast) with 3 stations at

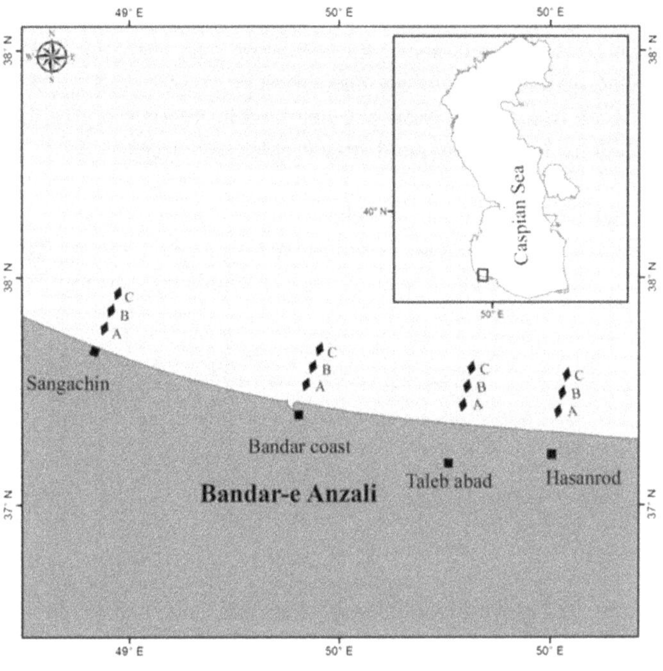

Fig. 1 Bandar-e Anzali Map from the southwestern Caspian Sea, showing the sampling sites

Table 1 Sampling transects and the location of stations in Bandar-e Anzali from the southwestern Caspian Sea

Transect	Station	Depth (m)	Longitude	Latitude
Hasan rod	A	Surface water (0)	49.40.55	37.27.21
	B	5	49.40.49	37.27.35
	C	10	49.40.36	37.27.51
Talebabad	A	Surface water (0)	49.32.53	37.28.04
	B	5	49.32.57	37.28.16
	C	10	49.33.01	37.28.39
Anzali coast	A	Surface water (0)	49.27.20	27.28.53
	B	5	49.27.25	37.29.10
	C	10	49.27.26	37.29.20
Sangachin	A	Surface water (0)	49.19.04	37.31.06
	B	5	49.19.19	37.31.22
	C	10	49.19.27	37.31.40

surface water, depths of 5 and 10 m (Fig. 1, Table 1). The sampling stations were fixed near shore. Sampling was performed seasonally, during spring 2012 to spring 2013.

In order to study phytoplankton, water samples were collected in Van Dorn water bottle (Ruttner) (Vollenweider 1974) and poured into plastic containers. Then, 7 ml of Lugol iodine solution was added to the containers (0.7 ml per each 100 ml of the sample) and after 1 h, 40 ml of 40% concentrated formaldehydes was added to the vessels. Therefore, the final solution contained 4% of formalin (APHA 2005). Three replicates were obtained at each station from surface water, with 5 and 10 m depth.

Phytoplankton samples were transported to the laboratory and stayed for 48 h. Then, the supernatant was siphoned in order to make the sample denser. The remaining was centrifuged for several times for 5 min at 3000 rpm to gain 30 ml volume of each sample (Centrifuge Model Number D-7200 tutilinge7).

To identify phytoplankton species, each sample was shaken well to obtain a homogenized sample. Thereafter, three replicates of each sample (1 ml) were pipetted from the 30 ml sample and poured into a Sedgewick-Rafter counting chamber under an invert microscope (Vollenweider 1974; Newell 1977). Species identification was performed using valid identification keys (Prescott 1962; Kasymov 2000; Tiffany and Britton 1971). Statistical analysis was done using the SPSS software (version 11.5). The data were tested to check for the normality. One-way analysis of variance (ANOVA) was used for the existence of significant differences ($P < 0.05$) between transects seasons and depths. Multivariate regression was used in this study. Phytoplankton phyla were considered as dependent variables and transect and depth were considered as independent variables.

Discriminate function analysis was applied to indicate how different abundances of phytoplankton phyla vary across transect. Shannon–Wiener Index was used to study species diversity. The values of this index are divided into five qualitative classes (Marques et al. 2009) (Table 4). Shanon Index is calculated by the following formula:

$$H' = -\Sigma\,(ni/N)\,\ln\,(ni/N),$$

N is the total population density, ni is the species density i.

Results

In this trail, 72 species from 59 genera, 45 families, 31 orders, and 5 main phytoplankton branches were identified (Table 2). Of these identified phytoplankton, 7 orders, 11 families, 16 genera, and 17 species were allocated to Green algae (Chlorophyta); 3 orders, 5 families, 10 genera and 6 species to blue-green algae (Cyanophyta); 5 orders, 6 families, 7 genera and 16 species to Dinoflagellata (Pyrrophyta); 1 order, 1 family, 3 genera, and 4 species to Euglenoids (Euglenophyta); 15 orders, 17 families, 23 genera and 29 species to diatoms (Bacillariophyta).

Among the phytoplankton species, 29 species (40%) were from diatoms (Bacillariophyta), 17 species (24%) from green algae (Chlorophyta), 16 species (22%) from Dinoflagellata (Pyrrophyta), 6 species (8%) from blue-green algae (Cyanophyta) and 4 species (6%) from Euglenophyta. The number of recorded species in spring, summer, autumn, and winter was 46, 50, 41 and 39, respectively.

Diatoms showed the most species diversity in winter (22 species) and autumn (21 species), while the lowest one was observed in spring (13 species). The highest species number of Pyrrophyta was detected in spring (15 species), Cyanobacteria in summer (6 species), Chlorophyta in spring (14 species) and Euglenophyta in summer (4 species) (Table 3).

Seasonally phytoplankton diversity in the southern Caspian Sea in Bandar-e Anzali was investigated based on Shannon-Diversity Index in the present study and exhibited the most diversity index in Autumn (2.532) and the least index in Spring (2.201) with an annual mean of this index (2.8) (Fig. 2). Eventually, Shannon index in Bandar-e Anzali is classified according to Table 4 in the moderate class.

During the first three seasons of the year, the highest abundance was discovered at 10 m depth while in winter it was observed at depths of 5 and 10 m. In different depths the most and the lowest abundance were recorded at depth of 10 and surface water, respectively. Transect and depth had significant effect on density of phytoplankton phyla ($P < 0.05$) (Table 5), but the interaction effect of depth by transect showed no significant effect ($P > 0.05$) (Fig. 3). The number of Cyanobacteria and Euglenophyta species is completely similar in different transects located at the southern Caspian Sea, whereas the species number of diatoms was variable at different transects. 12 species of diatoms, 9 species of Pyrrophyta and 13 species of other branches were revealed in Sangachin (Fig. 4). In fact, the main phytoplankton groups with most species number in Bandar-e Anzali were diatoms, green algae and Dinoflagellata. Moreover, Sangachin and Bandar coast transects had the highest and the least species diversity and cell abundance, respectively.

Discriminant Function Analysis was applied to better find any differences among transects based on phytoplankton cell abundance. As depicted in Fig. 5, DFA could indicate the discrimination between transect 1 (Hasan rod) and 3 (Bandar coast) as well as transect 2(Taleb abad) and 4(Sangachin). Transect 3 was well discriminated from other transects.

According to the results of DFA, among phytoplankton phyla isolated from the Caspian Sea, three phyla including Cyanophyta, Pyrrophyta and Euglenophyta had major role in discriminating the transects as their abundance among various transects showed significant differences ($P < 0.05$).

Table 2 Identified phytoplankton species from southern Caspian Sea in all seasons

Phylum	Order	Family	Genus	Species
Chlorophyta	Chlorellales	Oocystaceae	(1) *Gloeotaenium*	(1) *loitlesbergerianum*
	Chlorellales	Oocystaceae	(2) *Oocystis*	(2) *borgi*
				sp
				(3) *solitaria*
			(3) *Codotella*	sp
	Sphaeropleales	Hydrodictyaceae	(4) *Pediastrum*	(4) *tetras*
	Chlorellales	Botryococcaceae	(5) *Botryococcus*	sp
	Sphaeropleales	Scenedesmaceae	(6) *Scenedesmus*	(5) *quadricauda*
				(6) *abundans*
	Zygnematales	Zygnemataceae	(7) *Spirogira*	sp
	Sphaeropleales	Selenastraceae	(8) *Ankistrodesmus*	(7) *arcuatus*
				sp
	Desmidiales	Closteriaceae	(9) *Closterium*	(8) *moniliferum*
	Chlamydomonadales	Chlamydomonadacea	(10) *Chlamydomonas*	(9) *globosa*
				sp
				(10) *flusa*
				(11) *ovalis*
				(12) *monasovalis*
Chlorophyta	Chlorococcales	Chlorococcaceae	(11) *Selenstrum*	(13) *bibraianum*
	Ulotrichales	Ulotrichaceae	(12) *Binuclearia*	(14) *lauterbomii*
	Sphaeropleales	Scenedesmaceae	(13) *Coelastrum*	(15) *sphaericum*
	Chlorococcales	Chlorococcaceae	(14) *Scheroderia*	(16) *setigera*
	Chlorellales	Chlorellaceae	(15) *Chlorella*	sp
	Chlorococcales	Scenedesmaceae	(16) *Crucigenia*	sp
				(17) *quadrata*
Cyanophyta	Chlorococcales	Microcystaceae	(17) *Microcystis*	(18) -1*aeruginosa*
				sp
	Oscillatoriales	Oscillatoriaceae	(18) *Oscillatoria*	(19) -2*limosa*
				sp
	Nostocales	Oscillatoriaceae	(19) *Lyngbya*	(20) -3*limnetica*
				sp
	Chlorococcales	Chlorococcaceae	(20) *Aphanothece*	sp
	Nostocales	Nostocaceae	(21) *Nodularia*	sp
	Nostocales	Nostocaceae	(22) *Anabeanopsis*	(21) -4*arnoldii*
				sp
				(22) -5*rasiburaskii*
	Nostocales	Nostocaceae	(23) *Anabaena*	sp
				(23) -6*reniformis*
	Chlorococcales	Merismopediaceae	(24) *Merismopedia*	sp
	Chlorococcales	Microcystaceae	(25) *Gloeocapsa*	sp
	Nostocales	Oscillatoriaceae	(26) *Spirulina*	sp
Pyrrophyta	Peridiniales	Glenodiniaceae	(27) *Glenodinium*	(24) -1*cinctum*
				(25) -2*behningii*
				sp
				(26) -3*penardii*
	Peridiniales	Peridiniaceae	(28) *Peridinium*	(27) -4*subsalum*
				(28) -5*trochoideum*
				sp

Table 2 continued

Phylum	Order	Family	Genus	Species
				(29) -6*achromaticum*
	Prorocentrales	Prorocentraceae	(29) *Exuviaella*	(30) -7*cordata*
	Prorocentrales	Prorocentraceae	(30) *Prorocentrum*	(31) -8*paraximum*
				(32) -9*cordatum*
				(33) -10*scutelum*
				(34) -11*micans*
	Gonyaulacales	Gonyaulacaceae	(31) *Goniaulax*	(35) -12*digitale*
				sp
				(36) -13*polyedra*
				(37) -14*spinifera*
	Gymnodiniales	Gymnodiniacea	(32) *Gymnodinium*	(38) -15*variabile*
	Chattonellales	Vacuolariaceae	(33) *Gonyastomum*	(39) -16*depressum*
Euglenophyta	Euglenales	Euglenaceae	(34) *Euglena*	sp
				(40) -1*wangi*
				(41) -2*tuba*
	Euglenales	Euglenaceae	(35) *Phacus*	sp
	Euglenales	Euglenaceae	(36) *Trachelomonas*	sp
				(42) -3*planctoniea*
				(43) -4*slimilis*
Diatoms	Thalassiosirales	Thalassiosiraceae	(37) *Thalassiosira*	(44) -1*hustdti*
				sp
	Thalassiosirales	Skeletonemaceae	(38) *Rhizosolenia*	(45) -2*fragilissima*
				(46) -3*calcaravis*
	Thalassionematales	Thalassionemataceae	(39) *Thalassionema*	(47) -4*nitzschiodes*
	Bacillariales	Bacillariaceae	(40) *Nitzschia*	(48) -5*seriata*
				(49) -6*reversa*
				(50) -7*acicularis*
				sp
	Bacillariales	Bacillariaceae	(40) *Nitzschia*	(51) -8*tenirustris*
				(52) -9*sigmoidea*
	Thalassiosirales	Skeletonemaceae	(41) *Skletonema*	(53) -10*subsalum*
	Thalassiosirales	Skeletonemaceae		(54) -11*costatum*
Diatoms	Naviculales	Naviculaceae	(42) *Navicula*	sp
	Fragilariales	Fragilariaceae	(43) *Synedra*	(55) -12*ulna*
	Centrales	Chaetocerotaceae	(44) *Chaetoceros*	(56) -13*muelleri*
				sp
	Tribonematales	Tribonemataceae	(45) *Tribonema*	(57) -14*vulgar*
	Thalassiosirales	Stephanodiscaceae	(46) *Cyclotella*	(58) -15*menenghiniana*
	Cymbellales	Cymbellaceae	(47) *Cymbella*	(59) -16*tumidae*
				sp
	Coscinodiscales	Coscinodiscaceae	(48) *Coscinodiscus*	(60) -17*perforatus*
				(61) -18*gigas*
	Thalassiophysales	Catenulaceae	(49) *Amphora*	sp
	Naviculales	Diploneidaceae	(50) *Diplonois*	(62) -19*interupta*
				sp
	Achananthales	Cocconeidaceae	(51) *Cocconeis*	(63) -20*scutellum*
				sp
				(64) -21*placentula*

Table 2 continued

Phylum	Order	Family	Genus	Species
				(65) -22*husteli*
	Pennales	Fragilariaceae	(52) *Diatoma*	sp
	Melosirales	Melosiraceae	(53) *Melosira*	(66) -23*moniliformis*
				(67) -24*juergensii*
				(68) -25*granulata*
				(69) -26*varians*
	Fragilariales	Fragilariaceae	(54) *Fragilaria*	(70) -27*capucina*
	Chromulinales	Dinobryaceae	(55) *Dinobryon*	sp
	Naviculales	Pleurosigmataceae	(56) *Gyrosigma*	(71) -28*attenuatum*
	Naviculales	Pleurosigmataceae	(56) *Pleurosigma*	sp
				(72) -29*elongatum*
	Thalassiosirales	Stephonodiscacaeae	(58) *Stephonodiscus*	sp
	Surirellales	Surirellaceae	(59) *Surirella*	sp

The following equations obtained from DFA indicate the relationship between phytoplankton phyla and transects.

$$\begin{aligned}
\mathrm{DA1} &= -0.252\ \text{chlorophyta} + 0.495\ \text{cyanophyta} \\
&\quad + 0.846\ \text{pyrrophyta} + 0.295\ \text{euglenophyta} + 0.363\ \text{diatoms} \\
\mathrm{DA2} &= 0.755\ \text{chlorophyta} + 0.168\ \text{cyanophyta} \\
&\quad + 0.176\ \text{pyrrophyta} - 1.279\ \text{euglenophyta} + 0.280\ \text{diatoms} \\
\mathrm{DA3} &= 0.082\ \text{chlorophyta} + 0.998\ \text{cyanophyta} \\
&\quad - 0.461\ \text{pyrrophyta} - 0.392\ \text{euglenophyta} + 0.322\ \text{diatoms}
\end{aligned}$$

The following equations obtained from DFA indicate the relationship between phytoplankton phyla and depths.

$$\begin{aligned}
\mathrm{DA1} &= 1.092\ \text{chlorophyta} + 0.203\ \text{cyanophyta} \\
&\quad + 0.157\ \text{pyrrophyta} - 0.570\ \text{euglenophyta} + 0.459\ \text{diatoms} \\
\mathrm{DA2} &= -0.871\ \text{chlorophyta} + 0.817\ \text{cyanophyta} \\
&\quad + 0.502\ \text{pyrrophyta} + 0.604\ \text{euglenophyta} + 0.238\ \text{diatoms.}
\end{aligned}$$

The overall total average of cell abundance was approximately $43.55 \pm 5.10 \times 10^4$ cells/m^3. Actually, annual average dominance of cell abundance was discovered in diatoms with $14.99 \pm 16.97 \times 10^4$ cells/m^3 (Fig. 6). Because of high abundance of diatoms, the highest average abundance was detected in winter related to *Thalassionema nitzschiodes* at a depth of 10 m. The lowest and the highest cell abundance of diatoms were reported in summer and winter (5970 cells/m^3 and 359,370 cells/m^3, respectively) (Fig. 7) and their cell abundance showed significant differences at various seasons ($P < 0.05$).

Of Dinoflagellata, the most and least cell abundance was in spring (144,720 cells/m^3) and autumn (44,790 cells/m^3) with significant differences among seasons ($P < 0.05$) and the highest cell abundance was of *Exuviaella cordata* at a depth of 10 m (Fig. 7). By contrast, Cyanobacteria had their highest cell abundance in spring (25,792.5 cells/m^3) and their lowest was displayed in summer (187,890 cells/m^3). and highest cell abundance related to *Oscillatoria limosa* at a depth of 10 m. There were significant differences in cell abundance of this group among various seasons ($P < 0.05$) (Fig. 7). In spring and summer, the lowest (14,490 cells/m^3) and the most cell abundance (107,610 cells/m^3) of green algae were observed, respectively. And this index exhibited a significant difference between summer with other seasons ($P < 0.05$) whereas no significant

Table 3 Number of identified phytoplankton species from southern Caspian Sea in different seasons

	Seasons			
	Spring	Summer	Autumn	Winter
Chlorophyta				
1. *Ankistrodesmus arcuatus*	+	+	−	−
2. *Ankistrodesmus* sp.	+	+	−	+
3. *Binuclearia lauterbomii*	+	+	+	+
4. *Botryococcus* sp.	+	−	−	+
5. *Chlamydomonas globosa*	+	+	−	−
6. *Chlamydomonas flusa*	+	−	+	−
7. *Chlamydomonas monasovalis*	+	+	−	−
8. *Chlamydomonas ovalis*	+	+	−	+
9. *Chlamydomonas* sp.	+	+	+	+
10. *Chlorella* sp.	+	+	+	+
11. *Closterium moniliferum*	+	+	−	+
12. *Codotella* sp.	−	−	+	+
13. *Coelastrum sphaericum*	+	+	−	+
14. *Crucigenia quadrata*	+	+	−	−
15. *Crucigenia* sp.	+	+	−	−
16. *Gloeotaenium loitlesbergerianum*	+	+	−	−
17. *Oocystis borgi*	+	−	−	+
18. *Oocystis solitaria*	−	+	+	−
19. *Oocystis* sp.	+	+	−	+
20. *Pediastrum tetras*	−	+	−	+
21. *Scenedesmus abundans*	+	−	−	−
22. *Scenedesmus quadricauda*	+	+	+	−
23. *Scheroderia setigera*	−	−	+	−
24. *Selenstrum bibrajanum*	+	+	+	−
Total	20	18	9	12
Cyanophyta				
1. *Anabaena reniformis*	−	+	−	+
2. *Anabaena* sp.	+	+	+	+
3. Anabaenopsis arnoldii	−	+	+	−
4. Anabeanopsis raciborskii	−	+	−	+
5. Anabeanopsis sp.	+	+	+	+
6. Aphanothece sp.	+	+	+	−
7. Gloeocapsa sp.	+	+	+	−
8. Lyngbya limneticula	+	+	+	+
9. Lyngbya sp.	+	+	+	+
10. Merismopedia sp.	+	+	+	+
11. Microcystis aeruginosa	−	+	+	+
12. Microcystis sp.	+	+	−	−
13. Nodularia sp.	+	+	−	−
14. Oscillatoria limosum	+	+	+	+
15. Oscillatoria sp.	+	+	+	+
16. Spirulina sp.	+	−	+	+
Total	12	15	12	11
Pyrrophyta				
1. Exuviaella cordata				
2. Glenodinium behningii	+	+	+	+

Table 3 continued

	Seasons			
	Spring	Summer	Autumn	Winter
3. Glenodinium cinctum	+	–	–	–
4. Glenodinium penardii	+	–	–	–
5. Glenodinium sp.	+	–	+	+
6. Goniaulax digitale	+	–	–	+
7. Goniaulax polyedra	+	–	+	–
8. Goniaulax spinifera	–	–	+	–
9. Goniaulax sp.	+	–	+	+
10. Gonyastomum depressum	+	+	–	–
11. Gymnodinium variabile	+	–	+	+
12. Peridinium achromaticum	+	+	+	+
13. Peridinium subsalum	+	–	–	–
14. Peridinium trochoideum	+	–	–	–
15. Peridinium sp.	+	–	+	+
16. Prorocentrum cordatum	+	+	–	–
17. Prorocentrum micans	+	+	+	+
18. Prorocentrum paraximum	+	+	+	+
19. Prorocentrum scutellum	+	+	+	+
Total	17	7	11	10
Euglenophyta				
1. Euglena tuba	+	+	+	–
2. Euglena wangi	+	+	–	–
3. Euglena sp.	+	+	+	+
4. Phacus sp.	+	+	+	+
5. Trachelomonas planctoniea	–	+	–	–
6. Trachelomonas slimilis	–	+	–	+
7. Trachelomonas sp.	+	+	+	+
Total	5	7	4	4
Bacillariophyta				
1. Amphora sp.	–	–	–	+
2. Chaetoceros muelleri	+	–	+	+
3. Chaetoceros sp.	–	–	+	+
4. Cocconeis husteli	–	+	–	–
5. Cocconeis placentula	–	+	–	+
6. Cocconeis scutellum	–	+	–	–
7. Cocconeis sp.	–	–	+	+
8. Coscinodiscus gigas	–	+	+	+
9. Coscinodiscus perforatus	+	–	+	+
10. Cyclotella meneghiniana	+	+	+	+
11. Cymbella tumidae	–	+	+	+
12. Cymbella sp.	+	–	+	+
13. Diatoma sp.	+	–	–	–
14. Dinobryon sp.	+	–	–	–
15. Diplonois interupta	–	+	+	+
16. Diplonois sp.	–	+	+	–
17. Fragilaria capucina	–	+	–	–
18. Gyrosigma attenuatum	+	–	+	+
19. Melosira granulate	–	+	+	–

Table 3 continued

	Seasons			
	Spring	Summer	Autumn	Winter
20. Melosira juergensii	−	+	−	−
21. Melosira moniliformis	−	+	−	−
22. Melosira varians	−	−	+	+
23. Navicula sp.	+	+	+	+
24. Nitzschia acicularis	+	+	+	+
25. Nitzschia reversa	−	+	−	+
26. Nitzschia seriata	+	−	+	+
27. Nitzschia sigmoidea	+	+	+	+
28. Nitzschia tenirustris	−	+	−	+
29. Nitzschia sp.	+	−	+	+
30. Pleurosigma elongatum	−	+	+	−
31. Pleurosigma sp.	−	−	+	+
32. Rhizosolenia fragilissima	+	−	+	+
33. Rhizosolenia calcaravis	+	−	+	+
34. Skeletonema costatum	−	+	+	+
35. Skeletonema subsalum	−	+	+	+
36. Stephonodiscus sp.	−	−	+	−
37. Surirella sp.	−	−	+	+
38. Synedra ulna	+	−	+	+
39. Thalassionema nitzschiodes	+	−	+	+
40. Thalassiosira hustdti	+	−	+	+
41. Thalassiosira sp.	+	+	+	−
42. Tribonema vulgar	+	+	+	+
Total	19	22	31	30
Total of seasons	73	69	67	67

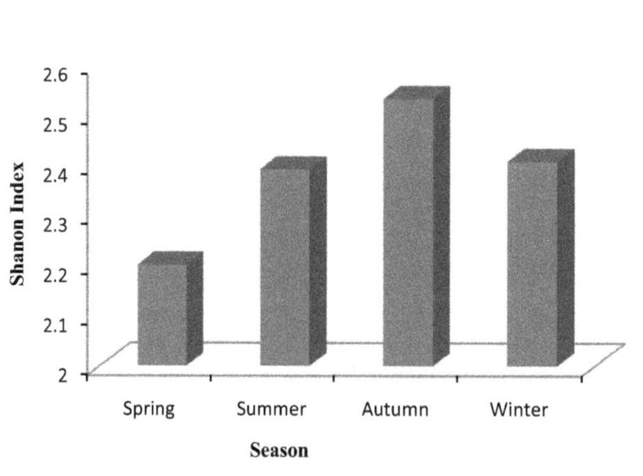

Fig. 2 Seasonality phytoplankton diversity in the southern Caspian Sea

differences were displayed among other seasons ($P > 0.05$) (Fig. 7). Euglenoids had less abundance than other phytoplankton groups and showed their lowest and the most abundance in autumn and summer (9570 cells/m^3 and 86,460 cells/m^3, respectively). Euglenoids abundance in summer revealed a significant difference

Table 4 Water quality classification, based on Shannon Diversity Index

Rating water quality	Shannon Index
Excellent	> 4
Good	3–4
Moderate	2–3
Poor	1–2
bad	0–1

Table 5 Multivariate tests (MANOVA) analyzing the effect of transect and depth on abundances of phytoplankton phyla in the southern Caspian Sea

Effect	Multivariate tests				
	Value	*F*	Hypothesis *df*	Error *df*	Sig.
Intercept					
Pillai's trace	0.819	115.691	5.000	128.000	0.000
Wilks' lambda	0.181	115.691	5.000	128.000	0.000
Hotelling's trace	4.519	115.691	5.000	128.000	0.000
Roy's largest root	4.519	115.691	5.000	128.000	0.000
Transect					
Pillai's trace	0.439	4.457	15.000	390.000	0.000
Wilks' lambda	0.575	5.231	15.000	353.753	0.000
Hotelling's trace	0.714	6.026	15.000	380.000	0.000
Roy's largest root	0.677	17.609	5.000	130.000	0.000
Depth					
Pillai's trace	0.687	13.508	10.000	258.000	0.000
Wilks' lambda	0.408	14.496	10.000	256.000	0.000
Hotelling's trace	1.220	15.496	10.000	254.000	0.000
Roy's largest root	0.983	25.369	5.000	129.000	0.000
Transect × depth					
Pillai's trace	0.202	0.925	30.000	660.000	0.584
Wilks' lambda	0.808	0.941	30.000	514.000	0.559
Hotelling's trace	0.227	0.957	30.000	632.000	0.533
Roy's largest root	0.167	3.668	6.000	132.000	0.002

with other seasons ($P < 0.05$), while there were no significant differences among other seasons ($P > 0.05$) (Fig. 7).

At the end, the highest cell abundance was observed in the spring, summer, autumn and winter seasons that belonged to species *Exuviaella cordata*, *Oscillatoria limosa*, *Thalassionema nitzschiodes* and *Thalassionema nitzschiodes* at a depth of 10 m, respectively.

The maximum and minimum average abundance of phytoplankton was observed in winter and spring respectively (Fig. 8).

Discussion

The main phytoplankton taxa in Caspian Sea are Bacillariophyta, Pyrrophyta and cyanophyta (Ghasemov and Bagherov 1983; Salmanov 1987; Ganjian et al. 2009; Ganjian 2007). In this study, Diatoms were the main group of phytoplankton, with the highest diversity of 29 species and 40% of the total taxa. The number of identified species was 46 in spring, 50 in summer, 41 in autumn, and 39 in winter. The highest and lowest

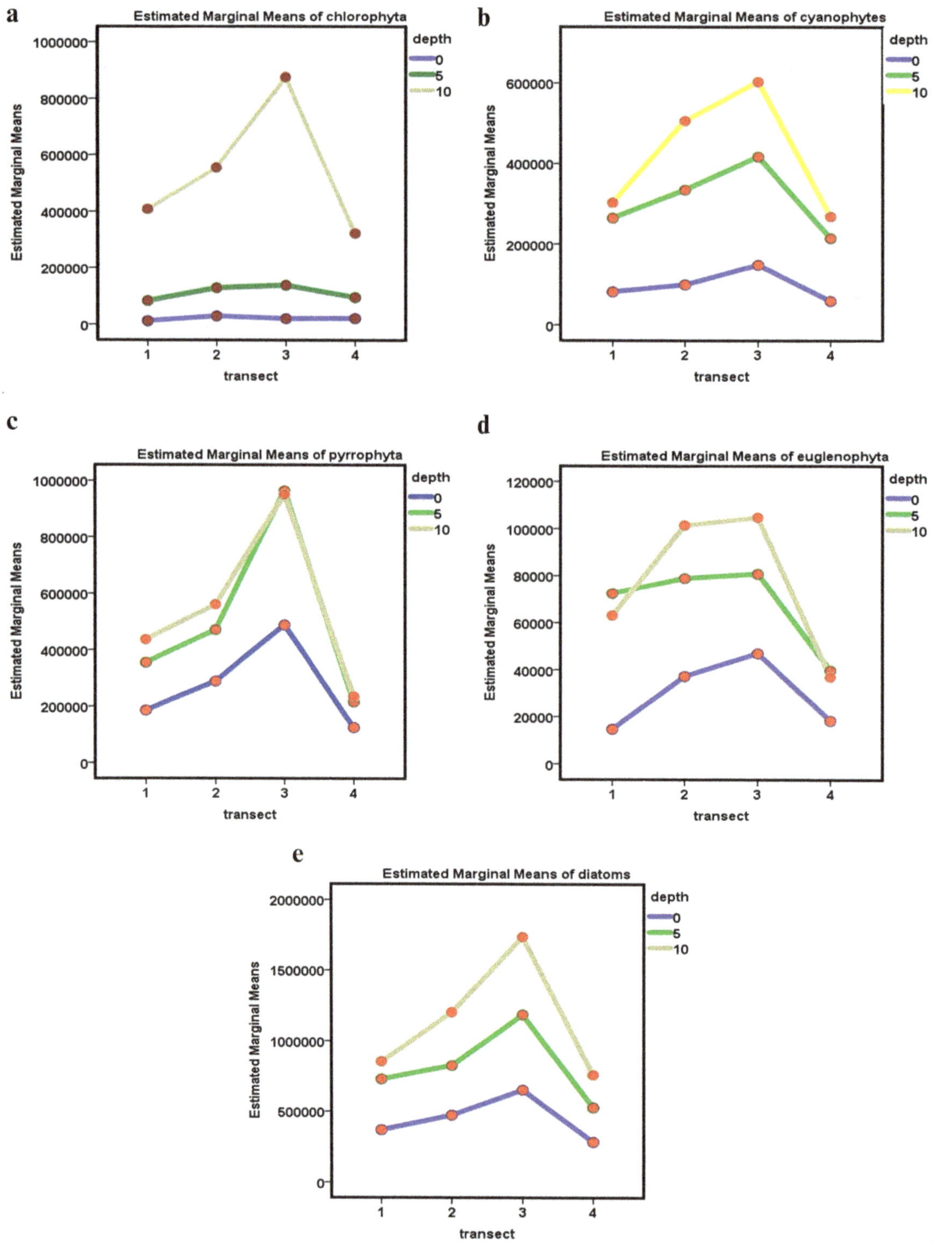

Fig. 3 Comparison between densities of phytoplankton phyla (**a** Chlorophyta, **b** Cyanophyta, **c** Pyrrophyta, **d** Euglenophyta, **e** Bacillariophyta) in different depths and transects

diversity was observed among phytoplankton groups in summer and autumn, respectively, which were similar to the finding in 2005–2006 recorded by Ganjian et al. 2010. They identified 101 species in the south of Caspian sea that the main phytoplankton group was Bacillariophyta (diatoms) with more than 71 species and 43% of the total taxa. The highest number of phytoplankton species was recorded in summer with 101 species in the southern part of the Caspian Sea. Temperature seems to be an important seasonal factor that controls the

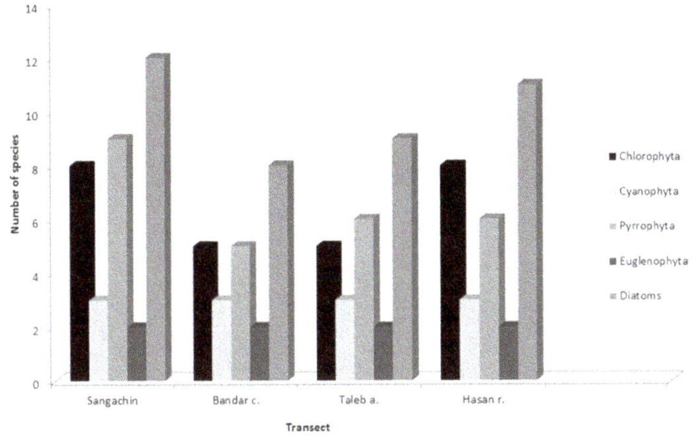

Fig. 4 Comparison between diversity of different phyla of phytoplankton in different transects

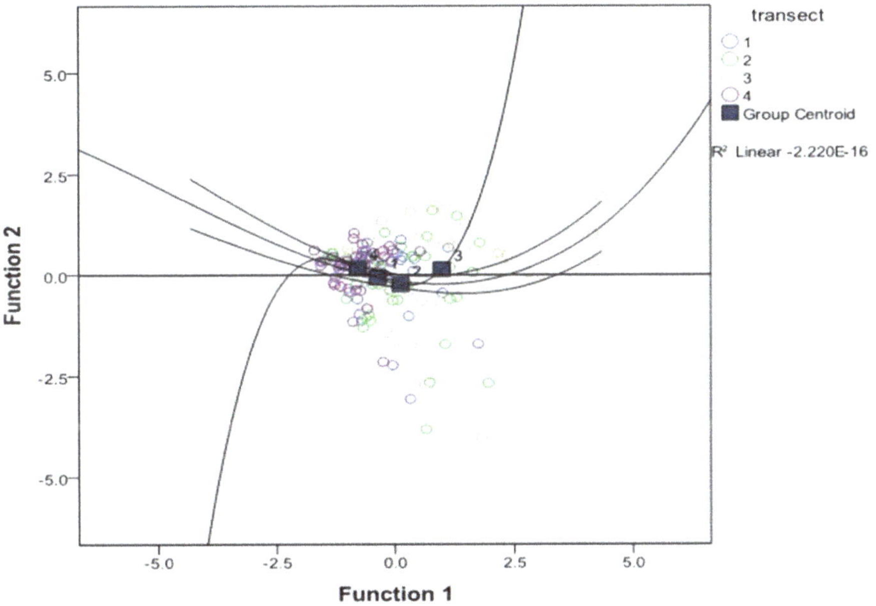

Fig. 5 Canonical Discriminant Function graph based on phytoplankton cell abundance at different transects

biological processes of phytoplankton communities (Vereshchaka and Anokhina 2014). Bat et al. (2011) reported Dinophyceae and Bacillariophyceae are the most dominant families for the number of species (respectively, 173 and 89 species) in the Black Sea. Also, Ozgur et al. (2010) published that a total of 129 taxa were recorded from neritic waters in Samsun Bay, including 76 Bacillariophyta, 2 Cyanobacteria, 1 Chlorophyta, 1 Euglenophyta and 48 other Phyla. Diatoms were dominant among phytoplankton groups except in November 2002 and June 2003.

Formerly diatoms were the most abundant in the Caspian Sea (Kosarev and Yablonskaya 1994). After the diatoms, chlorophytes and cyanophytes were the most abundant groups in the north of the Caspian Sea, while dinoflagellates dominated in the middle and southern parts of the Caspian Sea (Kosarev and Yablonskaya 1994; Ganjian 2007).

Pyrrophyta has the highest cell abundance in spring, which may be the environmental conditions during the spring lead to the growth. In the summer, cyanophyta had the highest density, which could be attributed to the increase in water temperature as seen in the Blanes Bay (Hense and Beckmann 2006), and the highest cell abundant of diatoms in the winter and autumn. Different researchers have reported that temperature was an important factor in fluctuations in the composition of phytoplankton and changes phytoplankton population

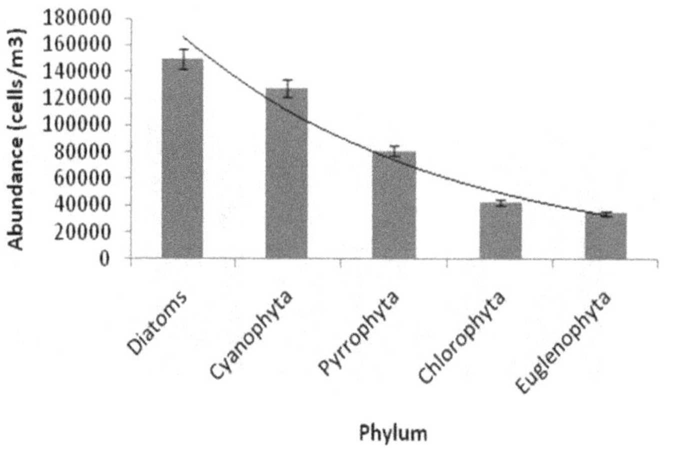

Fig. 6 The annual distribution changes in the average cell abundance (cells/m^3) of phytoplankton phyla

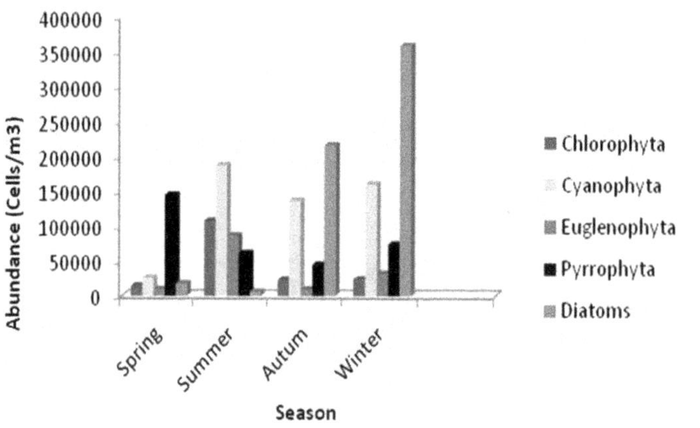

Fig. 7 Comparison between densities of different phyla of phytoplankton in different seasons

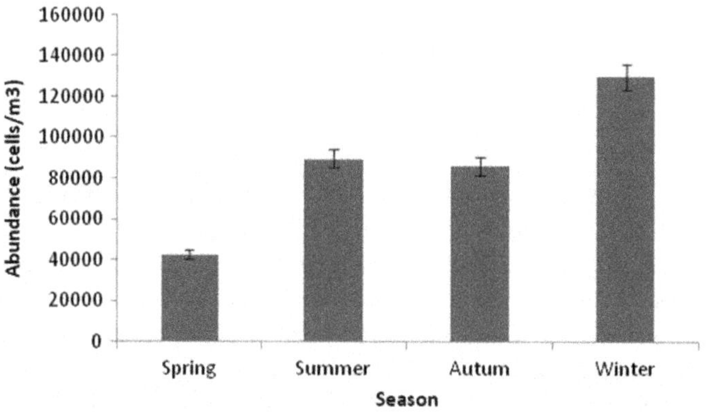

Fig. 8 The total average of cell abundance (cells/m^3) of phytoplankton in different seasons

with changing season and temperature differences, and diatoms were dominant species of the cold season in the Caspian Sea (Kideys et al. 2005; Resende et al. 2007; Bagheri et al. 2012). Compared to the previous findings, an increasing trend was observed in phytoplankton cell density. For example, the annual average density of phytoplankton was reported by Kosarev and Yablonskaya (1994) (1.4×10^4 cell L^{-1}), Kideys et al. (2005) (4.0×10^4 cell L^{-1}), and Bagheri et al. (2010) (2.10×10^5–3.90×10^5 cell L^{-1}) with the highest abundance of Pyrrophyta (51–56%) and Cyanophyta (22–29%). Dominant phytoplankton species were Prorocentrum cordatum of Pyrrophyta and *Oscillatoria* sp. of Cyanobacteria, and the present study (43.55×10^4 cell/m^3) showed that there is a large variation in the density of phytoplankton in the Caspian Sea.

Also in the same sea, Ozgur et al. 2010 showed that the annual average density of phytoplankton in the south Black Sea Samsun Bay was less than 0.01×10^6 in February 2003 and reached maximum value of 1.20×10^6 cells L^{-1} in July.

This survey showed that diatoms such as *Thalassionema nitzschiodes* and pyrrophyte *Exuviaella cordata* and cyanophyte *Oscillatoria limosa* were prevalent in Phytoplankton populations at the site.

Bagheri et al. (2014) showed that diatom communities were represented by a few dominant species such as *D. fragilissimus* and *T. nitzschioides* in the system.

In a previous research by Nasrollahzadeh Saravi et al. (2014) identified phytoplankton species, *Exuviaella cordata*, *Oscillatoria* sp., *Pseudonitzschia seriata*, *Thalassionema nitzschioides*, *Cerataulina pelagica* and *Chrysochromulina* sp. indicated the most abundance. Cyanophyta are living organisms from the natura ecosystem. Pointed out that warm (more than 20 °C), calm and stratified water in summer are desirable for the growth of *Oscillatoria* sp. (Chorus and Bartram 1999), *Oscillatoria* sp. is one of dwelling species of the Caspian Sea. However, this high abundance of the species in 2009 was introduced in the category of "harmful and potential bloom species "in the region.

In this study, dominant species had the highest density at a depth of 10 m. Also, the highest cell abundance of phytoplankton was observed in Bandar coasts transect. In this transect, due to the presence of breakwater, some species have come from the Anzali Wetland toward the sea, which increased the density of the phytoplankton compared with other transects.

In the Caspian Sea, various factors such as hydrological regimes, climate anomalies, increasing detergents, increasing the flow of fresh water through rivers, and increasing the level of nutrients due to the discharge of industrial and agricultural waste can affect the diversity and density of phytoplankton in coastal water of Caspian Sea (Kideys et al. 2008; Bilio and Niermann 2004).

Bagheri et al. (2014) theorized that hydro-biological changes in the Caspian Sea, Black Sea, and Baltic Sea during the 1990s and 2000s could be definitely correlated with climatical and hydrological characters.

Conclusion

There was no evidence of bloom in Bandar-e Anzali in southwestern of Caspian Sea. Although, species were observed that have a potential for blooming on this site, which should be controlled before reaching the bloom. As mentioned, increasing the detergents, fertilizers and nutrients resulting from industrial and agricultural sewages to the wetlands and coastal water of Caspian Sea and increase in deforestation are important elements for nutrient sources, and increase in these materials is an important factor in increasing the production of phytoplankton. The two top points were observed in the summer and winter. The top points were formed with *O. limosa* and *T. nitzschiodes* species from the phyla of Cyanophyta (29%) and diatom (34%), respectively. Opportunistic reproduction, toxin production and harmful characteristics of dominant species in the two seasons are the evidences of the under stress features of the Caspian Sea, which should be considered for environmental solutions.

Acknowledgements We are grateful to Laboratory of Department of the Environment, Marine Environment Division, Guilan, Iran. We wish to thank Mr. Armoodly in Fisheries Department of Iran. The Department of Marine Biology, Science and Research Branch, Islamic Azad University, Tehran, Iran is also gratefully acknowledged.

References

APHA (2005) Standard method for the examination of water and wastes water. American Publication Health Association, Washington, pp 10.1–10.47

Bagheri S, Mashhor M, Makaremi M, Mirzajani AR, Babaei H, Negarestan H, Wan-Maznah WO (2010) Distribution and composition of phytoplankton in the southwestern Caspian Sea during 2001–2002, a comparison with previous surveys. World J Fish Mar Sci 2(5):416–426

Bagheri S, Mashhor M, Turkoglu M, Makaremi M, Babaei A (2012) Temporal distribution of phytoplankton in the south-western Caspian Sea during 2009–2010: a comparison with previous surveys. J Mar Biol Assoc UK 92(6):1243–1255. https://doi.org/10.1017/S0025315412000094

Bagheri S, Turkoglu M, Abedini A (2014) Phytoplankton and nutrient variations in the Iranian waters of the Caspian Sea (Guilan region) during 2003–2004. Turk J Fish Aquat Sci 14:231–245. https://doi.org/10.4194/1303-2712-v14_1_25

Bat L, Sezgin M, Satilmis HH, Sahin F, Üstün F, Birinci-Özdemir Z, Gökkurt Baki O (2011) Biological diversity of the Turkish Black Sea Coast. Turk J Fish Aquat Sci 11:683–692. https://doi.org/10.4194/1303-2712-v11_4_04/trjfas

Bilio M, Niermann U (2004) Is the comb jelly to blame for it all? *Mnemiopsis leidyi* and the ecologial concerns about the Caspian Sea. Mar Ecol Prog Ser 269:173–183. https://doi.org/10.3354/meps269173

Boni AD (1925) Phytoplankton. Translated by Mohammad Reza Rahimi Bashar 2001 Publication of Green Code, p 218

Castro P, Huber ME (1991) Marine biology. McGraw-Hill Higher Education, New York, p 496

Chorus I, Bartram J (1999) Toxic cyanobacteria in water, A guide to their public health consequences, monitoring and management. E & FN Son, London, p 416P

Dumont HJ (1998) The Caspian Lake: history, biota, structure, and function. J Limnol Oceanogr 43:44–52

Falkowski PG, Raven JA (1997) Aquatic photosynthesis. Blackwell Science, Malden

Ganjian A (2007) Distribution, abundance and biomass of phytoplankton in the southern part of Caspian Sea (in Iranian waters), Thesis, Master of Science. School of Biological Sciences University Sciences Malaysia, p 144

Ganjian A, Wan Maznah WO, Khairun Y, Najafpour Sh, Najafpour GhD, Roohi A (2009) The assessment of biological indices for classification of water quality in southern part of Caspian Sea. World Appl Sci J 7(9):1097–1104

Ganjian A, Wan Maznah WO, Yahya K, Fazli H, Vahedi M, Roohi A, Farabi SMV (2010) Seasonal and regional distribution of phytoplankton in the southern part of the Caspian Sea. Iran J Fish Sci 9(3):382–401

Ghasemov, Bagherov (1983) The biology of Caspian Sea translated by Fathola pour fisheries research center of Gilan, p 184

Griffith MB, Hill BH, McCormick FH, Kauffman PR, Herlihy AT, Selle AR (2002) Comparative application of indices of biotic integrity based on periphyton, macroinvertebrata, and fish to southern rocky mountain streams. Ecol Ind 5(2):117–136

Grigorovich IA, Therriault TW, MacIsaac HJ (2003) History of aquatic invertebrate invasions in the Caspian Sea. Biol Invasions 5:103–115

Hense I, Beckmann A (2006) Towards a model of cyanobacteria life cycle-effects of growing and resting stages on bloom formation of N2-fixing species. In: Ecological modeling, I95. Elsevier, pp 205–218

Izaguirre I, Ofarrell I, Tell G (2001) Variation in phytoplankton composition and limnological features in a water-water ecotone of lower Parana Basin (Argentina). Freshw Biol 46:63–74

Kasymov A (2000) Methods of monitoring in Caspian Sea. In: QAPP-POLIQRAF, p 57

Kideys AE, Soydemir N, Eker E, Vladymyrov V, Soloviev D, Melin F (2005) Phytoplankton distribution in the Caspian Sea during March 2001. Hydrobiologia 543:159–168

Kideys AE, Roohi A, Eker-Develi E, Melin F, Beare D (2008) Increased chlorophyll levels in the southern Caspian Sea following an invasion of jellyfish. Res Lett Ecol 1–5

Kosarev AN, Yablonskaya EA (1994) The Caspian Sea. SPB Academic Publication, Moscow, p 259P

Marques JC, Salas F, Patricio J, Teixeiva H, Neto JM (2009) Ecological indicators for coastal and estuarine environmental assessment, a user's guide. WIT Press, University of Coimbra, Portugal, p 208. ISBN: 978-1845642099

Nasrollahzadeh Saravi H, Makhlough A, Eslami F, Leroy Suzanne AG (2014) Features of phytoplankton community in the southern Caspian Sea, a decade after the invasion of *Mnemiopsis leidyi*. Iran J Fish Sci 13(1):145–167

Newell GE (1977) Marine plankton. Hutchinson Co., London, p 320

Ozgur B, Arif G, Tufan K (2010) Temporal variations of phytoplankton in relation to eutrophication in Samsun Bay, Southern Black Sea. Turk J Fish Aquat Sci 10:363–372. https://doi.org/10.4194/trjfas

Prescott GW (1962) Algae of the western great lakes area. Brown company Publishing, Iowa, pp 1–933

Proshkina-Lavrenko AI, Makarova IV (1968) Plankton algae of the Caspian Sea. Leningrad, Nauka, p 287 **(in Russian)**

Resende P, Azeiteiro UM, Goncalves F, Pereira MJ (2007) Distribution and ecological preferences of diatoms and dinoflagellata in the west Iberian coasta zone (North Portugal). Acta Oecol 32:224–235. https://doi.org/10.1016/j.actao.2007.05.004

Salmanov MA (1987) The role of Microflora and phytoplankton in production process. Translated by Abolghasem shariati the science and industrial fishery centers in Mirza Kochagkhan, Rasht, p 349

Tiffany LH, Britton LE (1971) The algae of Illinois. Hanfer Publishing Company, New York, pp 1–407

Vereshchaka AL, Anokhina LL (2014) Composition and dynamics of the Black Sea Benthopelagic plankton and its contribution to the near-shore plankton communities. PLoS One 9(6):e99595. https://doi.org/10.1371/journal.pone.0099595

Vollenweider AR (1974) A manual on methods for measuring primary production in aquatic environmental. Blackwell Scientific publication Oxford, London, p 423

5

Horizontal and vertical variations in survival rates of juvenile bivalves, *Anadara kagoshimensis* (Tokunaga, 1906), *Ruditapes philippinarum* (Adams & Reeve, 1850) and *Atrina* sp., experimentally caged in the inner area of the Sea of Ariake, Japan

Takeo Kurihara · Kazumasa Hashimoto · Shoji Nakano · Yukihiko Matsuyama · Atsushi Ito · Masaei Kanematsu · Kazumaro Okamura

Abstract In the inner area of the Sea of Ariake, rearing experiments were conducted to examine how the survival rates of juveniles of three bivalve species (*Anadara kagoshimensis*, *Ruditapes philippinarum*, *Atrina* sp.) vary among sea areas and heights above the seabed. In three runs of the experiment, replicate samples of the three bivalves were kept in trays set at various heights above the seabed (range 0–40 cm) at one to six stations for 15–16 days. Results were as follows. (1) Survival rates varied among heights and stations more markedly for *Atrina* sp. than for *A. kagoshimensis* and *R. philippinarum*; (2) *Atrina* sp. always showed higher survival rates at greater heights above the seabed, whereas the other species did not always show this pattern. (3) Survival rates of *Atrina* sp. showed a different station-to-station variation pattern between experimental runs. (4) In statistical model selection, the spatial variations in the survival rate were not associated for either *A. kagoshimensis* or *R. philippinarum* with any of the environmental factors recorded (density of suspended solids, density of chlorophyll *a*, water temperature). In contrast, for *Atrina* sp. they were strongly and negatively associated with the spatial variations in density of suspended solids.

Keywords Survival rate · Spatial variations · Bivalves · Suspended solids

Introduction

Abundances of wild populations of many bivalves have decreased worldwide in recent years (Gosling 2015). This is also true for commercially important species in the Sea of Ariake, one of the largest basins in Japan. In this basin, the abundances and fishery catches of the ark shell *Anadara kagoshimensis* (Tokunaga, 1906), the clam *Ruditapes philippinarum* (Adams & Reeve, 1850), and the pen shell *Atrina* sp. decreased from 1970–1980s to 2000–2010s (Kumamoto Prefectural Fisheries Research Center 2006; Yoshino et al. 2007;

T. Kurihara (✉) · K. Hashimoto · S. Nakano · Y. Matsuyama · K. Okamura
Seikai National Fisheries Research Institute, Japan Fisheries Research and Education Agency, 1551-8 Tairamachi, Nagasaki 851-2213, Japan
e-mail: takeo@affrc.go.jp

A. Ito · M. Kanematsu
National Research Institute of Fisheries and Environment of Inland Sea, Japan Fisheries Research and Education Agency, 1760 Momoshima, Onomichi, Hiroshima 722-0061, Japan

Koga and Aramaki 2013; Ito 2017; Ministry of the Environment 2017; see "Materials and methods" for the taxonomic problems with *Atrina* sp.).

To restore these bivalve resources, it is necessary to elucidate how and why their densities vary spatially and temporally. The abundance of marine animals is in general strongly affected by the survival rate during their juvenile stage (e.g., Begon et al. 2006; Broitman et al. 2008; Haddon 2011). Indeed, this relationship is presented for *A. kagoshimensis* (Masaki and Onohara 2009), *R. philippinarum* (Nasu et al. 2008), and *Atrina* sp. (Ito 2017). Therefore, it is important to know how and why the juveniles of *A. kagoshimensis*, *R. philippinarum* and *Atrina* sp. change the survival rates.

The survival rates of juveniles of these bivalves in the Sea of Ariake are likely to vary horizontally (i.e., among sea areas) and vertically (i.e., between the seabed and water column). That is, juvenile *A. kagoshimensis* are likely to survive somewhat better near the mouth of the bay than in the inner area of the bay (Nakamuta et al. 2013) and in the water column compared to the seabed since the juveniles on spat collectors in the water column seem to survive better than the juveniles dropped from the collectors onto the seabed (Masaki and Onohara 2009). Juvenile *R. philippinarum* are likely to survive better in areas with coarser sediments (Kanzaki et al. 2017) and at the higher position on spat collectors (Nagamoto et al. 2017). Juvenile *Atrina* sp. were found to survive better in the east peripheral area (Suzuki et al. 2009) and in the water column than on the seabed (Matoba et al. 2016).

The patterns and causes for such variations in survival rates in the Sea of Ariake have, however, been studied using different methods among species and among study sites. For example, spat collectors were used in some surveys for *A. kagoshimensis* (Masaki and Onohara 2009) and *R. philippinarum* (Nagamoto et al. 2017) but not in any surveys reported for *Atrina* sp. (Matoba et al. 2016 and references therein); and *Atrina* sp. were kept in pocket nets in the water column and under fishery nets on the seabed (Matoba et al. 2016). As a result, it is difficult to determine whether the differences in survival rates are due to differences in methods or species-specific and/or site-specific characteristics. How and why the survival rates of each bivalve species horizontally and vertically vary in the Sea of Ariake should be studied to find locations suitable for culture and transplantation of the species, which have been sought by trial and error (Matoba et al. 2016; Nagamoto et al. 2017), and to elucidate what environmental factors are specific to such locations.

We conducted an in situ rearing experiment in the inner area of the Sea of Ariake, using a unified method to examine three problems: (1) how the survival rates of juvenile *A. kagoshimensis*, *R. philippinarum* and *Atrina* sp. vary among stations; (2) how the survival rates vary between the water column and the seabed; and (3) how the horizontal and vertical variations in the survival rates are correlated with turbidity, chlorophyll *a* concentration, and water temperature. These environmental factors have been found to correlate with the survival rates of many bivalve species (Gosling 2015). High turbidity is suggested to lead to the mortality of *Atrina* sp. in the Sea of Ariake (Matsuyama 2012; Ministry of the Environment 2017).

Materials and methods

The study site, the inner area of the Sea of Ariake, is characterized by the great tidal range (about 3.2 m on average; Matsukawa 2005). This causes strong tidal currents, which leads to cyclic suspension and deposition of sediments (Mdφ = 2–8). The water temperature and salinity at 5 m depth are 9 °C and 30 °C, respectively, in winter and 26 °C and 27 °C, respectively, in summer.

Three runs of the field experiment of the design in Table 1 were conducted in 2015. The ark shells *Anadara kagoshimensis* used in each run were wild specimens with shell length 14.5–27.8 mm collected in the Sea of Ariake. The clam *Ruditapes philippinarum* and pen shell *Atrina* sp. used were artificially produced seeds with shell length 12.6–14.7 mm and 8.8–37.9 mm, respectively. These specimens are regarded as juveniles, not sexually matured (Online Resource 1). How the differences in the source and body size of materials potentially affected species-specific results are examined in "Discussion".

The pen shells used were identified as "*Atrina pectinata* Lineage 2" (sensu Liu et al. 2011) by the analyses of the mtDNA COI region (Hashimoto et al. 2018). Their orthodox species name is, however, unclear owing to the taxonomic confusion of *Atrina* (Lemer et al. 2014). Therefore, we provisionally call the pen shells "*Atrina* sp".

Table 1 Design of each run of the rearing experiment

		Run 1	Run 2	Run 3
The number of specimens used	*Anadara kagoshimensis*	5	3	10
	Ruditapes philippinarum	5	3	10
	Atrina sp.	3	3	20
Shell length (mm)	*Anadara kagoshimensis*	27.8 ± 2.5	27.5 ± 2.4	14.5 ± 1.5
	Ruditapes philippinarum	14.7 ± 0.7	14.5 ± 0.7	12.6 ± 1.1
	Atrina sp.	37.9 ± 4.0	25.3 ± 2.8	8.8 ± 0.7
Preparation of bivalve specimens	*Anadara kagoshimensis*	Collected at Kashima (33.07°N, 130.19°E), the inner area of the Sea of Ariake in November 2014	Collected at Kashima (33.07°N, 130.19°E), the inner area of the Sea of Ariake in November 2014	Collected at Ooura (32.94°N, 130.23°E), the inner area of the Sea of Ariake in September 2015
	Ruditapes philippinarum	Artificially hatched from adult clams in November 2014	Artificially hatched from adult clams in November 2014	Artificially hatched from adult clams in May 2015
	Atrina sp.	Artificially hatched from adult pen shells in July 2014	Artificially hatched from adult pen shells in July 2014	Artificially hatched from adult pen shells in July 2015
Experimental stations		B, F	B	A, B, C, D, E, F
Heights of trays from the seabed (cm)		0, 40	0, 20, 40	0, 20
The number of trays for each station and height		1	2	2
Start of the experiment		27 May	5 September	7 October
Experimental period (days)		15	16	16

Experimental design differed among the three runs, depending on the availability of bivalves. Each run was conducted at some or all of the stations A to F in Fig. 1. The design of Run 1 was: two stations (B and F; Fig. 1) $\times$ two heights from the seabed (0 and 40 cm) $\times$ one rearing tray. Four trays were used in total. The trays were made of polypropylene with 35 cm width, 24 cm length, and 7 cm height. In each tray, anthracite (i.e., hard, compact coal) with a diameter of about 2 mm was filled to 5 cm height. Height of a tray was defined as the distance from the upper face of anthracite in the tray to the seabed. The heights of trays from the seabed was set < 1 m in each run, which is based on previous reports for *A. kagoshimensis* and *Atrina* sp. in the Sea of Ariake that survival rates sharply varied between specimens reared on the seabed and reared in the water column < 1 m distant from the seabed (Yurimoto et al. 2007; Matoba et al. 2016). Five *A. kagoshimensis*, five *R. philippinarum* and three *Atrina* sp. were embedded into the anthracite of each tray on 24 May. The trays were covered with a 2 mm $\times$ 2 mm mesh and were maintained in a bath with flowing seawater. The trays were then set at each height and station by divers on 27 May, using steel frames and stakes. They were recovered 15 days later to record the survival/death of each specimen.

The design of Run 2 was: one station (B in Fig. 1) $\times$ three heights from the seabed (0, 20 and 40 cm) $\times$ two trays. Six trays were used in total. Three specimens of each species were embedded into

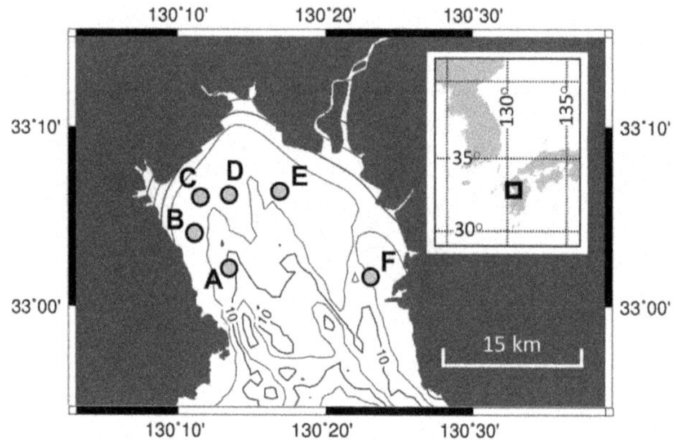

Fig. 1 Location of the study site, the inner part of the Sea of Ariake (rectangular area in a smaller map) and the experimental stations (gray circles in a larger map). The correspondence of these stations between the present and previous studies (e.g., Ministry of Agriculture, Forestry and Fisheries 2011) is as follows: "A" = "P6", "B" = "T14", "C" = "15", "D" = "O-5a", "E" = "SU-H", "F" = "T5"

anthracite in each tray on 5 September using the same method as in Run 1. The trays were set at the station on 7 September and were recovered 16 days later.

The design of Run 3 was: six stations (A to F in Fig. 1) × two heights from the seabed (0 and 20 cm) × two trays. In total, 24 trays were used. Ten *A. kagoshimensis*, 10 *R. philippinarum* and 20 *Atrina* sp. were embedded into anthracite in each tray on 5 October using the same method as in Run 1. The trays were set at the six stations on 7 October and were recovered 16 days later. Environmental factors were measured for only Run 3. Turbidity (FTU), chlorophyll fluorescence intensity, and temperature (°C) were measured for each station and height with an interval of 10 min throughout the run, using a data logger CLW-U (JFE Advantech Co., Ltd, http://ocean.jfe-advantech.co.jp/english/index.html). The records of turbidity and chlorophyll fluorescence intensity were converted to density of suspended solids (SS; mg/l) and chlorophyll *a* density (Chl-*a*; μg/l) with transformation formulae (Online Resource 2) that had been constructed before the experiments.

The data of each run were used to examine the horizontal and vertical variations in bivalve survival rates (except for Run 2, which lacked replications of stations and thus cannot be used to examine the horizontal variation in survival rates). The data of Run 3 were also used to examine how SS, Chl-*a* and water temperature spatially vary. Median of each environmental factor during Run 3 was compared between heights and among stations with a Friedman test. In addition, the data of Run 3 were used to analyze survival rates of each species with a generalized linear model with mixed effects (GLMM) based on a logit function and binomial distribution. The objective variable was survival rate for each tray. The explanatory variables were median SS, median Chl-*a*, median water temperature, height above the seabed, and station for each tray. The random factor was associated with the difference in uninvestigated factors (e.g., microenvironment) between two trays within each combination of height and station. The best combination of explanatory factors was selected with the Bayesian information criterion (BIC). The calculation was performed with R (R Core Team 2017) and the package, glmmML, ver. 1.0.2 (Broström 2017). Although the shell length of each species somewhat varied among experimental trays owing to random allocation, its effect on the tray-to-tray variation in survival rate was negligible (Pearson's correlation coefficient $r = -0.14$ and $p = 0.52$ for *A. kagoshimensis*; − 0.23 and 0.27 for *R. philippinarum*; 0.013 and 0.95 for *Atrina* sp).

Results

The horizontal variations in survival rate in Runs 1 and 3 (Figs. 2 and 3) were small for *Anadara kagoshimensis*, as its median survival rates were always 1.0 except for 0 cm height at stations D (0.90) and F (0.70) in Run 3. The variations were also small for *Ruditapes philippinarum*, as its survival rates were always 1.0 in Run 1 and were generally high (0.69–1.0) in Run 3 except for 0 cm height at station F (0.58). Compared to these two species, the horizontal variations in survival rate were greater for *Atrina* sp. Its survival rates were

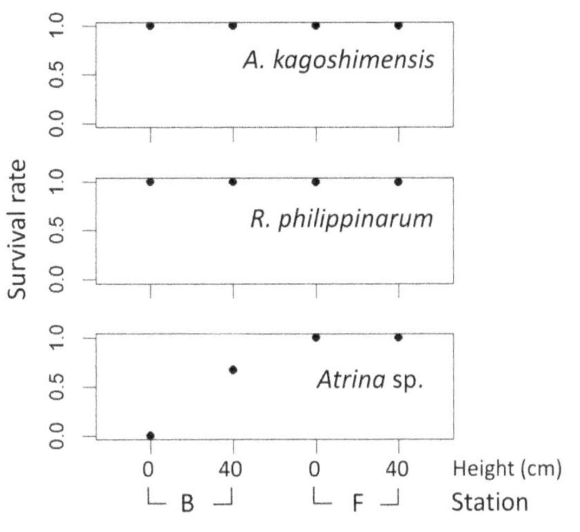

Fig. 2 Survival rates of *Anadara kagoshimensis*, *Ruditapes philippinarum*, and *Atrina* sp. for each combination of station and height in Run 1 of the rearing experiments

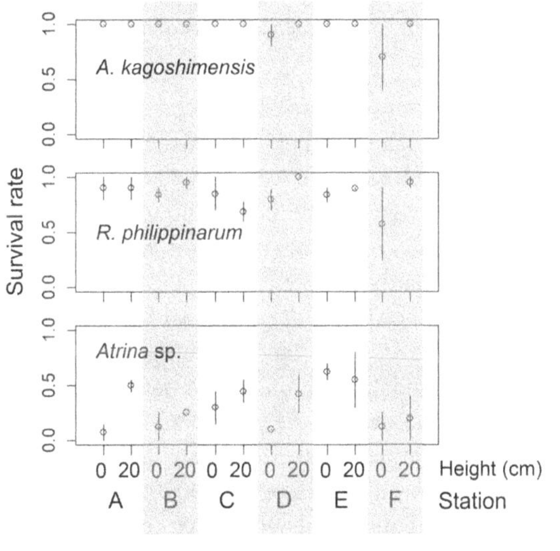

Fig. 3 Survival rates of *Anadara kagoshimensis*, *Ruditapes philippinarum*, and *Atrina* sp. for each combination of station and height in Run 3 of the rearing experiments. Circles denote median survival rates, and bars the range

much higher for station F (1.0) than station B (0.0–0.67) in Run 1, and for station E (0.55–0.63) than stations F and B (0.13–0.25) in Run 3.

Survival rates of *A. kagoshimensis* showed unclear vertical changes. The values were slightly higher for 20 cm height than 0 cm height in Run 2 (Fig. 4) and in Run 3 (stations D and F; Fig. 3), and the rates were the same between the heights in Run 1 (1.0; Fig. 2). Survival rates of *R. philippinarum* showed variable vertical changes. The rates were the same between the heights in Run 1 (1.0); higher for 0 cm height (1.0) than 20 cm height (0.83) in Run 2; and showed different vertical-variation patterns in Run 3, depending on the station (20 cm height > 0 cm height at stations B, D, E and F; 20 cm < 0 cm at station C; 20 cm = 0 cm at station A). In contrast to *A. kagoshimensis* and *R. philippinarum*, *Atrina* sp. showed lower survival rates nearer and on the seabed in general: at station B in Runs 1 and 2; and at most stations (stations A, B, C, D and F) in Run 3.

Figure 5 shows the horizontal and vertical variations in medians of SS, Chl-*a*, and water temperature in Run 3. Median SS was lower at stations A (25.2 and 16.5 mg/L for 0 and 20 cm heights, respectively) and E (20.5 and 19.3 mg/L) than the other stations (> 45.8 and > 31.2 mg/L) for each height; and 0 cm height (30.7–261.5 mg/L) than 20 cm height (32.5–125.6) for most stations. Median Chl-*a* was also lower at station A (5.4 and 5.0 μg/L for 0 and 20 cm heights, respectively) and E (2.4 and 2.2 μg/L) than the other stations

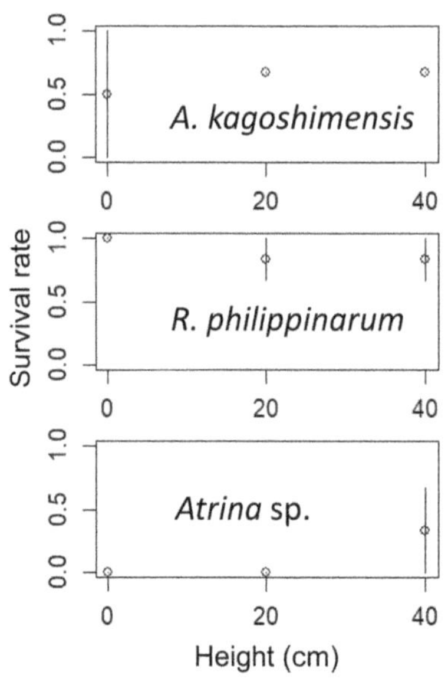

Fig. 4 Survival rates of *Anadara kagoshimensis*, *Ruditapes philippinarum*, and *Atrina* sp. in Run 2 of rearing experiments conducted at station B. Circles denote median survival rates, and bars the range

(> 8.9 and > 10.1 μg/L) for each layer, but less varied between layers (2.4–17.2 μg/L for 0 cm height and 2.2–14.5 μg/L for 20 cm height). Median water temperature was higher at stations A and F (> 22 °C) than the other stations (< 22 °C) and was nearly the same between heights. These spatial variations in environmental factors showed non-significant, but low, significance probabilities ($p = 0.103$–0.157 and $\chi^2 = 2.00$–2.67 for vertical variations; $p = 0.0752$–0.210 and $\chi^2 = 7.14$–10.00 for horizontal variations; Friedman test). The spatial variations were correlated significantly between Chl-*a* and SS (Spearman rank correlation $\rho = 0.713$, $p = 0.0276$ corrected with Holm's method) and between Chl-*a* and water temperature ($\rho = -0.768$, $p = 0.0107$) but not significantly between SS and water temperature ($\rho = -0.4965$, $p = 0.302$).

Of these spatial variations in environmental factors, the variations in median SS corresponded with the variations in *Atrina* sp. survival rate in that median SS was relatively low for the stations and height with the survival rates being relatively high (i.e., station E and 20 cm height). *Atrina* sp. survival rate was 0.45–0.55 for the lowest, median SS (16.5 mg/L), whereas it decreased to 0.00–0.25 for the highest, median SS (80.2 mg/L). Such a correspondence led to the selected statistical model that explains the variation in *Atrina* sp. survival rate with the variation in only SS (McFadden's pseudo R^2 for SS = 0.084; see Table 2 for the selected model and Online Resource 3 for every model examined). In contrast with *Atrina* sp., the other two species did not show a clear correspondence between their survival rates and environmental factors. This led to the selected statistical model that explains the variation in *R. philippinarum* survival rate with no environmental factors (Table 2 and Online Resource 4). GLMM did not converge for *A. kagoshimensis*. These species-specific correspondences between survival rate and SS are mirrored in Fig. 6 in that only *Atrina* sp. showed a sharp decrease in survival rate against SS.

Discussion

Although experimental design and replication differed among the three experimental runs, the patterns of horizontal and vertical variations in survival rates were similar among the runs in that the variations were great for only *Atrina* sp. in survival rates, as follows. Only *Atrina* sp. showed marked differences in survival rate among the experimental stations. In Run 1, *Atrina* sp. survived better at station F than station B. This concurs with Suzuki et al. (2009), who indicated relatively high survival rates of juvenile *Atrina* sp. in the east, inner area of the Sea of Ariake (including station F), based on their finding that the ratio of spat density to larval density was higher for the east sea area than a west inner sea area. In Run 3, however, juvenile *Atrina* sp. survived slightly better at station E than the other stations including station F. Hence, it is possible that the

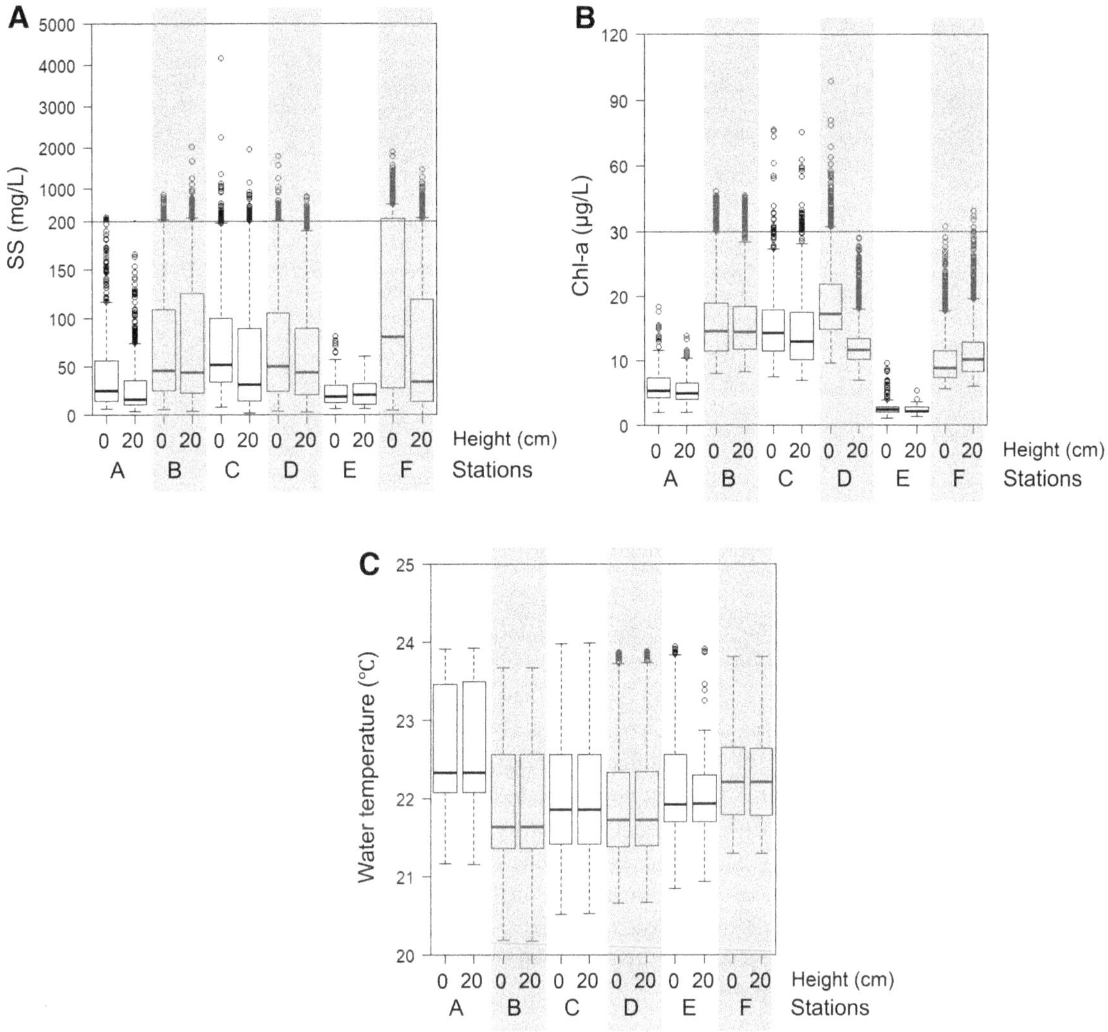

Fig. 5 Environmental factors in Run 3. **a** Suspended solids (SS, mg/L), **b** chlorophyll *a* concentration (µg/L), and **c** water temperature (°C). Suspended solids occasionally showed extremely high values and thus are displayed separately for 0–200 mg/L and 200–5000 mg/L at different scales. Likewise, chlorophyll *a* concentration is displayed separately for 0–30 µg/L and 30–120 µg/L

Table 2 Models showing the lowest BIC for (a) *Atrina* sp. and (b) *Ruditapes philippinarum*

Species	Model	BIC
Atrina sp.	$p = 1/(1 + \exp(-0.44 \pm 0.61 + (0.039 \pm 0.015)\ SS))$	75.91
R. philippinarum	$p = 1/(1 + \exp(-1.96 \pm 0.30))$	45.83

The estimates of parameters are presented as the most likely value ± standard error. See Online Resources 2 and 3 for more details of the results of the model selection

"*p*" denotes expected survival rates, *SS* the density of suspension solids

locations with *Atrina* sp. surviving well are variable and are affected by short-term fluctuations in some environmental factors. In addition, *Atrina* sp. of smaller body sizes was used in Run 3 than Run 1, which might have led to the different patterns in horizontal survival variation. For example, the tidal current is in general faster at station F than station E (Tsukamoto and Yanagi 2002), which possibly was more stressful for small specimens at station F than station E and led to the lower survival rates at station F than station E in Run 3.

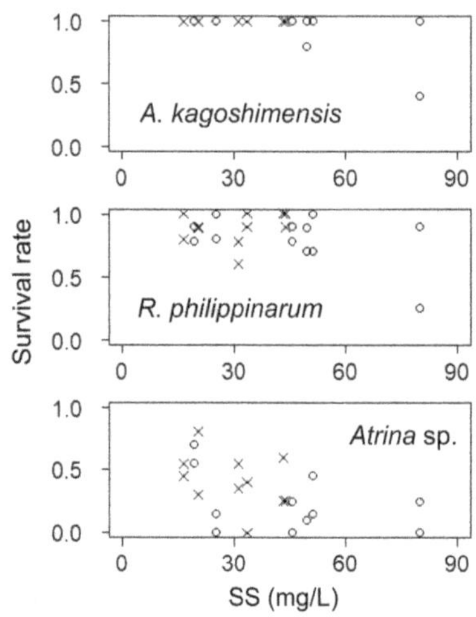

Fig. 6 Survival rates of *Anadara kagoshimensis*, *Ruditapes philippinarum*, and *Atrina* sp. plotted against suspended solids (SS) for each station and height in Run 3 of rearing experiments. Circles denote 0 cm height, and crosses 20 cm height

Atrina sp. survived better in trays higher in the water column in all runs, whereas neither *Anadara kagoshimensis* nor *Ruditapes philippinarum* showed this pattern. This accords with the previous experiments in the Sea of Ariake for *Atrina* sp. and *A. kagoshimensis* at least. *Atrina* sp. with shell length 7 cm showed higher survival rates (> 60%) about 1 m above the seabed than on the seabed (< 20%) during March–December 2015 (Matoba et al. 2016). In contrast to *Atrina* sp., *A. kagoshimensis* with shell length 3–4 cm showed similar survival rates, about 80%, 1 m above the seabed and on the seabed during February–October 2003 (Yurimoto et al. 2007). No such experiment comparing water column and sea bed has been reported for *R. philippinarum* in the Sea of Ariake to our knowledge. Although wild *A. kagoshimensis* and artificially produced *Atrina* sp. were used in our experiments, the difference between wild specimens and artificially hatched seeds is unlikely to result in the overestimation of the difference in survival rate between *A. kagoshimensis* and *Atrina sp.* This is because the survival rates of wild *A. kagoshimensis* are reported to be similar (Katayama et al. 1982) or lower (Takami and Yoshioka 1982, 1983), compared to those of artificially produced *A. kagoshimensis* in transplantation experiments. Therefore, if we had used artificially produced seeds for *A. kagoshimensis* as well as for *Atrina* sp., the difference in survival rates is expected to be either unchanged or larger.

In previous experiments rearing *A. kagoshimensis* and *Atrina* sp. near the seabed and in the water column in the Sea of Ariake, techniques for keeping bivalves differed between water column and seabed: pocket net and flat fishing net, respectively, for *Atrina* sp. (Matoba et al. 2016); and pocket net and resin-made basket, respectively, for *A. kagoshimensis* (Yurimoto et al. 2007). Our experiments showed that, regardless of such depth-dependent differences in equipment, *Atrina* sp. has higher survival rates when maintained in the water column than on the seabed, while *A. kagoshimensis* does not. Hence, the environmental factors show that *Atrina* sp. experience is likely to be less favorable near the seabed than in the water column.

One of the adverse environmental factors that *Atrina* sp. experience in some sea areas and near the seabed can be suspended solids. This is indicated from the result of model selection in Run 3, where the best model included the term SS and excluded the terms stations and height. Adverse effects of suspended solids are reported for many bivalve species (e.g., Bricelj et al. 1984; Navarro et al. 1992; Navarro and Widdows 1997) including the conspecific species *Atrina zelandica* (Ellis et al. 2002). *A. zelandica* showed a decrease in glycogen content and relative weight of soft tissue in the survey areas with abundant suspended solids. Suspended solids would adversely affect *Atrina* sp. through various pathways. Increased suspended solids can lead to decrease in clearance rate of particulates and filtration rate of water (e.g., Navarro et al. 1992; Navarro and Widdows 1997; Ellis et al. 2002), thereby causing decreased amount of food and oxygen acquired. Suspended solids can also lead to increased ejection of pseudofeces, thereby causing increased energy consumption (Madon et al. 1998). Such adverse effects would further increase, if suspended solids accumulate on the seabed and lead to an increased density of the suspended solids near the inhalant openings of *Atrina* sp.

Yurimoto et al. (2008) found that the glycogen content in its adductor muscle, an index of the condition, decreased in relation to the thickness of a high-density layer of suspended solids in a rearing experiment. Accumulation of suspended solids was found on the surface of the substrate used in our experimental trays (Kurihara, personal observation), which possibly increased the adverse effects of the solids for *Atrina* sp.

Why, of the three experimental species, only *Atrina* sp. showed clear negative correlations between its survival rate and the density of suspended solids might be related with endurance to hypoxia. *Atrina* sp. is more vulnerable to hypoxia than A. *kagoshimensis* and *R. philippinarum*: median lethal time, LT50, with dissolved oxygen < 0.4 mg/L is reported to be < 9 h for *Atrina* sp. (mean shell length 23 mm; Nagasoe et al. 2017), which is far shorter than 130 h for *A. kagoshimensis* (28 mm; estimated from the Fig. 1 of Miyamoto and Iwanaga 2012) and 70 h for *R. philippinarum* (31 mm; Yamada et al. 2016). Thus, if higher SS causes lower filtration rate (Navarro and Widdows 1997), thereby causing lower uptake of oxygen, *Atrina* sp. would suffer more greatly than *A. kagoshimensis* and *R. philippinarum.*

Future studies should examine how the patterns of spatial change in survival rate are affected by the body size and density of bivalves. It is likely that, for the bivalve species in the present study, smaller specimens are more vulnerable to various environmental stresses. This is reported for *R. philippinarum* exposed to physical disturbances due to typhoon (Mizuta et al. 2011) and for *Atrina* sp. exposed to hyposaline stress (Yamamoto and Handa 2011; Kurihara et al. 2017) and hypoxia (Nagasoe et al. 2017). Therefore, as we used juveniles for each species, interspecific differences in the responses to spatial variation in environmental factors might have been emphasized. Yet, even if we used larger specimens for each species, *Atrina* sp. would probably show great spatial variations in survival rates. This is suggested by Matoba et al. (2016) and Yoshida et al. (2007) who showed that large *Atrina* sp. (shell length > 200 mm in Matoba et al. 2016 and about 100 mm in Yoshida et al. 2007) survived better in pocket nets in the water column (90% per 10 months) than in cages on the seabed (20% per 10 months); and on the tidal flats (20–30% per 10 months) than in the subtidal area (0% per < 10 months). The densities of *A. kagoshimensis* and *Atrina* sp. in our experimental trays (35.7–119 m^{-2} and 35.7–238 m^{-2}, respectively) were higher than those in their natural habitats (< 50 m^{-2} and < 80 m^{-2}, respectively; Yoshino et al. 2007; Suzuki et al. 2009). Whether the present experimental results for such high densities differ from the results for natural, low densities should be examined in future.

We found that survival rates were lower for *Atrina* sp. than for *A. kagoshimensis* and *R. philippinarum* on the seabed in the inner area of the Sea of Ariake. This suggests that the environmental factors are especially unfavorable for *Atrina* sp. at present. This notion is in accordance with the abundance and the status of fisheries for the three species in the inner area of the Sea of Ariake in recent years. The density of *A. kagoshimensis* averages about 100 m^{-2} in the intertidal area (Nakamuta et al. 2013) and 200 m^{-2} in the subtidal area (Okamura, personal data for 2015). The density of *R. philippinarum* attains 1000–5000 m^{-2} in the intertidal area (Kanzaki et al. 2017; no information for the subtidal area). These two species are still harvested by fishermen (Ministry of the Environment 2017). In contrast, the densities of juvenile *Atrina* sp. have been < 1 m^{-2} in most of the subtidal area of the inner part of the Sea of Ariake between 1997 and 2011 (Koga and Aramaki 2013). It has been legally prohibited to catch *Atrina* sp. therein since 2012 owing to the low abundance (Ito 2017). It is necessary to investigate the environmental factors especially unfavorable for *Atrina sp.*

Acknowledgements We thank Junichi Uchikawa (Kumamoto Prefectural Fisheries Research Center) for providing the bivalve samples. We thank Kengo Suzuki (Hokkaido National Fisheries Research Institute) for commenting on the present study. We thank Nishimura Syokai Co., Ltd., Japan Justice Co., Ltd., and Minefuku Kaiji for their support in our field experiments. The present study was made as a part of contract work with Ministry of the Environment, Japan.

References

Begon M, Townsend CR, Harper JL (2006) Ecology: From Individuals To Ecosystems, 2nd edn. Blackwell, Malden, p 738

Bricelj VM, Malouf RE, Quillfeldt C (1984) Growth of juvenile *Mercenaria mercenaria* and the effect of resuspended bottom sediments. Mar Biol 84:167–173

Broitman BR, Blanchette CA, Menge BA, Lubchenco J, Krenz C, Foley M, Raimondi PT, Lohse D, Gaines SD (2008) Spatial and temporal patterns of invertebrate recruitment along the west coast of the United States. Ecol Monogr 78:403–421

Broström G (2017) Generalized linear models with clustering. https://cran.r-project.org/web/packages/glmmML/glmmML.pdf. Accessed 4 Feb 2018

Ellis J, Cummings V, Hewitt J, Thrush S, Norkko A (2002) Determining effects of suspended sediment on condition of a suspension feeding bivalve (*Atrina zelandica*): results of a survey, a laboratory experiment and a field transplant experiment. J Exp Mar Biol Ecol 267:147–174

Gosling E (2015) Marine Bivalve Molluscs, 2nd edition. NY: John Wiley & Sons, Ltd. 524 pages

Haddon M (2011) Modelling and quantitative methods in fisheries, 2nd edn. Taylor & Francis Group, New York

Hashimoto K, Yamada K, Nagae A, Matsuyama Y (2018) Lineage specific detection of the scaly form of the pen shell *Atrina* spp. by a loop-mediated isothermal amplification method. Fish Sci. https://doi.org/10.1007/s12562-018-1231-4

Ito S (2017) Reviving the pen shell, *Atrina* spp. diving apparatus fishery in the inner part of Ariake Sound. Bull Saga Pref Ariake Fish Res Develop Cent 28:147–166

Kanzaki H, Tsukuda M, Tsujo K (2017) Distribution of Manila clam, *Ruditapes philippinarum*, on the tidal flat at the mouths of Tara and Itoki Rivers in Ariake Sound. Bull Saga Pref Ariake Fish Res Develop Cent 28:119–123

Katayama S, Miyake Y, Ikegami T, Ikeda Z (1982) Experiments on aquaculture of *Anadara kagoshimensis* in Okayama Prefecture, Japan. In: Nansei National Fisheries Research Institute (eds) Rearing experiments for causes of high mortality of *Anadara kagoshimensis*. Volume I. Hiroshima, p 2–15 (in Japanese)

Koga H, Aramaki H (2013) History of the pen shell, *Atrina* spp. fishery and the factors relevant to the fishing ground formation in Saga Ariake Bay—especially, some consideration about the factors of big catch in 2009 fiscal fishing period. Bull Saga Pref Ariake Fish Res Develop Cent 26:13–24

Kumamoto Prefectural Fisheries Research Center (2006) Guidance for resource management of the Asari clam *Ruditapes philippinarum* in Kumamoto Prefecture, version II. https://www.pref.kumamoto.jp/common/UploadFileOutput.ashx?c_id=3&id=9666&sub_id=1&flid=13700.Accessed 24 Nov 2018 (in Japanese)

Kurihara T, Nakano S, Matsuyama Y, Hashimoto K, Yamada K, Ito A, Kanematsu M (2017) Survival time of juvenile pen shell *Atrina pectinata* (Bivalvia: Pinnidae) in hyposaline water. Int Aquat Res. https://doi.org/10.1007/s40071-017-0183-0

Lemer S, Buge B, Bemis A, Giribet G (2014) First molecular phylogeny of the circumtropical bivalve family Pinnidae (Mollusca, Bivalvia): evidence for high levels of cryptic species diversity. Mol Phylogenet Evol 75:11–23. https://doi.org/10.1016/j.ympev.2014.02.008

Liu J, Li Q, Kong LF, Zheng XD (2011) Cryptic diversity in the pen shell *Atrina pectinata* (Bivalvia: Pinnidae): high divergence and hybridization revealed by molecular and morphological data. Molec Ecol 20:4332–4345

Madon SP, Schneider DW, Stoeckel JA, Sparks RE (1998) Effects of inorganic sediment and food concentrations on energetic processes of the zebra mussel, *Dreissena polymorpha*: implications for growth in turbid rivers. Can J Fish Aquat Sci 55:401–413

Masaki K, Onohara T (2009) Relationship between juvenile ark shell, *Scapharca kagoshimensis*, development and weather conditions in the inner area of the Ariake Sound. Bull Saga Pref Ariake Fish Res Develop Cent 24:13–18

Matoba T, Hirose M, Nagamoto A, Yoshida M, Shinohara N (2016) Research on the causes for the mortality of the pen shell *Atrina* sp. in Fukuoka sea area of the Sea of Ariake, IV. Bull Fukuoka Fish Mar Tech Res Cent 26:1–8

Matsukawa Y (2005) Physical process. In: The Oceanographic Society of Japan (ed) Restoration of ecosystem in the Sea of Ariake. Koseisha Koseikaku, Tokyo, pp 3–11

Matsuyama Y (2012) Abnormal mortality of the pen shell *Atrina* spp. and efforts for recovery of the resource. In: Ohshima Y (ed) Current status and issues of the Ariake Bay, a potentially highly productive Coastal Sea. Koseisha Koseikaku, Tokyo, pp 53–62

Ministry of Agriculture, Forestry and Fisheries (2011) Dissolved oxygen. http://www.maff.go.jp/kyusyu/seibibu/isahaya/pdf/011_6_2_4_youzonnsannso_1.pdf. Accessed 3 Feb 2018

Ministry of the Environment (2017) Ariakekai Yatsushirokai Sogo Tyosa Hyoka Iinkai Hokokusho. https://www.env.go.jp/council/20ari-yatsu/rep061221/index.html Accessed 17 April 2018; (in Japanese)

Miyamoto Y, Iwanaga C (2012) Biochemical responses to anoxia and hypoxia in the ark shell *Scapharca kagoshimensis*. Plankton Benthos Res 7:167–174

Mizuta K, Hirano K, Higano J, Tamaki A (2011) Effects of typhoons on artificially-made culture grounds for the manila clam *Ruditapes philippinarum* along Konagai coast of Isahaya Bay, Kyushu, Japan. Aquac Sci 59:75–88

Nagamoto A, Matoba T, Shinohara N (2017) Efficiency of preventive measures against burial of bivalve spat collector using gravel bag in Fukuoka area of the Sea of Ariake, Japan. Bull Fukuoka Fish Mar Tech Res Cent 27:1–8

Nagasoe S, Tokunaga T, Matsuyama Y (2017) Effect of hypoxic conditions on the survival of juvenile pen shell *Atrina* cf. *japonica* (an East Asian species). Suisan Zosyoku (Aquacult Sci) 65:125–132

Nakamuta H, Fujisaki H, Yoshida K (2013) Abnormal mortality of the ark shell, *Scapharca kagoshimensis*, occurred from the autumn to winter in 2011. Bull Saga Pref Ariake Fish Res Develop Cent 26:33–48

Nasu H, Ikushima N, Tobase N, Nakahara Y (2008) The present status and a problem of the fishery for the clam *Ruditapes philippinarum* in Kumamoto Prefecture. Rep Kumamoto Pref Fish Res Cent 8:89–99

Navarro JM, Widdows J (1997) Feeding physiology of *Cerastoderma edule* in response to a wide range of seston concentrations. Mar Ecol Prog Ser 152:175–186

Navarro E, Iglesias JIP, Ortega MM (1992) Natural sediment as a food source for the cockle *Cerastoderma edule* (l): effect of variable particle concentration on feeding, digestion and the scope for growth. J Exp Mar Biol Ecol 156:69–87

R Core Team (2017) R: A language and environment for statistical computing. R foundation for statistical computing, Vienna, Austria. https://www.R-project.org/. Accessed 16 Dec 2017

Suzuki K, Yurimoto T, Watanabe Y, Koshiishi Y, Kimoto K, Yoshida M, Fujisaki H, Fujii A, Nasu H (2009) Distribution of planktonic larvae and spats of the pen shell *Atrina* spp. in Ariake Bay, southern Japan for years 2003–2005. Bull Jpn Soc Fish Oceanogr 73:161–171

Takami T, Yoshioka S (1982) Rearing experiments of *Anadara kagoshimensis* in Yamaguchi Prefecture. In: Nansei National Fisheries Research Institute (ed) Rearing experiments for causes of high mortality of *Anadara kagoshimensis*. Volume I. Hiroshima, p 16–28 (in Japanese)

Takami T, Yoshioka S (1983) Rearing experiments of *Anadara kagoshimensis* in Yamaguchi Prefecture. In: Nansei National Fisheries Research Institute (ed) Rearing experiments for causes of high mortality of *Anadara kagoshimensis*. Volume II. Hiroshima, p 9–19 (in Japanese)

Tsukamoto H, Yanagi T (2002) Tide and tidal current in Ariake Bay. Umi To Sora 78:31–38

Yamada K, Miyamoto Y, Nakano T, Okamura K (2016) Inter- and intraspecific variation in anoxic survival among three bivalve species in intertidal and subtidal areas along the coast of Japan. Plankton Benthos Res 11:49–56

Yamamoto K, Handa T (2011) Effect of low salinity in ambient water on ventilation in the pen-shell *Atrina* (*Servatrina*) *lischkeana*. Suisan Zosyoku (Aquacult Sci) 59:535–540

Yoshida M, Ito T, Chikushi Y (2007) The study of transplantation of the pen-shell (*Atrina pectinata japonica*). Bull Fukuoka Fish Mar Tech Res Cent 17:1–5

Yoshino K, Yamamoto K, Hayami Y, Hamada T, Kasagi T, Ueno D, Ohgushi K (2007) Benthic fauna of the inner part of Ariake Bay: long-term changes in several ecological parameters. Plankton Benthos Res 2:198–212

Yurimoto T, Nasu H, Tobase N, Maeno Y (2007) Growth, survival and feeding of ark shell *Scapharca kagoshimensis* with hung and settled culture in Ariake Bay, Japan. Suisan Zosyoku (Aquacult Sci) 55:535–540

Yurimoto T, Tanaka K, Nasu H, Matsuoka K (2008) Influence of resuspended sediments and their surface accumulation on a pen shell *Atrina pectinata* in Ariake Bay, West Japan. Suisan Zosyoku (Aquacult Sci) 56:335–342

6

An overview of Cladoceran studies conducted in mine water impacted lakes

Jaakko Johannes Leppänen

Abstract Mine-fed waters have been rigorously studied, but most of the ecological research on mine water has been conducted in riverine systems. Lakes, however, are known to recover from pollution more slowly than riverine systems and, thus, the impacts of mine water on lakes are equally interesting. One of the most important biological components in lakes are the Cladocera, an order of crustacean zooplankton. Cladocerans are regarded as excellent indicators of environmental change and, for example, genus Daphnia is one of the most used test organisms in ecotoxicology. While in vitro tests regarding pollutants and cladocerans have been reviewed multiple times, the literature regarding the community level responses to mine pollution in natural settings is still to be better explored. The main aim of this paper is to screen and compile the current literature related to cladoceran communities and mine-induced water pollution. In addition, the applicability of cladocerans as a bioindicators group in mine water studies will be explored. This review shows that cladocerans have been studied in many cases of mining related pollution and most of the research has been conducted in North America, central Europe and Brazil. Acidity, turbidity and metals pollution are nearly equally important in shaping cladoceran communities in mining impacted lakes. The most tolerant taxa to mining pollution are *Bosmina* spp. and *Chydorus sphaericus*. The group clearly has potential as community level bioindicator/biomonitor in mining pollution studies, but challenges remain. Namely, the lack of data regarding the most sensitive taxa is a major problem when indicator value of any single species is assessed.

Keywords Mining · Pollution · AMD · Tailings · Zooplankton · Water flea

Introduction

Although the reservoirs of high grade deposits are continuously decreasing (Watling 2015), the total mineral production has actually doubled in the past 30 years (Reichl et al. 2016). The increased exploitation of low grade ores results in higher amount of mine waste (Hudson-Edwards and Dold 2015) and, in many cases, subsequently larger quantities of potentially harmful mine water. Thus, there is a growing need for monitoring tools, which are suitable for reliable assessment of ecological conditions in mine water impacted lakes. Potentially harmful mine waters are known to exhibit a wide range of acid to alkaline conditions (pH − 2.5 to 11.7) and based on chemical composition, mine water can be characterized in a plethora of ways (e.g., saline, radioactive, corrosive or toxic; Nordstrom 2011). However, the environmental impacts of mine water are even

J. J. Leppänen (✉)
Environmental Change Research Unit (ECRU) and Helsinki Institute of Sustainability Science, Faculty of Biological and Environmental Sciences, University of Helsinki, P.O. Box 65, 00014 Helsinki, Finland
e-mail: jaakko.leppanen@helsinki.fi

more diverse. This is mainly due, but not limited to, the variability in species sensitivity to harmful substances and to the physical and chemical characteristics of the receiving environment (e.g., Kozlova et al. 2008). Environmental impacts of mine waters on lakes have been acknowledged for centuries (e.g., Wolkersdorfer and Bowell 2005) but because the effluent is usually released into lotic (flowing) waters, this has gained most of the attention in research. While riverine systems are known to recover fast when pollution stops (Yount and Niemi 1990), lakes tend to recover slowly because of longer residence times. In some cases, the recovery may be further delayed by sedimentary contamination (Rogalski 2015) and mine water induced meromixis, especially in pit lakes (e.g., Schultze et al. 2017).

Bioindicators/biomonitors are attractive tools for pollution researchers. The idea is that the biota is the best indicator of ecological changes as it provides the data concerning the actual biological impacts (Zhou et al. 2008). In mine water studies, the popular bioindicator species are riverine insect larvae (Cain et al. 1992; Salmelin et al. 2017) and acidophilic algae (e.g., Valente and Gomes 2007). In lakes, one of the most promising indicator groups for mine pollution is the Cladocera; an order of microscopic aquatic Crustacea, which are excellent indicators of environmental change (Jeppesen et al. 2011). Cladocera colonize almost every type of fresh waters and they are usually highly abundant. The group is of vital importance to the wellbeing of aquatic systems due to their position in food webs. Cladocerans transfer energy from primary producers towards higher trophic levels, such as juvenile fish (Sterner 2009).

There is a large body of knowledge regarding the effects of harmful substances on cladocerans from in vitro testing (Sarma and Nandini 2006; Suhett et al. 2015). In fact, the applicability of cladocerans to mining contamination assessment is further highlighted in Canadian metal mining effluent regulations (Canadian Government 2002), which include *Daphnia magna* monitoring tests to be conducted on a monthly basis as a part of mine water monitoring programs. Despite the obvious advantages of in vitro studies (standardized protocols, low cost), the comparability of in vitro tests with changes in situ is not well understood. Whereas the results of ecotoxicological laboratory tests using Cladocera have been reviewed several times (e.g., Sarma and Nandini 2006; Persoone et al. 2009; Bownik 2017), the impact of mine water on natural cladoceran communities has not been reviewed.

This contribution focuses on the effects of mine-impacted waters on lentic cladoceran communities and populations in an attempt to compile relevant studies. The second goal is to assess the suitability of cladocerans as a community-level bioindicator tool in studies of mine-related lake pollution.

Scientific bibliographic databases (BioOne, EBSCO, Web of Science, Google Scholar) were used to find the research literature considering the impacts of mine water release or tailings dumping on cladoceran communities (search terms cladocera AND mining, cladocera AND AMD, cladocera and mine water, cladocera AND tailings, cladocera AND acid mine drainage). Because the mine water impact is difficult to differentiate from other stressor sources in the catchment and airborne pollution, only the cases where point sources are clearly identified are reviewed here. In addition, all in vitro tests are omitted from this review whereas mining pit lakes and man-made reservoirs are included and are discussed together with natural waterbodies.

Mine water pollution as a multi-stressor problem

The mine waters only rarely contain a singular component, which can be deemed as a single factor which induces ecological impacts. Mine waters can simultaneously exhibit high metals concentrations, elevated salinity and high concentrations of solids and these different constituents and characteristics affect each other. For example, pH and redox potential can be regarded as the most important parameters which control the solubility, and thus the availability, of trace metals to biota. The processes involved are precipitation, adsorption, oxido-reduction and complexation (Bourg and Loch 1995). Toxicity tests for Cladocera are being designed to discriminate the toxicity due to metals and low pH (Lopes et al. 1999) and the bioavailability of some metals can be assessed using biotic ligand models (BLM), where local water characteristics are taken into account (Niyogi and Wood 2004). However, all BLM tools have pronounced applicability ranges for many parameters (e.g., pH, DOC, Ca, Mg, Na, SO; Vink and Verschoor 2010). Further, in some cases, mine effluent affects these parameters and subsequently hampers the analysis of the mine water impact. Thus, the issue is complicated by the synergic or antagonistic effects of mine water components (e.g., metals; Smirnov 2017). In addition, the varying effects of contaminants also depend on the route of exposure (e.g., dietary exposure, direct exposure; Sofyan et al. 2009). Even though the environmental impacts of elevated salinity and dissolved solids have not been studied as intensively as acid drainage in mine water research, they constitute an important and potentially critical component in mine water toxicity. Salinity (sulfate, sodium, calcium) induces osmotic stress to fresh water biota and may turn the water body meromictic (Molenda 2014). The meromixis is a result of the incomplete mixing of dense (saline) effluent and overlying freshwater which may lead to permanent anoxia in deep bottoms. The elevated concentrations of solids may contribute to habitat degradation due to diminished water clarity and elevated siltation. Further, elevated concentrations of suspended minerals (turbidity) are harmful to cladocerans (McCabe and O'Brien 1983; Cuker 1987; Hart 1987; Kirk and Gilbert 1990). In addition, the accompanying impacts (changes in water clarity, trophic status, acidification, predator community or food availability) may induce additional difficulties in mine pollution research. Due to the complex character of mine pollution, the most convenient way to present the current research is to organize the literature based on mined material rather than pollutant.

Cladocera and metal mining

Bauxite

Lake Batata, located in Amazonia, Brazil has been impacted by bauxite mining tailings, which are high in very fine solid particles and in aluminum and iron oxides. The tailings were dumped into the lake for 10 years (18 million m^3/year), which has resulted in greater variation in cladoceran density and brood size when compared to a non-impacted area (Bozelli 1996). In addition, cladocerans collected from impacted site showed smaller body sizes but higher weights which were explained by ingestion of tailings material (Maia-Barbosa and Bozelli 2005). Moreover, a 7-year study indicated that the highest impact of tailings was detected during the low water period and that *Diaphanosoma birgei* and *Bosmina hagmanni* dominated during the turbid phases (Garrido et al. 2003) suggesting high tolerance to turbidity, whereas *Moina minuta* was negatively impacted by high turbidity. Thus, *M. minuta* would be interesting species if the impacts of suspended solids pollution were to be assessed in the future studies in the region. The impact of seasonal water level fluctuations, which also result in changes in turbidity, is characteristic in these studies. Therefore, it is difficult to postulate the most important stressor, because other season-related environmental conditions (e.g., temperature, food availability) probably vary simultaneously with the turbidity. The ingestion of tailings particles by cladocerans is potentially highly important component when ecological impacts of tailings are considered because species exhibit morphological differences of feeding apparatus, which has been suggested to contribute to the ability of selective (e.g., ability to reject inedible particles) feeding (Smirnov 2017).

Copper and gold

Doig et al. (2015) investigated the impact of acid and metal-contaminated effluent originating from Cu–Zn mine on *Bosmina longirostris* abundance in a small boreal lake in Manitoba, Canada. The authors detected severe reduction in community size, which was related both to acidity and zinc contamination. Similarly, Kerfoot et al. (1999) detected a pronounced decline in *Bosmina* production in oligo-mesotrophic (pH 6.5–9.1) lake, Keweenaw Waterway in Michigan USA, due to copper pollution (maximum copper sediment flux 579 μg/cm^2/yr) inflicted by historical (1848–1920) mining activities. Different cladoceran species exhibit high variation in copper tolerance (12-fold differences in copper tolerance have been reported; Bossuyt and Janssen 2005). *B. longirostris* is regarded as more sensitive to copper pollution than many other cladoceran species (Koivisto et al. 1992; Koivisto and Ketola 1995; Bossuyt and Janssen 2005) which may explain the above mentioned deleterious impact. Leppänen et al. (2017b) reported decreased diversity and species richness in Lake Kirkkojärvi, southern Finland, which has been affected by increased input of mineral matter and minor metal pollution, originating from tailings impoundment of Au–Cu mine around 1960–1970, but the lake has never acidified due to rapid dilution. The most pronounced impact was detected during the input of fine mineral tailings. *Polyphemus pediculus*, *Eubosmina coregoni*, *Daphnia* sp., *Alonella excisa*, *A. quandranqularis*, *Alonopsis elongata* and *Disparalona rostrata* were negatively affected. In addition, the cladoceran community in the shallow embayment of the same lake is unharmed by low volume but continuous high-Cu low pH effluent (since 1970s) because the pollution is most intense during early spring, when the cladoceran communities are mostly absent (Leppänen et al. 2017a). The lucky mismatch is further reflected in increased abundance of *B. longirostris*, which is especially sensitive to copper pollution but thrives in eutrophicated systems, such as Lake Kirkkojärvi.

Iron

Holopainen et al. (2008) reported potassium to be the most harmful element to the planktonic ecosystem in Russian lakes impacted by alkaline mine water released from iron mine. Species number was low (< 7) and the impacted lakes (pH 7.7–8.0, EC 49–391 µS/cm) were dominated by *B. longirostris* whereas in the non-impacted reference lake (pH 6.7–6.8, EC 24–28 µS/cm) the dominating species was *Daphnia cristata*. The elevated conductivity of the impacted lake may explain the dominance of *B. longirostris,* a species known to inhabit high conductivity waters (Zawisza et al. 2016). In contrast, Bradbury and Megarad (1972) found that the cladoceran community in Shagawa Lake (Minnesota, US) shifted from a *B. longirostris* dominated system towards higher abundance of *Chydorus sphaericus* due to onset of anthropogenic eutrophication and the effluent (hematite rich effluent) originating from iron mine. The authors suggest that the reason for species shift is related to cyanobacterial blooms, but still, this trend is very interesting, because both species are regarded to prefer eutrophicated waters (e.g., Nevalainen and Luoto 2013) and to tolerate cyanobacterial blooms (Tõnno et al. 2016). However, the species differ in their tolerance on heavy metals (Bossuyt and Janssen 2005) and solids pollution (elevated conductivity; Zawisza et al. 2016). However, no metals concentrations data are available from Lake Shagawa. Changes in species assemblages are also reflected in species richness as Winegardner et al. (2017) reported the pronounced decrease in cladoceran species richness to correlate with iron mining derived metal pollution in Quebec, Canada. In species level, *C. sphaericus* was able to remain relatively unharmed by pollution. Moreira et al. (2016) compared zooplankton communities in two artificial lakes, in Mata Porcos, Brazil, which receive mining effluent from two different mines. The cladoceran community in the lake which was affected by kaolinite mining (pH 6.8–0.7.1, EC 12.1–16.2 µS/cm, barium 24.0–53.3 µg/L, manganese 66.3–248.7 µg/L, zinc 0.0–12.5 µg/L) was dominated by *B. longirostris*, whereas the cladoceran assemblage was more diverse (*Eubosmina tubicen, Bosminopsis deitersis, Alonella clathratula, Ilyocryptus spinifer*) in the lake which was affected by iron mine effluent (pH 5.4–6.4, EC 2.9–18.1 µS/cm, barium 4.2–6.9 µg/L, manganese 12.4–109.7 µg/L, zinc 4.2–68.9 µg/L). Here, despite some differences in water chemistry between two lakes, the reported concentrations are low in terms of toxicity (ECOTOX 2017) and no clear reason behind the community differences can be drawn from the water chemistry alone.

Nickel, tin and zinc

Saline and metal-contaminated mine water originating from Talvivaara Terrafame Ni–Zn–Cu mine (central Finland) resulted in nearly total disappearance of *Daphnia cucullata* and clear dominance of *B. longirostris* and turned the receiving lakes meromictic with maximum concentration of 1200 mg/L of sulfate in surface water. Whereas the metals concentration in lake water has remained relatively low, the salinity-induced stress and the stratification of the waterbody (and resulting physical changes in lake) probably contributed more to the community change (Leppänen et al. 2017c). The cladoceran production greatly diminished and clear changes in species composition were detected in Datun Lake, China, which was impacted by mineral tailings and elevated arsenic concentrations originating from tin mining. Particularly, *Eubosmina longispina* and *Alona guttata/rectangula* were adversely affected as the production declined over 80% (Chen et al. 2016). Wilk-Woźniak et al. (2011) studied plankton communities in fishponds contaminated by heavy metal runoff from Zn-Pb mine "Matylda" in south Poland and found an extremely poor cladoceran community. The maximum proportion of cladocerans among zooplankton was only 3% (total density) and the only reported species was *E.coregoni*.

Uranium

Ferrari et al. (2009) reported *Bosmina sp.* as the only species in waste water of a uranium mine in Poços de Caldas Plateau region, Brazil, with elevated conductivity (2415 µS/cm) pH of 3.9 and high uranium concentration (4.3 mg/L). However, also the low impact reference lake was noted as being unsuitable for cladocerans as only 2 species (*Diaphanosoma* and *Bosmina*) were recorded. The reference lake is characterized by fluctuation in the water properties (e.g., SO_4^{2-} 12.4–253.4 mg/L, hardness 41.49–323.4 mg/L) (Ferrari et al. 2017) which may explain the low cladoceran diversity. In addition, *Bosmina* –group species (*Bosminopsis deitersi* and *Bosmina* sp.) were detected to be the only cladoceran taxa in a newly formed acidic uranium pit lake in the same region with mean pH 3.8, electrical conductivity of 2391 μS/cm, sulfate concentration of 1413 mg/L and uranium concentration of 3 mg/L (Ferrari et al. 2015). Interestingly, Melville (1995) reported a clear shift (the loss of *Bosmina*) in the cladoceran community and pronounced dominance of *Ceriodaphnia reticulata* in a small boreal lake in Canada, which was affected by uranium mine effluent resulting in elevated concentrations of uranium (0.54 mg/L) and sulfate (1140 mg/L) in lake water. These pronounced differences in dominant species in high U and high sulfate lakes suggest that the reasons behind the cladoceran changes are complicated and in addition to water chemistry are probably also related to regional (e.g., climatic) variables.

Cladocera and fuel minerals mining

Black coal, lignite and oil sands

Canton and Ward (1981) reported extremely poor cladoceran community (sole species was *B. longirostris* < 1 individual/L) in a coal mine pond (Pond 1) in NW Colorado USA, which had received mine water (pH 7.4, total dissolved solids (TDS) 3811 mg/L, nitrate 16.34 mg/L) for a decade. In the older pond (Pond 2), with lower contamination levels (pH 7.4, TDS 1770 mg/L, nitrate 0.55 mg/L), the cladoceran community covered more species (*C. sphaericus, D. pulex, B. longirostris*). The cladoceran community of the control site (Pond 4) (pH 7.6, TDS 547 mg/L, nitrate 0.14 mg/L) was devoid of *B. longirostris* and the dominating species was *C. reticulata*, followed by *D. pulex, C. sphaericus and A. guttata.* Whereas the extremely low cladoceran abundance in the most polluted pond (Pond 1) may result either from nitrate or solids pollution, the nitrate levels in other two ponds (Ponds 2 and 4) are well below safe levels (Camargo et al. 2005) suggesting impact due to solids pollution. Because many lignite deposits in central Europe are located in sulfide-rich areas, the abandoned mines once filled with water are prone to acidification and metal pollution (Miller et al. 1996). In Lusatia (eastern Germany) on its own, the number of acidic pit lakes exceeds 400 (Nixdorf et al. 1998). In central Germany lignite mining district, approximately 50% were earlier or currently are acidic (Schultze et al. 2010). Mining pit lakes are occasionally affected by varying dilution regimes or pollution gradients. For example, a strong pH gradient in mining pit lake Lake Senftenberger See (Germany) is reflected in cladoceran species richness along this gradient (Belyaeva and Deneke 2007) with *C. sphaericus* being the only species present along the whole pH range (pH 3–7). The pH tolerance of *C. sphaericus* was also demonstrated by Klapper and Schultze (1995) as they noticed *C. sphaericus* in lignite pit lakes in Germany at the pH of 2.9, whereas *Daphnia* was detected only in lakes with the pH of 6 or higher. In addition, Wollmann et al. (2000)reported *C. sphaericus* from acidic pit lakes Felix (pH 3.6) and L117 (pH 2.8) in Germany. Similarly, Sienkiewicz and Gasiorowski (2016) found that *C. sphaericus* was the only species present in a highly acidic mining lake (Lake TR-33, Poland) before neutralization and subsequent normalization of the cladoceran community. Moser and Weisse (2011) compared zooplankton communities between two meromictic mining lakes in Austria, one of which was neutralized 25 years ago whereas the other had remained acidic (pH 2.6). The acidic lake was devoid of cladocerans, whereas *Daphnia* and *Bosmina* were recorded occasionally in the neutralized lake during the 2-year period. However, the characteristics of the neutralized lake varied inter-annually, resulting in fluctuation between pH 4.3 and 8.0 probably hampering the successful establishment of a rich cladoceran community. The mining pollution impacts on cladocerans, even when pollutants are highly toxic, are not always so straightforward. Kurek et al. (2012) studied cladoceran community changes in six lakes in Alberta, Canada, that have been affected for decades by PAHs (polycyclic aromatic hydrocarbons) input originating from the oil sands mining. However, the major shifts in cladoceran communities occurred before the oil sands mining started and, thus, the effect of PAHs contamination is uncertain.

Cladocera and other mines

Apatite

Vandysh (2004) concluded that the pollution induced by apatite-nepheline tailings and mine water since the 1930s induced a clear shift in cladoceran community in Belaya River watershed located on the Kola peninsula, NW Russia. The author suggested that the tailings-derived turbidity and also the eutrophication resulted in the disappearance of *D. cristata*, *Holopedium gibberum*, *Leptodora kindti* and *Bythotrephes cedestromii*. The most common species in polluted areas was *Bosmina obtusirostris* (syn. *Bosmina (Eubosmina) coregoni*). The loss of large sized species is particularly interesting as the large bodied species are valuable food items for fish but also more vulnerable to fish predation than small bodied taxa (Brooks and Dodson 1965), adding another component to the ecological changes in Belaya River watershed.

Basalt

El-Bassat and Taylor (2007) noted cladoceran species being extremely rare (*B. longirostris*, *Pleuroxus* sp. and *Alona* sp. were detected) in alkaline (pH 7.8–8.5), saline (EC 28,000 μS/cm) and eutrophic Abp Zaabal Lake, which occupies a former basalt mine, in Cairo, Egypt. It must be noted, however, that Abp Zaabal Lake is highly impacted by waste water input unrelated to mining, which prevents any generalizations regarding the role of initial pit lake chemistry on cladoceran community.

Diamonds

The effluent (showing elevated concentrations of Ca^{2+}, K^{+}, SO_4^{2-}, Na^{+}, Mg^{2+}, HCO_3^{-}) originating from the Ekati mine, a diamond mine located in northern Canada, was reported to inflict distinct changes in cladoceran communities in the chain of lakes along the pollution gradient. The species, which usually dominates softwater low-Ca lakes in the region, *Holopedium glacialis*, was substituted by *Daphnia longiremis* and *Daphnia middendorfiana* due to elevated pH and Ca enrichment (Griffiths et al. 2018).

Gypsum

Alkaline (pH 7.6–0.7.9) gypsum mine pit lakes in Croatia, with very high concentrations of solids (TDS > 2000 mg/L), calcium (~ 1300–1400 mg/L) and sulfate (~ 400–1300 mg/L) were dominated by *Chydorus* sp. and *C. reticulata,* whereas a non-gypsum karst lake (pH 7.9, EC < 482 μS/cm, calcium 240 mg/L, sulfate 12.8 mg/L) was dominated by *B. longirostris* and *Ceriodaphnia quadrangula.* The community difference was suggested to originate from the reduced feeding efficiency in gypsum lakes (Stankovic et al. 2011).

Hematite

Coard et al. (1983) detected the total, but relatively temporal loss of planktonic taxa and subsequent dominance of a chydorid species in a lake located in Cornwall, England, which was affected by hematite–clay input due to mining activities in the early twentieth century. The loss of planktonic taxa may be related to increased concentrations of solids and subsequently decreased feeding efficiency for planktonic filter feeder species (e.g., Arruda et al. 1983; Stankovic et al. 2011).

Sand

Sand and gravel extraction pit lakes in, e.g., Hungary (Vad et al. 2012) and Croatia (Stankovic and Ternjej 2009) harbor relatively rich and diverse cladoceran communities. In addition, one species (*Macrothrix laticornis*), seldom detected in Belgium in the early twentieth century, has been discovered to exploit newly dug sand pits and has thus been able to become relatively widespread (Dumont 1989).

Sulfur

A pit lake, formed due to the abandonment of a sulfur mine in Tarnobrzeg (Poland) is characterized by slightly alkaline pH (7.2–8.6), elevated hardness (1000–2000 mg $CaCO^3$), and the lake is meromictic. *Chaoborus flavicans,* which is an efficient predator and may affect the cladoceran community (Jäger et al. 2011), was detected. The general cladoceran abundance was low and planktonic species *D. cucullata* and *B. longirostris* dominate the cladoceran community (Wilk-Woźniak and Zurek 2006). This particularly interesting case would deserve long-term monitoring study or paleolimnological study to examine the roles of meromixis, water chemistry and predation on cladoceran community structure.

Cladocera in mine water research

The available literature regarding the impacts of mine water on cladoceran communities in lake environments is still rather scarce. Moreover, only 11 of the reviewed studies included an adequate temporal dimension to allow the analysis of the community dynamics, which constitutes vital information regarding impact assessments or restoration planning. There seems to be few regional hotspots where cladoceran communities and mining pollution have been studied in natural waters (Table 1). This regional variation in cladoceran studies is probably related to the number of lakes in different regions. Most important stressors in reviewed papers were acidity, metals and salinity or solids. In most cases, species shifts or changes in productivity are reported. In addition to direct toxic impacts inflicted by high aquatic concentrations of harmful elements, also low concentrations of harmful elements (or changes in lake chemistry regarding non-toxic constituents) or other ecological changes may result in distinct impacts in cladoceran populations. The stress induced by exposure to elevated salinity has been reported to interfere with cladoceran reproduction (Elphick et al. 2011; Van Dam et al. 2014), whereas cladoceran growth is negatively affected by changes in food availability (mineral ingestion instead of edible particles; Maia-Barbosa and Bozelli 2005) and in food quality (e.g., increased abundance of cyanobacteria; Lundstedt and Brett 1991). Declining clutch sizes and inhibited growth are reflected in cladoceran productivity. The most sensitive taxa disappear first, which is reflected in diversity. Because cladocerans rely on sedimentary egg banks as temporal refugia to bypass seasonal or ecological "hardships" (e.g., winter or crowding), the sedimentary pollution may also have a pronounced impact on cladoceran communities (Rogalski 2015).

The most tolerant species regarding mine water impact belong to the *Chydorus* and *Bosmina*—groups. Namely, *C. sphaericus* is clearly able to thrive in acidic lakes, whereas *Bosmina* sp. does not exhibit special tolerance to any particular pollutant but seems to tolerate solids pollution relatively well. Comprehensive autecological analysis has not been published for either *B. longirostris* or *C. sphaericus* but according to (Błędzki and Rybak 2016) *Chydorus* cf. *spahericus* and *Bosmina* spp. are wide spread and abundant in many types of water bodies both in littoral and pelagial zone with high stress tolerance.

The applicability of cladocera as bioindicators/biomonitors

As defined by Markert et al. (2003), bioindicators are organisms (or, e.g., communities) that contain information on the quality of the environment, whereas biomonitors are organisms (or, e.g., communities) that contain information on the quantitative prospects of the environment. Thus, biomonitors are always bioindicators as well. In general, cladocerans fulfill many of the characteristics that are deemed important for any bioindicator organism. Namely, cladocerans are sensitive to pollution (Von Der Ohe and Liess 2004) and have short life cycles (Smirnov 2017), resulting in rapid responses to environmental stress. Recently, cladocerans have been used to study community level responses to environmental variables in many types of environments, e.g., dystrophic lakes in central Europe (Zawisza et al. 2016), subarctic lakes in Finland (Leppänen et al. 2016) and crater lakes in western Uganda (Rumes et al. 2011). The potential of cladocerans as a community level indicator tool in mining pollution studies is highlighted in the results of this review as the community shifts are clearly visible in most of the studies which span from pre- to post impact conditions. In addition, most of the results indicate distinct differences between impacted and reference sites. However, whether the overall conditions of an ecosystem can be determined by studying only one group of organisms is

Table 1 General characteristics of mine pollution and impacts on cladoceans

Country	Mine	Pollution type	Study type	Pollution characteristics	Dominant species during impact	Other notes	Citation
Austria	Lignite	Pit lake	Descriptive	Impacted lake low pH, reference lake neutralized. Both meromictic, and internannual pH variation	No cladocerans	*Daphnia* and *Bosmina* only occasionally detected in reference lake	Moser and Weisse (2011)
Brazil	Uranium	Effluent	Comparison	High U, low pH, high conducitivty in impacted site	*Bosmina* sp.	*Bosmina* and *Diaphanosoma* detected in reference site	Ferrari et al. (2009)
Brazil	Uranium	Effluent	Descriptive	High U, low pH, high conducitivty, high sulfate in impacted site	*Bosmina* sp. and *Bosminopsis deitersi*	Low density	Ferrari et al. (2015)
Brazil	Bauxite	Tailings	Comparison	Fine particulate tailings of Al and Fe oxides	*Diaphanosoma polyspina, B. deitersi, Bosmina hagmanni, Ceriodaphnia cornuta, Moina minuta*	Density and brood size oscillate more in the impacted area	Bozelli (1996)
Brazil	Bauxite	Tailings	Comparison	Fine particulate tailings of Al and Fe oxides	*Diaphanosoma birgei, B. deitersi, B. hagmanni, C. cornuta, M.minuta*	Size and weight affected by tailings in impacted area	Maia-Barbosa and Bozelli (2005)
Brazil	Kaolinite and Iron	Effluent	Descriptive	Effluents with near neutral pH, varying elemental concentrations originating from two mines	*Bosmina longirostris* and *Eubosmina tubicen*	More diverse community in Iron mine reservoir	Moreira et al. (2016)
Canada	Copper	Effluent	Continuous	Low pH and elevated metals concentration	*B. longirostris*	Community reduction due to pollution	Doig et al. (2015)
Canada	Iron	Effluent	Continuous	Elevated metals concentrations	*B. longirostris, E. longispina*	Decreased cladoceran species richness	Winegardner et al. (2017)
Canada	Uranium	Effluent	Continuous	High U, elevated salinity and metals concentration	*Ceriodaphnia reticulata*	Loss of cladoceran diversity	Melville (1995)
Canada	Oil sands	Effluent	Continuous	PAHs contamination	*Daphnia*	Uncertain effects	Kurek et al. (2012)
Canada	Diamonds	Effluent	Comparison	elevatod Ca, K, Mg, Na, SO concentrations in effluent	*Daphnia middendorfiana* and *Daphnia longiremis*	*Holopedium glacialis* dominates the unimpacted lakes	Griffiths et al. (2018)
China	Tin	Effluent	Continuous	Arsenic contamination	*Bosmina* spp.	Cladoceran production decline	Chen et al. (2016)
Egypt	Basalt	Pit lake	Descriptive	Saline, eutrophic pit lake	*B. longirostris*	Very low cladoceran diversity and species richness	El-Bassat and Taylor (2007)
Finland	Gold	Effluent	Comparison	Low pH and elevated metals concentrations	*B. longirostris*	Masked by eutrophication	Leppänen et al. (2017a)

Table 1 continued

Country	Mine	Pollution type	Study type	Pollution characteristics	Dominant species during impact	Other notes	Citation
Finland	Gold	Tailings	Continuous	Mineral tailings	*B. longirostris*	Decline in cladoceran diversity and species richness	Leppänen et al. (2017b)
Finland	Nickel	Effluent	Continuous	Meromictic, elevated salinity and metals concentrations	*B. longirostris*	Low cladoceran diversity	Leppänen et al. (2017c)
Germany	Lignite	Pit lake	Comparison	Low pH	*C. sphaericus*	Species shift along pH gradient	Belyaeva and Deneke (2007)
Germany	Lignite	Pit lake	Descriptive	Low pH	*C. spahericus*		Wollmann et al. (2000)
Germany	Lignite	Pit lake	Comparison	Low pH	*C. spahericus*	Daphnia only in pH 6 and above lakes	Klapper and Schultze (1995)
Poland	Zinc	Effluent	Comparison	Elevated metals concentrations	*Bosmina coregoni*	Extremely poor cladoceran community	Wilk-Wozniak et al. (2011)
Poland	Sulfur	Pit lake	Descriptive	High hardness, meromixis	*Daphnia cucullata* and *B. longirostris*		Wilk-Wozniak and Zurek (2006)
Poland	Lignite	Pit lake	Continuous	Low pH	*C.spahericus*	Cladoceran production and diversity increased due to neutralization	Sienkiewicz and Gasiorowski (2016)
Russia	Iron	Effluent	Comparison	Alkaline effluent, elevated EC and Na, K, Li, Ca, Mg concentrations	*Bosmina sp.*	*Daphnia* dominated reference lake	Holopainen et al. (2008)
Russia	Apatite	Effluent	Comparison	Apatite-nepheline effluent and tailings	*B. coregoni*	Sensitive species disappeared	Vandysh (2004)
UK	Granite	Tailings	Continuous	Hematite-clay	*Chydorids*		Coard et al. (1983)
USA	Copper	Effluent	Continuous	Copper pollution		Decline in *Bosmina* production	Kerfoot et al. (1999)
USA	Iron	Tailings	Continuous	Hematite and limonite particles	*C. sphaericus*		Bradbury and Megarad (1972)

Table 1 continued

Country	Mine	Pollution type	Study type	Pollution characteristics	Dominant species during impact	Other notes	Citation
USA	Coal	Tailings	Comparison	High concentration of solids, high N	*B. longirostris*	Very low diversity and richness	Canton and Ward (1981)

Study type denotes whether a study is descriptive without comparison to non-impacted reference, comparative (e.g., non-impacted lake versus impacted lake) or continuous with adequate temporal dimension to allow the assessment of community changes the central question in every bioindicator application (Markert et al. 2003) but due to the central location in food webs, the cladoceran community has the potential to reflect the surrounding environment in great detail.

The suggested endpoints for community level bioindicator studies are food web structure shifts or changes in species diversity (Harwell et al. 1987). With cladocera, the food web structure shifts (e.g., change in predator regime) are reflected in size structure in cladoceran community (Brooks and Dodson 1965) and to morphological changes among some species (Korosi et al. 2013). The increased dominance of the most tolerant taxa is reflected in diversity and richness indices due to the disappearance of rare or sensitive species from the community. Distinct changes in species composition (e.g., total loss of macrophyte associated species) may indicate the destruction of habitat (e.g., pronounced decline in macrophyte cover). However, it is challenging to determine whether a community has factually changed due to mine water input and what criteria should be used to detect the change. Namely, the natural variability of pre-disturbance community, disturbance impact and the post-disturbance community dynamics must be assessed. This can be achieved either by long term monitoring (e.g., Stow et al. 1998; Lindenmayer and Likens 2009), which are very rare, or paleolimnology (e.g., Smol 1992; Saros 2009). Luckily, historical cladoceran community can be reconstructed using paleolimnological methods (Korhola and Rautio 2001) and, in fact, the group is deemed a powerful tool for environmental reconstruction (Tolotti et al. 2016). Another highly important information, i.e., the exact point (or time) of community change, can be determined by applying statistical methods, such as broken-stick zonation (Bennett 1996), where significant changes in community are pinpointed or analysis of similarities, ANOSIM (Clarke 1993), which can be used as a test of significant difference between pre- and post-pollution communities. The perpetuity of the shift can be assessed by post-disturbance research (e.g., if the community recovers to its pre-disturbance form or is shifted to a new steady state). The information regarding pre-disturbance community (e.g., Battarbee 1999) and the possible recovery dynamics are of vital importance also in restoration and management purposes (European Union 2000). It must be noted that after the disturbance, the new community structure depends on source populations for recolonization and, thus, the niche availability (e.g., whether the most important niches are re-occupied or not) should be taken into account in addition to actual post-disturbance species composition.

Unfortunately, the reviewed literature does not allow the reliable identification of most vulnerable species, which could be used as early warning tools in mining pollution research. Partly, this is a result of incomplete reporting of cladoceran community. In many studies, only the dominant species were recorded and potentially very interesting data regarding most sensitive species were not available. In addition to community level considerations, widely distributed and highly resistant taxa *C. sphaericus* and *Bosmina* spp. may be successfully used as individual indicators. However, to further explore the potential of these two taxa as mine water indicators, more research (e.g., morphological changes due to pollution, population impacts, and life history changes) with *Bosmina* and *Chydorus* is needed. Luckily, some of this work has already been started (e.g., sedimentary toxicity test for *C. spahericus* has been developed; Dekker et al. 2005) but both of those groups are deemed challenging in terms of taxonomy (Belyaeva and Taylor 2009; Adamczuk 2016) which will hamper the applicability of *Bosmina* and *Chydorus* as global or regional indicators until the taxonomic issues are clarified. Moreover, as pointed out by Markert et al. (2003) bioindicator community or species level responses should be compared and "calibrated" to determine the indicator applicability of same species or species groups across regions and ecosystems. In addition to toxicological aspect, cladocerans can also be used in biomagnification studies (Stewart et al. 2008) and food web impact research (Draves and Fox 1998) in mine water studies. An interesting species, worth mentioning, is halophylic *Daphnia exilis*, which has been detected in reservoirs impacted by copper mine pollution in Chile (Heine-Fuster et al. 2010) and could thus be used as an indicator of saline mine drainage in the natural distribution area of this species.

Further reading

Besides the literature reviewed here, there are additional important resources for anyone working with cladocerans and mine pollution: The current role of cladocerans in ecotoxicological research in general has recently been reviewed by Sarma and Nandini (2006) and Suhett et al. (2015). In addition, the most important cladoceran species in toxicology research is *Daphnia*, which was extensively reviewed by Lampert (2011). Adamczuk (2016) reviewed *Bosmina longirostris*, one of the most important species when pollution impacts on cladoceran species composition are studied. Moreover, extensive data on mortality and reproduction in response to increasing concentrations of toxicants are available from databases, such as ECOTOX (2017). Detailed contribution on cladoceran physiology, where deeper look into the impact of xenobiotics is provided, was presented by Smirnov (2017). In addition to mine water impact, mining industry has also other effects on cladoceran communities. The lakes in the Sudbury mining region have been rigorously studied and the impacts of pollution originating from smelting and roasting processes on cladocerans have been assessed in many papers (e.g., Labaj et al. 2014, 2015; Thienpont et al. 2016). These studies would be a great benefit for anyone interested in cladocerans and mining pollution.

Conclusion

Clearly, there is a need to conduct further research in the field. The most interesting and necessary aspect is related to most sensitive species. Thus, it would be highly preferable if researchers published full species composition data even in cases, where only selected taxa are being studied in greater detail. Cladoceran potential as community level indicators is another issue which deserves more attention in research. For example, cladoceran community studies utilizing spatial and temporal aspects along pollution gradient could hold great potential in this sense. Finally, ecotoxicological research regarding the most tolerant taxa (*Chydorus sphaericus* and *Bosmina longirostris*) in natural setting is of utmost importance, but this should be preceded by clarification of the taxonomical issues of those species.

Acknowledgements This work was funded by Tellervo and Juuso Walden Foundation.

References

Adamczuk M (2016) Past, present, and future roles of small cladoceran *Bosmina longirostris* (O. F. Müller, 1785) in aquatic ecosystems. Hydrobiologia 767:1–11

Arruda JA, Marzolf GR, Faulk RT (1983) The role of suspended sediments in the nutrition of zooplankton in turbid reservoirs. Ecology 64:1225–1235

Battarbee RW (1999) The importance of paleolimnology to lake restoration. Hydrobiologia 395(396):149–159

Belyaeva M, Deneke R (2007) Colonization of acidic mining lakes: *Chydorus sphaericus* and other Cladocera within a dynamic horizontal pH gradient (pH 3 – 7) in lake Senftenberger See (Germany). Hydrobiologia 594:97–108

Belyaeva M, Taylor DJ (2009) Cryptic species within the *Chydorus sphaericus* species complex (Crustacea: Cladocera) revealed by molecular markers and sexual stage morphology. Mol Phylogenet Evol 50:534–546

Bennett K (1996) Determination of the number of zones in a biostratigraphical sequence. New Phytol 132:155–170

Błędzki LA, Rybak JI (2016) Freshwater crustacean zooplankton of Europe. Springer, Swizerland

Bossuyt BTA, Janssen CR (2005) Copper toxicity to different field-collected cladoceran species: intra- and inter-species sensitivity. Environ Pollut 136:145–154

Bourg ACM, Loch JPG (1995) Mobilization of heavy metals as affected by pH and redox conditions. In: Salomons W, Stigliani WM (eds) Biogeodynamics of pollutants in soils and sediments. Environmental science. Springer, Berlin, pp 87–102

Bownik A (2017) Daphnia swimming behavior as a biomarker in toxicity assessment: a review. Sci Total Environ 601–602:194–205

Bozelli RL (1996) The influence of bauxite tailings on the cladoceran populations of Lake Batata, Amazonia, Brazil. Int Rev Hydrobiol 81:621–634

Bradbury JP, Megarad RO (1972) Stratigraphic record of pollution in Shagawa Lake, Northeastern Minnesota. Geol Soc Am Bull 83:2639–2648

Brooks JL, Dodson SI (1965) Predation, body size and composition of plankton. Science 150:28–35

Cain DJ, Luoma SN, Carter JL, Fend SV (1992) Aquatic insects as bioindicators of trace element contamination in cobble-bottom rivers and streams. Can J Fish Aquat Sci 49:2141–5154

Camargo JA, Alonso A, Salamanca A (2005) Nitrate toxicity to aquatic animals: a review with new data for freshwater invertebrates. Chemosphere 58:1255–1267

Canadian Government (2002) Metal mining effluent regulations. SOR/2002-222

Canton SP, Ward JV (1981) Benthos and zooplankton of coal strip mine ponds in the mountains of northwestern Colorado, U.S.A. Hydrobiologia 85:23–31

Chen G, Shi H, Tao J, Chen L, Liu Y, Lei G, Liu X, Smol P (2016) Industrial arsenic contamination causes catastrophic changes in freshwater ecosystem. Sci Rep 5:17419. https://doi.org/10.1038/srep17419

Clarke KR (1993) Non-parametric multivariate analysis of changes in community structure. Aust J Ecol 18:117–143

Coard MA, Cousen SM, Cuttler AH, Dean HJ, Dearing JA, Eglinton TI, Greaves AM, Lacey KP, O'Sullivan PE, Pickering DA, Rhead MM, Rodwell JK, Simola H (1983) Paleolimnological studies of annually-laminated sediments in Loe Pool, Cornwall, U.K. Hydrobiologia 103:185–191

Cuker BE (1987) Field experiment on the influences of suspended clay and P on the plankton of a small lake. Limnol Oceanogr 32:840–847

Dekker T, Greve GD, Ter Laak TL, Boivin ME, Veuger B, Gortzak G, Dumfries S, Lücker SMG, Kraak MHS, Admiraal W, van der Geest HG (2005) Development and application of a sediment toxicity test using the benthic cladoceran *Chydorus sphaericus*. Environ Pollut 140:231–238

Doig LE, Schiffer ST, Liber K (2015) Reconstructing the ecological impacts of eight decades of mining, metallurgical, and municipal activities on a small boreal lake in Norther Canada. Integr Environ Assess Manag 11:490–501

Draves JF, Fox MC (1998) Effects of a mine tailings spill on feeding and metal concentrations in juvenile perch (*Perca fluviatilis*). Environ Toxicol Chem 17:1626–1632

Dumont HJ (1989) The non-marine Cladocera in Belgium. In: Wouters K, Baert, L (eds) Proceedings of the symposium "Invertebrates of Belgium". Brussel, 25–26 November 1988, pp 137–142

ECOTOX (2017) The ecotoxicology database. https://cfpub.epa.gov/ecotox/ecotox_home.cfm. Accessed 17 May 2017

El-Bassat RA, Taylor WD (2007) The zooplankton community of Lake Abo Zaabal, a newly-formed mining lake in Cairo, Egypt. Afr J Aquat Sci 32:185–192

Elphick JR, Davies M, Gilron G, Canaria EC, Lo B, Bailey HC (2011) An aquatic toxicological evaluation of sulfate: the case for considering hardness as a modifying factor in setting water quality guidelines. Environ Toxicol Chem 30:247–257

European Union (2000) Directive 2000/60/EC of the European Parliament and of the Council of 23 October 2000 establishing a framework for community action in the field of water policy. Off J Eur Communities L327:1–72

Ferrari CR, Wisniewski MJ, Roque CV, Ronque LB, Campos MB, Rodgher S, Nascimento MRL, de Azevedo H (2009) Preliminary assessment of the zooplankton community composition in a region under the influence of a uranium mine (Caldas, southeastern Brazil). International Nuclear Atlantic Conference—INAC 2009. Rio de Janeiro, RJ, Brazil, September 27 to October 2, 2009 (ISBN 978-85-99141-03-8)

Ferrari CR, de Azevedo H, Wisniewski MJS, Roddgher S, Roque CV, Nascimento MRL (2015) An overview of an acidic uranium mine pit lake (Caldas, Brazil): composition of the zooplankton community and limnochemical aspects. Mine Water Environ 34:343–351

Ferrari CR, de Azevedo H, Nascimento F, Rodgher S, Almeida T, Bruschi AL, Nascimento MRL, Bonifacio RL (2017) Effects of the discharge of uranium mining effluents on the water quality of the reservoir: an integrative chemical and ecotoxicological assessment. Sci Rep 7:13919. https://doi.org/10.1038/s41598-017-14100-w

Garrido AV, Bozelli RL, De Esteves FA, Alves LS (2003) Long-term patterns of the planktonic cladoceran community of Batata Lake, Amazonia, Brazil. Acta Limnol Bras 15:41–53

Griffiths K, Thienpont J, Jeziorski A, Smol JP (2018) The impact of Ca-rich diamond mining effluent on downstream communities in softwater lakes of the Northwest Territories, Canada. Can J Fish Aquat Sci. https://doi.org/10.1139/cjfas-2017-0469

Hart RC (1987) Population dynamics and production of five crustacean zooplankton in a subtropical reservoir during years of contrasting turbidity. Freshw Biol 18:287–318

Harwell MA, Harwell CC, Weinstein DA, Kelly JR (1987) Anthropogenic stresses on ecosystems: issues and indicators of response and recovery. ERC-153. Ecosystems Research Center, Ithaca

Heine-Fuster I, Vega-Retter C, Sabat P, Ramos-Jiliberto R (2010) Osmoregulatory and demographic responses to salinity of the exotic cladoceran *Daphnia exilis*. J Plankton Res 32:1405–1411

Holopainen IJ, Holopainen A-L, Huitu E, Rahkola-Sorsa M, Zingel P (2008) The pelagic food web in forest lakes affected by alkaline mining waste in NW Russia. Est J Ecol 57:214–228

Hudson-Edwards KA, Dold B (2015) Mine waste characterization, management and remediation. Minerals 5:82–85

Jäger IS, Hölker F, Flöder S, Walz N (2011) Impact of *Chaoborus flavicans* on the zooplankton in a mesotrophic lake—a three year study. Internat Rev Hydrobiol 96:191–208

Jeppesen E, Nõges P, Davidson T, Haberman J, Nõges T, Blank K, Lauridsen T, Søndergaard M, Sayer C, Laugaste R, Johansson L, Bjerring R, Amsinck S (2011) Zooplankton as indicators in lakes: a scientific—based plea for including zooplankton in the ecological quality assessment of lakes according to the European Water Framework Directive (WFD). Hydrobiologia 676:279–297

Kerfoot WC, Robbins JA, Weider LJ (1999) A new approach to historical reconstruction: combining descriptive and experimental paleolimnology. Limnol Oceanogr 44:1232–1247

Kirk KL, Gilbert JJ (1990) Suspended clay and the population dynamics of planktonic rotifers and cladocerans. Ecology 71:1741–1755

Klapper H, Schultze M (1995) Geogenically acidified mining lakes—living conditions and possibilities of restoration. Int Rev Hydrobiol 80:639–653

Koivisto S, Ketola M (1995) Effects of copper on life-history traits of Daphnia pulex and Bosmina longirostris. Aquat Toxicol 32:255–269

Koivisto S, Ketola M, Walls M (1992) Comparison of 5 cladoceran species in short- and long-term copper exposure. Hydrobiologia 248:125–136

Korhola A, Rautio M (2001) Cladocera and other Branchiopod crustaceans. In: Smol JP, Birks HJB, Last WM (eds) Tracking environmental change using lake sediments, vol 4. Zoological indicators. Kluwer Academic Publishers, Dordrecht, pp 5–41

Korosi JB, Kurek J, Smol JP (2013) A review on utilizing *Bosmina* size structure archieved in lake sediments to infer historic shifts in predation regimes. J Plankton Res 53:444–460

Kozlova T, Wood CM, McGeer JC (2008) The effect of water chemistry on the acute toxicity of nickel to cladoceran Daphnia pulex and the development of a biotic ligand model. Aquat Toxicol 91:221–228

Kurek J, Kirk JL, Muir DCG, Wang X, Evans MS, Smol JP (2012) Legacy of a half century of Athabasca oil and development recorded by lake ecosystems. Proc Natl Acad Sci USA 110:1761–1766

Labaj AL, Jeziorski A, Kurek J, Smol JP (2014) Long-term trends in Cladoceran assemblages related to acidification and subsequent liming of Middle Lake (Sudbury Canada). Water Air Soil Pollut 225:1868. https://doi.org/10.1007/s11270-014-1868-2
Labaj AL, Kurek J, Jeziorski A, Smol JP (2015) Elevated metal concentrations inhibit biological recovery of Cladocera in previously acidified boreal lakes. Freshw Biol 60:347–359
Lampert W (2011) Daphnia: development of a model organism in ecology and evolution. Excellence in ecology, book 21. International Ecology Institute, Oldendorf
Leppänen J, Siitonen S, Weckström J (2016) The stability of cladoceran communities in sub-arctic NW Finnish Lapland lakes. Polar Biol 40:2211–2223
Leppänen J, Weckström J, Korhola A (2017a) Paleolimnological fingerprinting of the impact of acid mine drainage after 50 years of chronic pollution in a southern Finnish lake. Water Soil Air Pollut 228:227. https://doi.org/10.1007/s11270-017-3417-2
Leppänen J, Weckström J, Korhola A (2017b) Mining pollution triggered a regime shift in cladoceran community of Lake Kirkkojärvi. J Paleolimnol, S Finland. https://doi.org/10.1007/s10933-018-0030-3
Leppänen J, Weckström J, Korhola A (2017c) Multiple mining impacts induce widespread changes in ecosystem dynamics in a boreal lake. Sci Rep 7:10581. https://doi.org/10.1038/s41598-017-11421-8
Lindenmayer DB, Likens GE (2009) Adaptive monitoring: a new paradigm for long-term research and monitoring. Trends Ecol Evol 24:482–486
Lopes I, Goncalves F, Soares AMVM, Ribeiro R (1999) Discriminating the ecotoxicity due to metals and to low pH in acid mine drainage. Ecotox Environ Safe 44:207–214
Lundstedt L, Brett MT (1991) Differential growth rates of three cladoceran species in response to mono- and mixed-algal cultures. Limnol Oceanogr 36:159–165
Maia-Barbosa PM, Bozelli RR (2005) Length-weight relationships for five cladoceran species in an Amazonian lake. Braz Arch Biol Technol 48:303–308
Markert BA, Breure AM, Zechmeister HG (2003) Definitions, strategies and principles for bioindication/biomonitoring. In: Markert BA, Breure AM, Zechmeister HG (eds) Bioindicators & biomonitors. Principles, concepts and applications. Trace metals and other contaminants in the environment 6. Elsevier, Kidlington, pp 3–39
McCabe GD, O'Brien WJ (1983) The effects of suspended silt on feeding and reproduction of Daphnia pulex. Am Midl Nat 110:324–337
Melville GE (1995) Changes in pelagic crustacean zooplankton of high-boreal Island Lake, Saskatchewan, associated with uranium mining. Environ Monit Assess 34:45–58
Miller GC, Lyons WB, Davis A (1996) Understanding the water quality of pit lakes. Environ Sci Technol 30:118A–123A
Molenda T (2014) Impact of saline mine water: development of a meromictic reservoir in Poland. Mine Water Environ 33:327–334
Moreira FWA, Leite MGP, Fujaco MAG, Mendoca FPC, Campos LP, Eskinazi-Sankt'Anna EM (2016) Assessing the impacts of mining activities on zooplankton functional diversity. Acta Limnol Bras. https://doi.org/10.1590/S2179-975X0816
Moser M, Weisse T (2011) The most acidified Austrian lake in comparison to a neutralized mining lake. Limnologica 41:303–315
Nevalainen L, Luoto TP (2013) Limnological deterioration forces community and phenotypic changes in Cladocera: tracking eutrophication of Mallusjärvi, a lake in southern Finland. Boreal Env Res 18:209–222
Nixdorf B, Mischke U, Leßmann D (1998) Chrysophytes and chlamydomonads: pioneer colonists in extremely acidic mining lakes (pH < 3) in Lusatia (Germany). Hydrobiologia 369(370):315–327
Niyogi S, Wood CM (2004) Biotic ligand model, a flexible tool for developing site-specific water quality guidelines for metals. Environ Sci Technol 38:6177–6192
Nordstrom DK (2011) Mine waters: acidic to circumneutral. Elements 7:393–398
Persoone G, Baudo R, Cotman M, Blaise C, Thompson KCL, Moreira-Santos M, Vollat B, Törökne A, Han T (2009) Review on the acute Daphnia magna toxicity test—evaluation of the sensitivity and the precision of assays performed with organisms from laboratory cultures or hatched from dormant eggs. Knowl Manag Aquat Ecosyst 393:1. https://doi.org/10.1051/kmae/2009012
Reichl C, Schatz M, Zsak G (2016) World mining data. Volume 31. Minerals Production. International organizing committee for the world mining congresses. Federal Ministry of Science, Research and Economy, Vienna
Rogalski MA (2015) Tainted resurrection: metal pollution is linked with reduced hatching and high juvenile mortality in *Daphnia* egg banks. Ecology 96:1166–1173
Rumes B, Eggermont H, Verschuren D (2011) Distribution and faunal richness of Cladocera in western Uganda crater lakes. Hydrobiologia 676:39. https://doi.org/10.1007/s10750-011-0829-7
Salmelin J, Leppänen MT, Karjalainen AK, Vuori K-M, Gerhardt A, Hämäläinen H (2017) Assessing ecotoxicology of biomining effluents in stream ecosystems by in situ invertebrate bioassays: a case study in Talvivaara, Finland. Environ Toxicol Chem 36:147–155
Sarma SSS, Nandini S (2006) Review of recent ecotoxicological studies on cladocerans. J Environ Sci Health B 41:1417–1430
Saros JE (2009) Integrating neo- and paleolimnological approaches to refine interpretations of environmental change. J Paleolimnol 41:243–252
Schultze M, Pokrandt K-H, Hille W (2010) Pit lakes of the Central German lignite mining district: creation, morphometry and water quality aspects. Limnologica 40:148–155
Kozlova T, Wood CM, McGeer JC (2008) The effect of water chemistry on the acute toxicity of nickel to cladoceran Daphnia pulex and the development of a biotic ligand model. Aquat Toxicol 91:221–228
Kurek J, Kirk JL, Muir DCG, Wang X, Evans MS, Smol JP (2012) Legacy of a half century of Athabasca oil and development recorded by lake ecosystems. Proc Natl Acad Sci USA 110:1761–1766

Smirnov N (2017) Physiology of the Cladocera. Elsevier Academic Press, London

Smol JP (1992) Paleolimnology: an important tool for effective ecosystem management. J Aquat Ecosyst Health 1:49–58

Sofyan A, Shaw JR, Birge WJ (2009) Metal trophic transfer from algae to cladocerans and the relative importance of dietary metal exposure. Environ Toxicol Chem 25:1034–1041

Stankovic I, Ternjej I (2009) New data on Croatian crustacean plankton fauna (Copepoda & Cladocera) in Cingi-Lingi Lake. Nat Croat 18:181–184

Stankovic I, Ternjej I, Mihaljevic Z, Furac L, Kerovec M (2011) Crustacean plankton community (Crustacea: Copepoda and Cladocera) in gypsum karst lakes and their relation to abiotic parameters. Hydrobiologia 666:145–153

Sterner RW (2009) Role of zooplankton in aquatic ecosystems. In: Likens GE (ed) Encyclopedia of inland waters. Academic Press, Oxford, pp 678–688

Stewart RA, Saiki MK, Kuwabara JS, Alpers CN, Marvin-DiPasquale M, Krabbenhoft DP (2008) Influence of plankton mercury dynamics and trophic pathways on mercury concentrations of top predator fish of a mining-impacted reservoir. Can J Fish Aquat Sci 65:2351–2366

Stow CA, Carpenter SR, Webster KE, Frost TM (1998) Long-term environmental monitoring: some perspectives from lakes. Ecol Appl 8:269–276

Suhett AL, Santangelo JM, Bozelli RL, Steinberg CEW, Farjalla VF (2015) An overview of the contribution of studies with cladocerans to environmental stress research. Acta Limnol Bras 27:145–159

Thienpont JR, Korosi JB, Hargan KE, Williams T, Eickmeyer DC, Kimpe LE, Palmer MJ, Smol JP, Blais JM (2016) Multi-trophic level response to extreme metal contamination from gold mining in a subarctic lake. Proc R Soc Lond B Biol Sci 283:1836. https://doi.org/10.1098/rspb.2016.1125

Tolotti M, Milan M, Szeroczynska K (2016) Subfossil Cladocera as a powerful tool for paleoecological reconstructions. Adv Oceanogr Limnol. https://doi.org/10.4081/aiol.2016.6467

Tõnno I, Agasild H, Kõiv T, Freiberg R, Nõges P, Nõges T (2016) Algal diet of small-bodied crustacean zooplankton in a cyanobacteria-dominated eutrophic Lake. PLoS ONE 11(4):e0154526. https://doi.org/10.1371/journal.pone.0154526

Vad CF, Horvàth Z, Kiss KT, Àcs È (2012) Microcrustacean (cladoceran, copepod) communities in artificial lakes in the region of the north hungarian mountains, with special reference to the adventive species. Acta Zool Acad Sci Hung 58:47–61

Valente T, Gomes CL (2007) The role of two acidophilic algae as ecological indicators of acid mine drainage sites. J Iber Geol 33:283–294

Van Dam RA, Harford AJ, Lunn SA, Gagnon MM (2014) Identifying the cause of toxicity of a saline mine water. PLoS ONE 9(9):e106857. https://doi.org/10.1371/journal.pone.0106857

Vandysh OI (2004) Zooplankton as an indicator of the state of lake ecosystems polluted with mining wastewater in the Kola Peninsula. Russ J Ecol 35:134–140

Vink J, Verschoor A (2010) Biotic ligand models: availability, performance and applicability for water quality assessment. Deltares, Netherlands

Von Der Ohe PC, Liess M (2004) Relative sensitivity distribution of aquatic invertebrates to organic and metal compounds. Environ Toxicol 23:150–156

Wolkersdorfer C, Bowell R (2005) Contemporary Reviews of Mine Water Studies in Europe, Part 3. Mine Water Environ 24:58–76. https://doi.org/10.1007/s10230-005-0074-2

Watling HR (2015) Review of biohydrometallurgical metals extraction from polymetallic mineral resources. Minerals 5:1–60

Wilk-Woźniak E, Zurek R (2006) Phytoplankton and its relationships with chemical parameters and zooplankton in the meromictic Piasecno reservoir, South Poland. Aquat Ecol 40:165–176

Wilk-Woźniak E, Pociecha A, Ciszewski D, Aleksander-Kwaterczak U, Walusiak E (2011) Phyto- and zooplankton in fishponds contaminated with heavy metal runoff from a lead-zinc mine. Oceanol Hydrobiol Stud 40:77–85

Winegardner AK, Salter N, Aebischer S, Pienitz R, Derry AM, Wing B, Beisner BE, Gregory-Eaves I (2017) Cladoceran diversity dynamics in lakes from a northern mining region: responses to multiple stressors characterized by alpha and beta diversity. Can J Fish Aquat Sci 74:1654–1667

Wollmann K, Deneke R, Nixdorf B, Packroff G (2000) Dynamics of planktonic food webs in three mining lakes across a pH gradient (pH 2-4). Hydrobiologia 433:3–14

Yount JD, Niemi GJ (1990) Recovery of lotic communities and ecosystems from disturbance—a narrative review of case studies. Environ Manage 14:547–569

Zawisza E, Zawiska I, Correa-Metrio A (2016) Cladocera community composition as a function of physicochemical and morphological parameters of dystrophic lakes in NE Poland. Wetlands 36:1131–1142

Zhou Q, Zhang J, Fu J, Shi J, Jiang G (2008) Biomonitoring: an appealing tool for assessment of metal pollution in the aquatic ecosystem. Anal Chim Acta 606:135–150

7

The East River tidal strait, New York City, New York, a high-nutrient, low-chlorophyll coastal system

Yaqin Li · Shannon L. Meseck · Mark S. Dixon · Gary H. Wikfors

Abstract The East River tidal strait, located between New York Harbor and Western Long Island Sound, is characterized by high suspended silt concentrations with low organic content kept in suspension by intense tidal currents. Inorganic nutrients, including nitrate, nitrite, ammonia, and phosphate, were high even during the summer. Dissolved inorganic nitrogen (DIN) concentrations generally were above 20 μM and did not likely limit phytoplankton growth. Despite high nutrient concentrations, median chlorophyll *a* concentration was only 1.53 μg l^{-1}, making the East River tidal strait a high-nutrient, low-chlorophyll (HNLC) area, likely a result of suspended silt blocking light penetration into the surface water. There were times at which the ratio of mixed layer to depth of the euphotic zone was generally greater than what has been suggested for phytoplankton to produce net primary production. The high-nutrient East River tidal strait is likely one of the sources of nutrients fueling summer phytoplankton production and consequent hypoxia in the Western Long Island Sound as silt settles from surface water in the lower turbulence conditions of the western narrows of Long Island Sound, thereby allowing light penetration and subsequent consumption of dissolved nutrients by phytoplankton.

Keywords High-nutrient · Low-chlorophyll · East River tidal strait · Western Long Island Sound · Water quality

Introduction

The East River tidal strait (referred to simply as the 'East River' in some references and maps) is a water body with a length of about 25 km connecting New York Harbor to the east to Western Long Island Sound to the west (Fig. 1); thus, it is not a true river. As a "tidal strait," the East River receives little fresh water input compared to conventional rivers. The East River tidal strait is divided into upper and lower sections by the Hell Gate sill where the tidal flow is restricted through a narrow passage (Marmer 1935). The upper East River tidal strait, from the Hell Gate to Willets Point near Western Long Island Sound, ~ 13.5 km in length, is shallower and wider than Hell Gate. The lower East River tidal strait, ~ 12 km, from Hell Gate to the Battery

Y. Li (✉) · S. L. Meseck · M. S. Dixon · G. H. Wikfors
NOAA, National Marine Fisheries Service, Northeast Fisheries Science Center, Milford Laboratory, Milford, CT 06460, USA
e-mail: judy.yaqin.li@noaa.gov

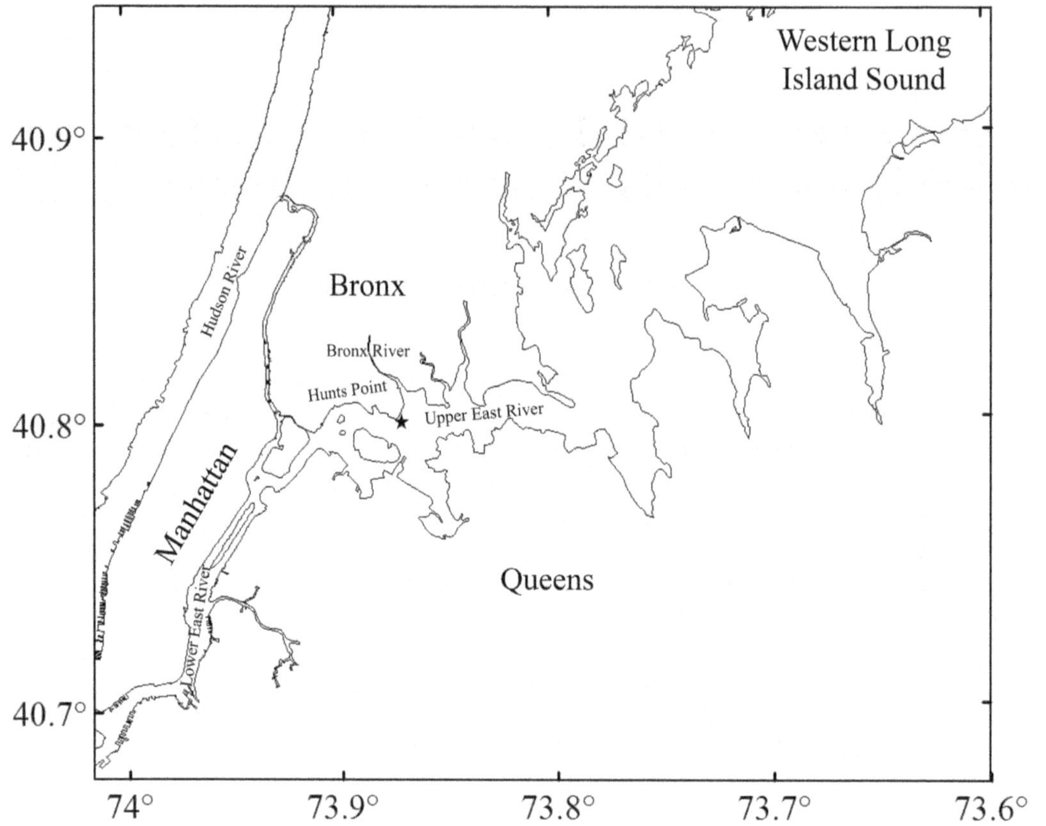

Fig. 1 Map of the study site located at 40°48′5.99″N, 73°52′17.76″W, off Hunts Point, Bronx, New York

where the Hudson River enters the New York Harbor, is narrower and deeper (Bowman 1976). The flow in the East River tidal strait is tidally dominated, with a semidiurnal pattern resulting from the superposition of the Long Island Sound and New York Harbor tidal waves (Bowman 1976; Blumberg and Pritchard 1997). In the lower East River tidal strait, the tidal current is swift and can reach as high as 3 m s^{-1} near Hell Gate (Bowman 1976). In the upper East River tidal strait, the current is slower but still reaches 1 m s^{-1}. Within the East River tidal strait, there is a two-layered flow with a near-surface transport toward Long Island Sound and transport at depth toward the New York Harbor. Overall net water transport is thought to be toward the Sound (Bowman 1976; Browning et al. 2014). In the upper East River tidal strait, the water is turbid with high concentration of inorganic materials in the total suspended seston, but very low particulate organic matter or chlorophyll (Galimany et al. 2013). Little information, however, is available concerning phytoplankton, light availability, and other relevant environmental variables to understand if the system is able to produce net primary production.

Most coastal areas, particularly those that are highly populated, are subject to high nutrient loading (e.g., Varekamp et al. 2014). With high nutrients, high phytoplankton abundance is common (e.g., Nixon 1995). In the turbid coastal environment, however, light can become the factor limiting phytoplankton growth and primary productivity (Grobbelaar 1985; Cloern 1987; Irigoien and Castel 1997; Domingues et al. 2011b; Gameiro et al. 2011). Phytoplankton may be limited by the light part of the year or year-round (Kromkamp et al. 1995; Kocum et al. 2002; Domingues et al. 2011b). In nutrient-rich, turbid environments, primary productivity may be directly proportional to the availability of light (Underwood and Kromkamp 1999; Domingues et al. 2011a). Thus, in very turbid environments, high nutrients do not necessarily lead to high phytoplankton abundance (Irigoien and Castel 1997). Phytoplankton may adapt to low light conditions, whether caused by high turbidity or low incident light (Grobbelaar 1985; McMinn et al. 2010), with increased photosynthetic efficiency. Estuaries, such as the Chesapeake Bay, Delaware Bay, and San Francisco Bay (Cole and Cloern 1984; Harding et al. 1986), have areas within them that are typically classified as high nutrient low chlorophyll-a (HNLC) attributed to high turbidity.

We characterized nutrient and seston conditions in the upper East River tidal strait. We report here phytoplankton, inorganic nutrients, euphotic zone depth, phytoplankton potential maximum photosynthetic efficiency, and other environmental variables. As coastal areas tend to exhibit high temporal variability over a short (day) and long term (e.g., seasonal and inter-annual), we conducted high-frequency measurements of some environmental variables over spring–autumn seasons. Our results suggest that phytoplankton were at times not able to achieve net primary production despite high nutrient availability because of low light penetration. Our findings, combined with existing knowledge of tidal currents, offer insights into the nutrient dynamics of western Long Island Sound, an area in which symptoms of eutrophication prevent achievement of water-quality standards.

Materials and methods

Study site

The sampling station at the upper East River tidal strait is located at 40°48′5.99″N, 73°52′17.76″W, off Hunts Point, Bronx, New York (Fig. 1). It is in the highly industrialized section of New York City. The site is within close proximity to the Hunts Point Wastewater Treatment Plant, one of the largest wastewater treatment plants in New York City. The upper East River tidal strait reportedly has varied depth. The depth near Hunts Point, at Clarkson Point/College Point is between 15 m on the south Channel and 24 m at the center of the channel (Blumberg and Pritchard 1997). The Department of Environmental Protection of the State of New York has conducted a year-round, water-quality survey in the area in recent years. They reported that a station (E14) near this study site is well mixed and has a mean depth of 6.5 m (http://www.nyc.gov/html/dep/html/harborwater/harbor_water_sampling_results.shtml).

Water chemistry measurements

Water samples were collected weekly from April 2012 to October 2012 from 0.5 m below the surface for major nutrients (including nitrate plus nitrite, ammonium, and phosphate), chlorophyll *a*, and total suspended solids. Chlorophyll *a* was measured fluorometrically. Sample filtration and extraction followed Li et al. (2009), and fluorescence intensity was measured by a Perkin Elmer Plate reader following Li et al. (2013). In short, a known amount of water (generally 500 ml) was filtered through a GF/F glass fiber filter immediately after collection. Filters were kept on ice until transported to the lab where they were stored at -20 °C until analyses. Immediately prior to analyses, filters were sonicated in 90% acetone with a Tekmar Sonic Disruptor. The sonicated solutions were kept refrigerated overnight to allow for pigment extraction. Samples then were centrifuged at 8000×*g* for 10 min. A 300-μl supernatant sub-sample was pipetted into a 96-well plate well. Samples were read at 690 nm by a PerkinElmer Luminescence Spectrometer LS 50B (PerkinElmer inc., Waltham, MA, USA). The regression between fluorescence and chlorophyll *a* was calibrated with a chlorophyll *a* standard from Sigma-Aldrich Co. LLC. (St. Louis, MO, USA).

Nutrient samples were taken with protocols detailed by Meseck et al. (2012). Briefly, 120 ml of water was filtered through a pre-combusted (450 °C overnight), 25-mm GF/F filter placed in a low-pressure, syringe-filter holder. The filtered water was stored in a cooler with ice until analysis at the laboratory.

Total ammonia, phosphate, silicate, and nitrate plus nitrite were determined using a Quattro Auto Analyzer (Seal Analytical, WI, USA). Detailed descriptions for quantifying total ammonia, nitrate plus nitrite, phosphate, and silicate using this instrument can be found in Methods of Seawater Analysis (Hansen and Koroleff 1999). Briefly, total ammonia was determined using the Berthelot reaction (detection limit, 0.05 μM). For nitrate plus nitrite, a cadmium column was used to reduce nitrate to nitrite, and then the formation of the red azo dye was used (detection limit 0.005 μM). The formation of β-silicomulybic acid was used to determine silicate (detection limit 0.1 μM), and a phosphomolybdate hetropoly acid reduced to a blue compound was used to determine phosphate (detection limit 0.01 μM).

Urea determinations were done following the method of Goeyens et al. (1998). Briefly, the method involves the formation of imidazolone–thiosemicarbazide complex at room temperature for 72 h. The detection limit is 0.10 uM with 1.6% reproducibility.

Total suspended solids (TSS) samples were collected on 25 mm pre-weighed and combusted Gf/F filters. Samples were put immediately at – 20 °C until analysis. Samples were then dried at 60 °C and weighted.

Continuous, high-frequency measurements of environmental variables

Temperature, salinity, dissolved oxygen, chlorophyll *a*, pH, turbidity, and under-water PAR (photosynthetically available radiation) were measured in situ by sensors attached to a YSI 6600 V2-4 Sonde (YSI Inc., Yellow Springs, OH, USA) suspended about 0.5 m below the surface from a moored raft. All of the sensors attached to the Sonde took a reading every 15 min. The data collections were from April to October 2012, with gaps plotted on Fig. 4. Sensor models used were as follows: temperature and salinity (combined), 6360; dissolved oxygen, ROX™; chlorophyll *a*, 6025; pH, 6561; turbidity, 6136; under-water PAR, 2π quantum sensors (model: LI-192) by LI-COR® (LI-COR®, Lincoln, Nebraska USA). The top PAR sensor was 0.5 m below the water surface and the bottom one was at 1 m below the surface. Sensor calibration and maintenance followed the recommendation of the manufacturer. Non-photochemical quenching of chlorophyll *a* fluorescence during the day resulted in chlorophyll *a* calculated values underestimated from the in situ fluorescence (Dandonneau and Neveux 1997; Chekalyuk and Hafez 2008). Accordingly, data corrections were made following Li et al. (2012).

Phytoplankton identifications

Each 100-ml water sample was fixed in 2% acidic Lugol's solution (Throndsen 1978) and stored in an amber glass bottle until analyses. A 50-ml sub-sample was settled following the standard Utermöhl settling chamber method (Hasle 1978) and examined with a Zeiss Axio inverted microscope (Carl Zeiss Microscopy, LLC, Thornwood, NY, USA).

Photosynthetic variables

The FIRe (Fluorescence Induction and Relaxation) System (by Satlantic Inc., Halifax, Canada) was used for variable chlorophyll fluorescence measurements twice a month from May to October 2012. FIRe is based upon the Fast Repetition Rate Fluorometry (FRRF) technique. Water samples collected from the study site (0.5 m below the surface) were kept in the dark at ambient water temperature for at least 20 min before measurements. F_v/F_m, maximum quantum efficiency or maximum quantum yield of PSII, was calculated through the built-in instrument protocol.

Euphotic zone depth calculations

The depth of euphotic zone, Zeu, taken as the depth to which 1% of the surface light penetrates, was calculated two ways: first from the measurements of the two under water PAR sensors following the equation:

$$I_z = I_0 \times e^{-kz}$$

in which k is the vertical extinction coefficient, and z equals the vertical distance between I_z and I_0, which represent irradiance at depth z and the surface, respectively. A daily median Zeu was then calculated based on data from 10:00 am to 2:00 pm.

Secondly, Zeu was calculated from Secchi disk depth measured twice monthly at the site. Zeu was 3.5 times of the secchi disk depth based on Holmes (1970).

Statistical analyses

To test the influence of tide upon the water variables, correlations were calculated between the individual variables and the tidal heights using Statgraphics Plus. Tidal heights were based upon NOAA tidal predictions at station Kings Point, approximately 10 km east of the study site. The NOAA table provided height at only 4 time points on each date; therefore, the gaps between these time points were filled assuming a sine function for the tidal heights. To focus on the tide and minimize the other variations, such as day-to-day or seasonal variations, the sampling time series was broken into 5 seasonal periods based upon water temperature. Pearson correlations between water variables and tidal heights were calculated for each of the 5 periods (Table 1).

Results

Total suspended solids (TSS) from April to October ranged from 7.73 to 46.93 with a median of 19.30 mg l^{-1} (Fig. 2). April to June had higher TSS than the rest of the months, with a median of 27.65 mg l^{-1}. Median TSS from July to October was 15.83 mg l^{-1}.

In terms of dissolved nutrients, the highest nitrate concentration was 34.71 μM, with a median of 14.01 μM (Fig. 3). There were 3 occasions during July–August when one or more dissolved nitrogen species was below the method detection limit. The highest ammonia was 43.19 μM, with a median of 23.90 μM. The total inorganic nitrogen (TIN) (nitrate, nitrite and ammonia combined) ranged from 15.23 to 73.84 μM from April to October. This range of TIN is high, particularly for the summer when most temperate, coastal areas experience undetectable or low TIN. Urea was also fairly high, with a maximum of 27.19 μM; the concentration never went below the detection limit. There also were fairly high concentrations of dissolved phosphate and silica in the water, ranging from 2.58 to 14.17 μM and 21.94 to166.72 μM for phosphate and silicate, respectively.

Seasonal changes in temperature and other environmental variables at the site are presented in Fig. 4. 96-point (24 h) moving averages were used to remove diel fluctuations. Temperature showed a seasonal warming from April to late August and a gradually decreasing trend afterwards. Salinity ranged from 22.11 to 27.43. The relatively small fluctuation in salinity is consistent with the fact that the upper East River tidal strait has limited freshwater input. Chlorophyll *a* ranged from 0.2 to 4.6 μg l^{-1} with a median value of 1.53 μg l^{-1}. Dissolved Oxygen (DO) ranged from 2.15 to 9.66 mg l^{-1}.

Table 1 Correlations between tidal height and other variables

Time period	Statistics	Variables				
		Temperature	Salinity	pH	DO (mg l^{-1})	Chl *a*
25/4/12 to 31/5/12	Pearson correlations	0.014	0.025	− 0.118	0.048	− 0.108
(Temperature increased)	*P* value	0.107	0.107	<0.001	0.005	<0.001
	Covariances	0.016	0.016	− 0.006	0.031	0.052
31/5/12 to 19/6/12	Pearson correlations	0.165	− 0.337	− 0.236	− 0.048	− 0.197
(Temperature increased)	*P* value	< 0.001	< 0.001	< 0.001	0.002	< 0.001
	Covariances	0.156	− 0.235	− 0.014	− 0.019	− 0.150
20/6/12 to 10/7/12	Pearson correlations	0.341	− 0.437	− 0.318	0.051	− 0.143
(Temperature stable)	*P* value	< 0.001	< 0.001	< 0.001	0.022	< 0.001
	Covariances	0.206	− 0.181	− 0.0083	0.011	− 0.134
12/7/12 to 26/7/12	Pearson correlations	0.539	− 0.441	0.009	0.455	− 0.109
(Temperature increased)	*P* value	< 0.001	< 0.001	0.73	< 0.001	< 0.001
	Covariances	0.222	− 0.195	0.001	0.360	− 0.085
29/8/12 to 27/9/12	Pearson correlations	0.008	− 0.334	0.106	0.263	− 0.022
(Temperature decreased)	*P* value	0.680	0.083	< 0.001	< 0.001	0.430
	Covariances	0.006	− 0.029	0.007	0.095	− 0.018

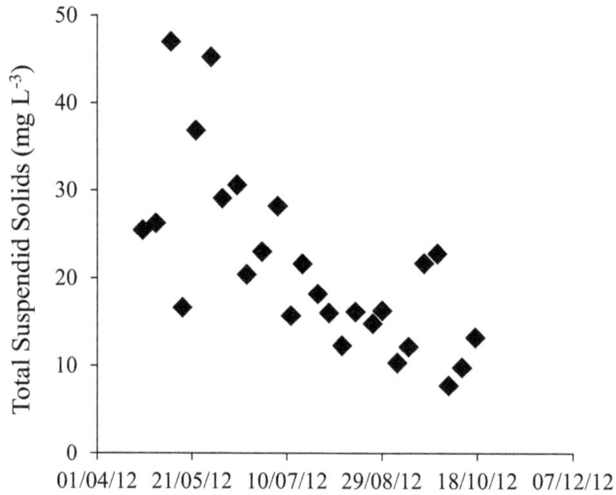

Fig. 2 Total suspended solids at the site

In addition to the seasonal changes in the environmental variables, there was also short-term variability revealed by plotting raw, un-averaged data (Figs. 5, 6). Water temperature, pH, and dissolved oxygen generally exhibited two highs and two lows daily.

Water temperature was positively correlated with tidal height from 31 May to 26 July (Table 1). Salinity was negatively correlated with the tidal height except for the period from 25 April to 31 May (Table 1) when rain in the spring may have upset the correlation. A negative rather than positive relationship between salinity and tidal height is consistent with water circulation in the East River tidal strait, which includes a net flow from New York Harbor toward the Western Long Island Sound. New York Harbor has lower salinity (influenced by the Hudson River) than western Long Island Sound. Correlation between pH and tidal height was not consistent between different time periods (Table 1); a negative correlation in the spring and early summer was replaced by a positive correlation in the summer to early fall. Dissolved oxygen showed consistent positive correlation with tidal height (Table 1). Chlorophyll *a* had a consistent negative correlation with tidal height (Table 1).

To focus specifically on the diel patterns of dissolved oxygen (DO) and pH, data from 7 June to 13 June were selected when day-to-day variation was small. DO and pH showed the same diel variation with two daily peaks and two valleys (Fig. 6).

The phytoplankton community was dominated by diatoms, which made up more than 70% of the community, numerically. The most common species were: *Skeletonema costatum*, *Thalassiosira* spp., *Thalassiosira nordenskioldii*, *Thalassionema nitzschioides*, *Asterionella japonica*, and *Chaetoceros* spp. In addition to diatoms, the green algae *Euglena* spp. and *Eutreptiella* spp. were frequently in the water. The dinoflagellate *Scrippsiella* sp. was present in low numbers. The phytoplankton maximum photosynthetic efficiency of PSII, F_v/F_m, at the site ranged from 0.359 ± 0.04 to 0.552 ± 0.051 (Fig. 7).

The euphotic zone depth ranged from 1.3 to 7.9 m, with a median of 3.8 m (Fig. 8). The values calculated from the measurement of underwater irradiance and Secchi disk depth agree well. It appears that light penetration was the lowest in June and July, with a median euphotic zone depth of 3.1 m. Light penetration increased late in the season in September and October, with a median euphotic zone depth of 6.1 m.

Discussion

Suspended particular matter in the East River tidal strait was remarkably high. The highly turbid condition was attributable to inorganic matter, as the chlorophyll level at the site was fairly low with a median value of 1.53 μg l^{-1}. This is in agreement with observations by Galimany et al. (2013) at the same site the previous year; this earlier study reported high suspended matter with low organic content. Galimany and co-workers attributed the high suspended solids at the site to tidal turbulence and Bronx River interactions with the East

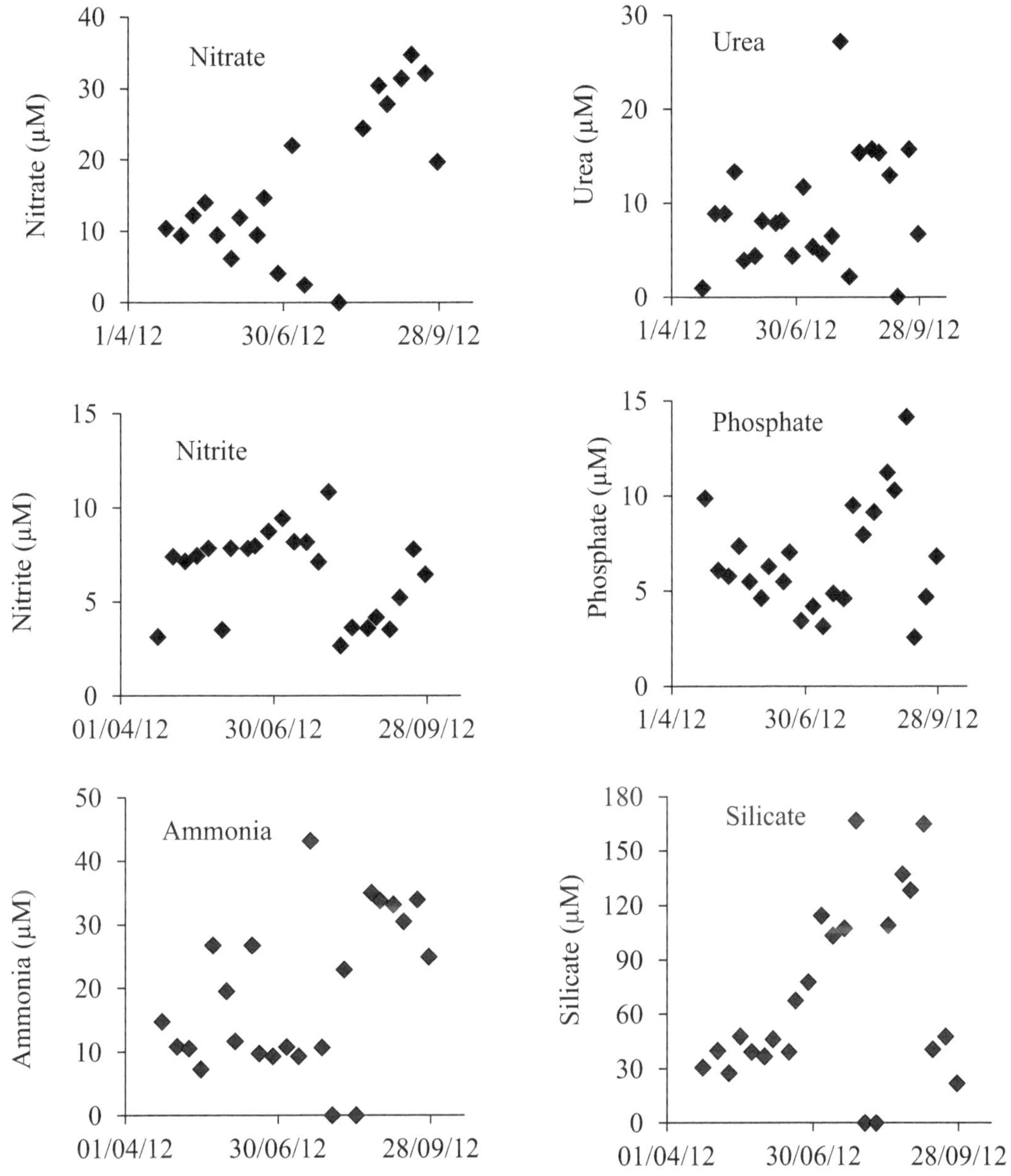

Fig. 3 Nutrients at the site from April 2012 to October 2012

River tidal strait. The swift tidal water movements at the East River tidal strait near our study site were documented by Bowman (1976), who reported a peak current speed of 3 m s^{-1} at the Hell Gate in the lower East River tidal strait and 1 m s^{-1} at the upper East River tidal strait close to our study site. Turbulent mixing of surface water at our study site was visually apparent at all times.

The phytoplankton community at the East River tidal strait site we studied was dominated by diatoms. The species composition was similar to that present in the adjacent western Long Island Sound (Capriulo et al. 2002; Lopez et al. 2014). One noticeable difference was the frequent presence of *Euglena* spp. and *Eutreptiella* spp. at the Hunts Point site. Although both *Euglena* and *Eutrepiella* were reported in the Long Island Sound, they were not in high numbers and not frequent (Capriulo et al. 2002; Liu and Lin 2008). Euglenoids can be associated with highly turbid conditions in an estuary (Pourafrasyabi and Ramezanpour 2014), possibly because this group is known to be mixotrophic (Yamane et al. 2001), and therefore, less dependent upon light than many other phytoplankters.

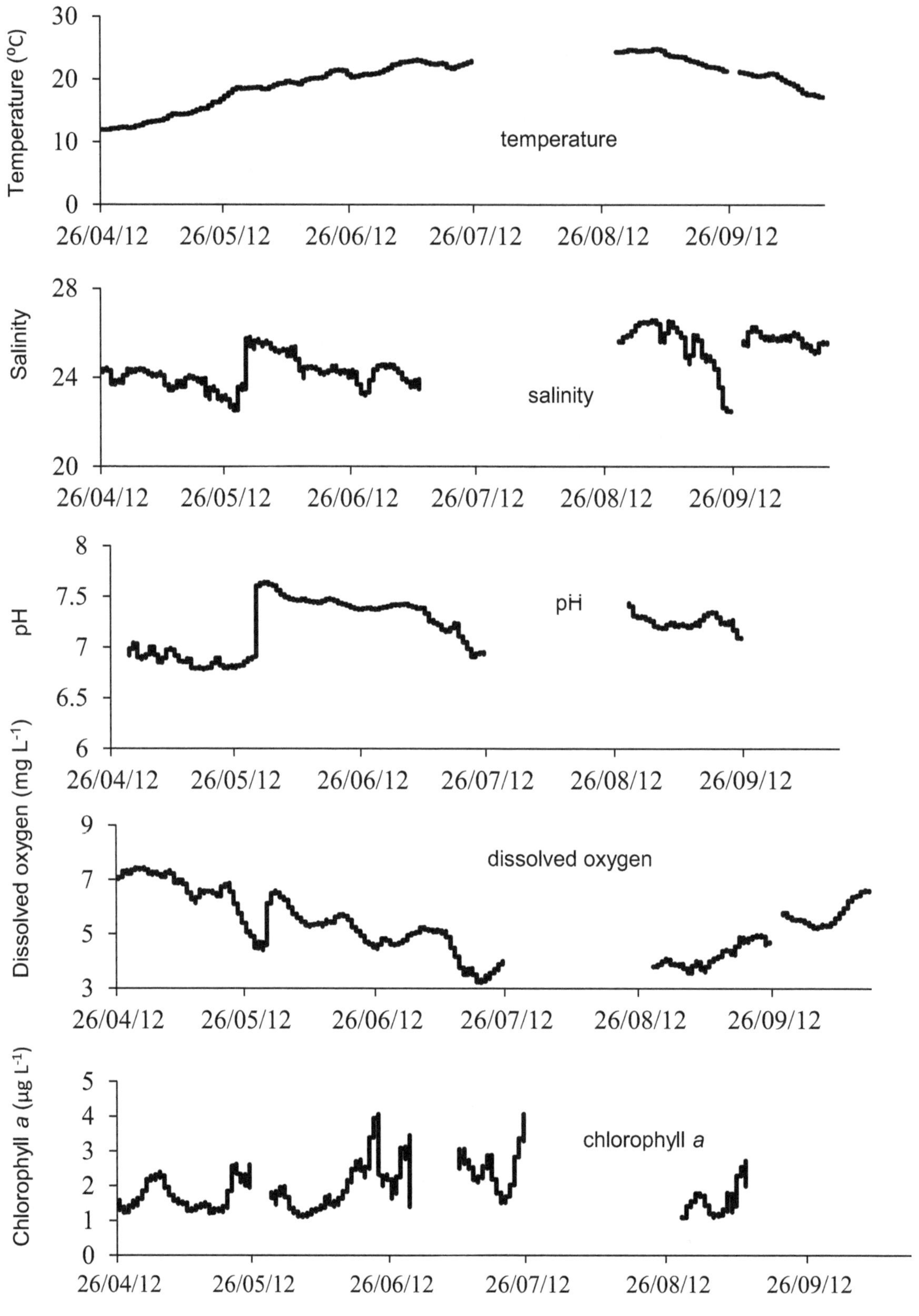

Fig. 4 Seasonal change in water variables at the site measured every 15 min (96 readings daily). Data presented are 96-point moving average. Thus, the long-term change is shown without the diel signals

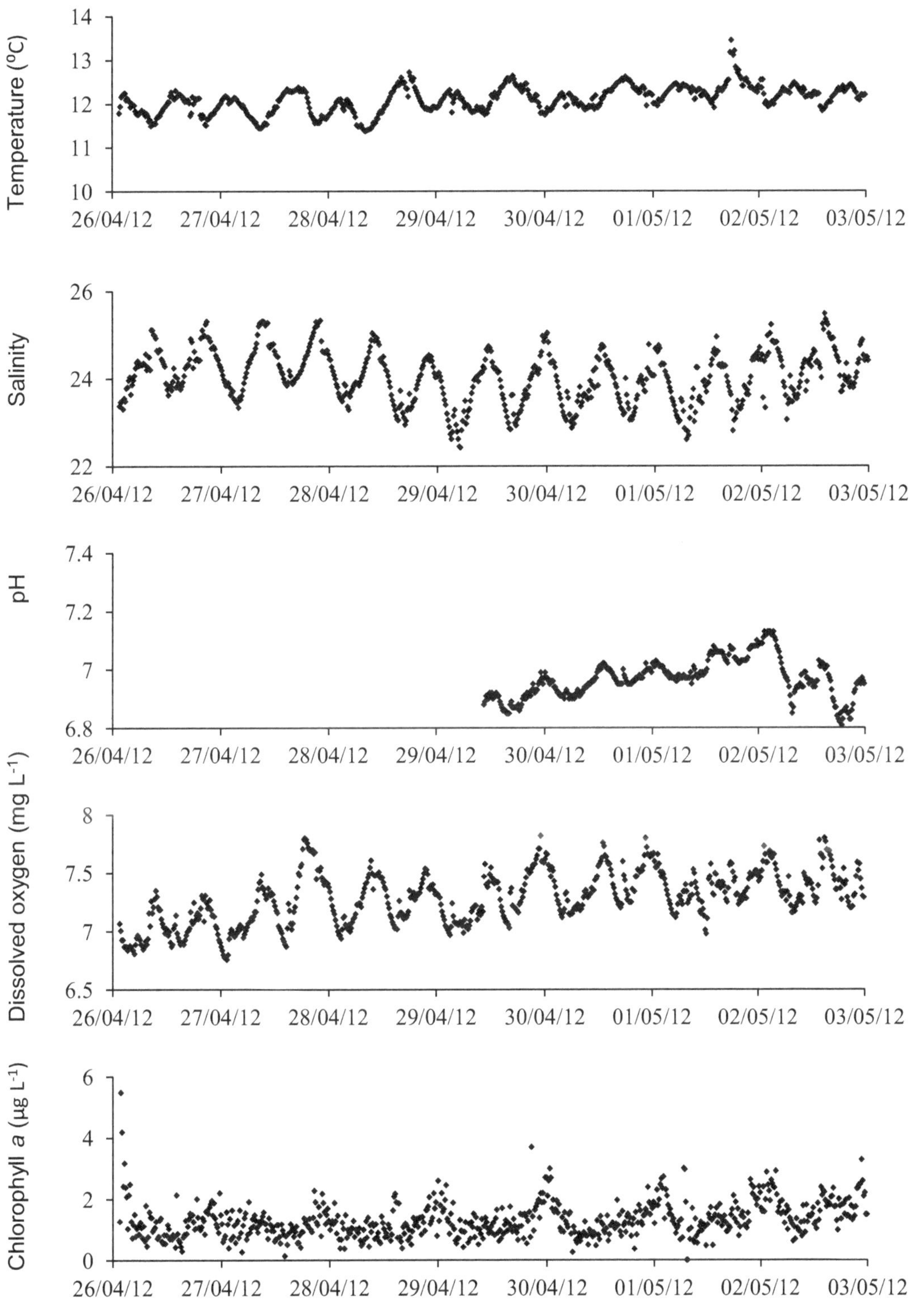

Fig. 5 Time series change in water variables at the site measured every 15 min. Data presented are from 26/04/2012 to 3/05/2012

Phytoplankton F_v/F_m in the East River tidal strait was lower compared to measurements in the Long Island Sound (personal communication with Matt Lyman, CT DEEP). F_v/F_m, a measure of photosynthetic efficiency of PSII, has been used as an indicator of general physiological stress. In the South Pacific Ocean, as well as in a HNLC (high nutrients low chlorophyll) equatorial upwelling region, reduction in Fv/Fm was related to iron limitation (Behrenfeld and Zbigniew 1999; Johnson et al. 2010). Goto et al. (2008) related changes in Fv/Fm

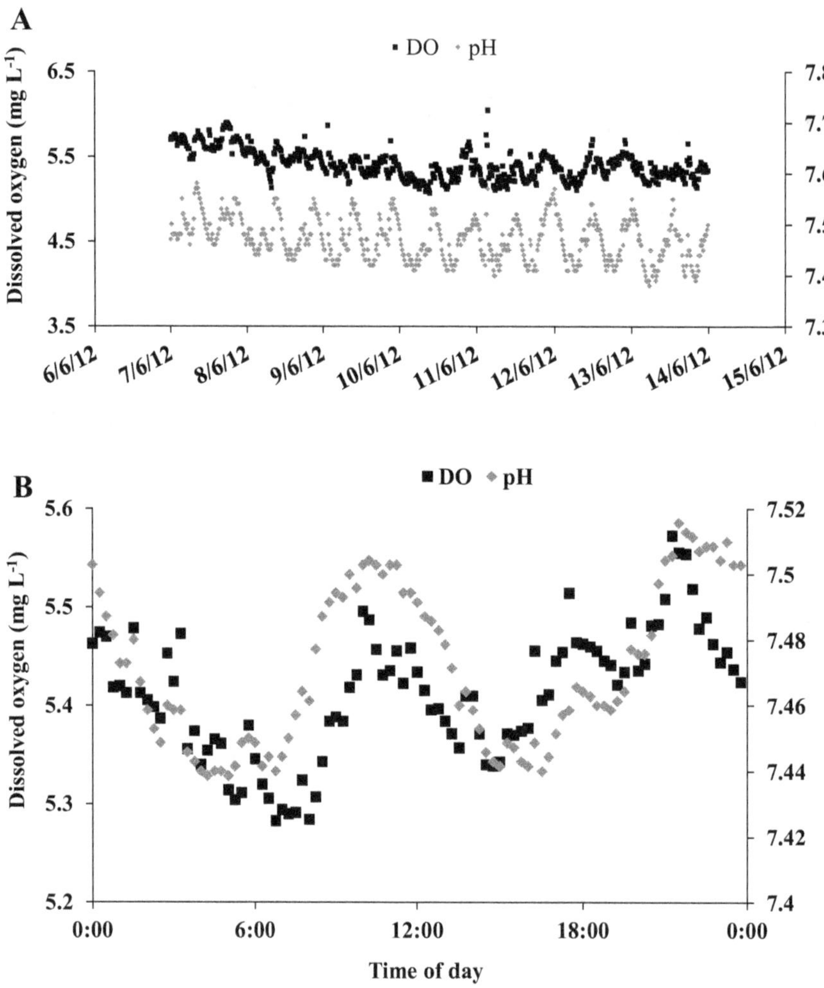

Fig. 6 Dissolved oxygen and pH at the site from June 7 2012 to June 13 2012. **a** Time series plots of data with 15 min-interval. **b** Mean diel cycle for the time period

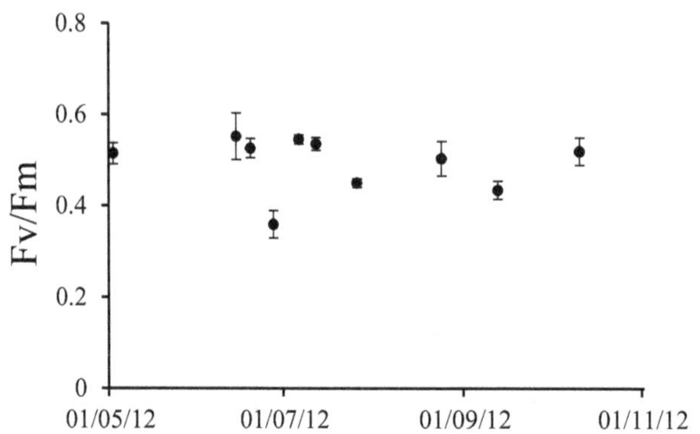

Fig. 7 Fv/Fm over time. Vertical bars represent one standard error of the mean

to the photo-damage of phytoplankton near the surface of lakes. Thus, low Fv/Fm in the East River Tidal strait likely indicated that the phytoplankton community in general was stressed.

In turbid estuaries, the ratio of mixed layer depth to euthotic zone depth (Z_m/Z_{eu}) has been used to indicate if phytoplankton bloom is possible (e.g., Irigoien and Castel 1997). In San Francisco Bay, it was possible for phytoplankton to produce net primary production and for a bloom to initiate when Z_m/Z_{eu} was less than 6

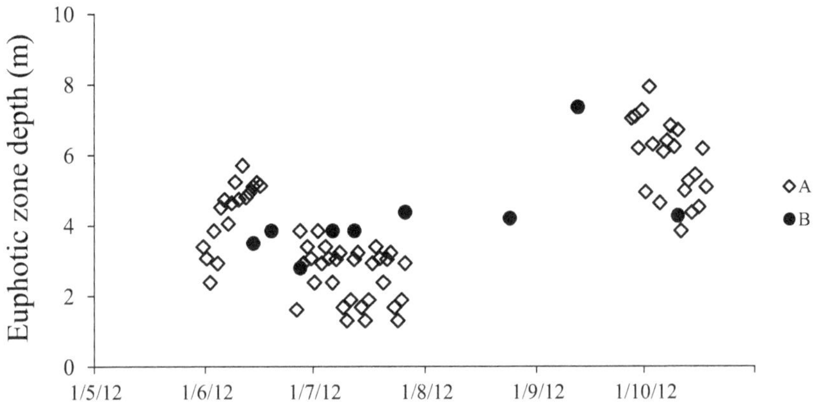

Fig. 8 Euphotic zone depth. **a** Daily median calculated from the under water PAR measured by the 2 sensors; **b** values calculated from the Secchi disk depth

(Cole and Cloern 1984; Alpine and Cloern 1988). It generally is considered to be possible for a net primary production when Z_m/Z_{eu} is 5–6 or lower (Cole and Cloern 1984; Grobbelaar 1985; Irigoien and Castel 1997). For the upper East River tidal strait, in the deep central channel (24 m depth), the Z_m/Z_{eu} was great than 6 for the sample period; therefore, it is likely that the central channel would not have net primary production to initiate a phytoplankton bloom. In the south channel with a depth of 15 m, Z_m/Z_{eu} was close to 5 in June/July and it is barely low enough for a bloom to initiate; whereas in September/October Z_m/Z_{eu} was 2.5 suggesting that there was enough light for a net primary production.

The high suspended solids and shallow euphotic zone at the study site had a noticeable effect on the growth of aquacultured macroalga. Kim et al. (2014) reported that the growth of *Gracilaria tikvahiae* McLachlan at the site declined from just 0.5 m below surface to 2.5 m depth, with maximum growth near the surface. The higher nitrogen and lower carbon content at the depth implied slower photosynthetic rates beneath 0.5 m depth.

The present study provides evidence that phytoplankton community photosynthesis was light-limited in the East River tidal strait. Whether or not phytoplankton photosynthesis and growth were sufficient to maintain the population on the site cannot, however, be ascertained from the data. Other factors contributing to the answer are: phytoplankton respiration rates, loss to grazing by zooplankton and other grazers, and gain or loss through water exchange (Wofsy 1983; Cole et al. 1992). In two previously studied environments, the turbid upper Hudson River and San Francisco Bay, phytoplankton respiration alone was shown to exceed photosynthesis in well-mixed areas (Alpine and Cloern 1988; Lewis 1988; Cole et al. 1992). In these systems, phytoplankton relied upon external input (immigration) to maintain biomass.

In contrast to most high-nutrient estuaries, the East River tidal strait has low chlorophyll concentrations. Our results suggested that the phytoplankton at the site were not able to produce net primary productivity at least part of the year because of the low light penetration in the water. We speculate, therefore, that the low chlorophyll at the site was caused by low in situ phytoplankton growth rates rather than loss of biomass through grazing, sinking, or horizontal transport.

Driven largely by water column photosynthesis and respiration, dissolved oxygen (DO) and pH of a productive coastal system usually exhibited one peak and one valley over a diel cycle (e.g., Li et al. 2012). In contrast, the East River tidal strait exhibited two peaks and two valleys in DO and pH over a diel cycle suggesting that tidal rather than biological activities (photosynthesis and respiration) played a major role. The effect of tidal influence on DO and pH is also witness by the significant, although weak, correlation between tidal height and DO, tidal height and pH, for most of the time periods (Table 1).

The Western Long Island Sound (WLIS), adjacent to the East River tidal strait, has exhibited seasonal hypoxia in the bottom water for decades (Parker and O'Reilly 1991; O'Donnel et al. 2014). The most common belief for the cause of hypoxia is that phytoplankton in the WLIS respond to the high loading of nitrogen and phosphorous from various sources. Seasonal stratification is thought to allow greater residence of phytoplankton in the upper euphotic zone. Thus, hypoxia develops when bacteria consume and respire the supply of phytoplankton and oxygen, particularly in the summer. The relaxation of stratification in the fall allows re-oxygenation of bottom waters. There are also indications that phytoplankton in the WLIS may be nutrient-limited, particularly in the summer. For example, Anderson and Taylor (2001) reported that phytoplankton production in the WLIS responded to the input of ammonia from rain events, and the severity of hypoxia also was coupled with ammonia pulses related to rain events.

Although there is a net flow of surface water from the East River tidal strait toward the WLIS (Bowman 1976; Browning et al. 2014), the East River tidal strait cannot be a significant source of phytoplankton biomass to WLIS because chlorophyll *a* levels in the East River tidal strait are much lower than those observed in the WLIS. The East River tidal strait with high nutrient concentrations, however, certainly can provide net nutrient input to the WLIS where light penetrations was higher to support phytoplankton growth. As the net tidal flow fluctuates on a weekly basis following the lunar cycle, nutrient input to the WLIS likely also fluctuates on a weekly basis. Coincidentally, hypoxia in WLIS was shown to oscillate on a weekly basis (Anderson and Taylor 2001). Thus, there is likely cause and effect between the nutrient input from East River tidal strait and the dynamics of hypoxia in the WLIS.

Acknowledgements The following individuals are appreciated for participating the field work: Genevieve Bernatchez, Eve Gallimony, Julie Rose, Marguerite Petit, and Aynur Lok. Matt Lyman of the CT Department of Environmental Protection and Energy (DEEP) is acknowledged for making available the Long Island Sound Monitoring Program data. We thank LTJG Erick Estela for providing the ArcView map of the study site. This study was supported by the NOAA Office of Aquaculture and NOAA Fisheries Northeast Fisheries Science Center. Brian Gionado was very accommodating for our field work at his construction site and he is appreciated for making one of his trailers available to us for fluorescence measurements in the dark.

References

Alpine AE, Cloern JE (1988) Phytoplankton growth rates in a light-limited environment, San Francisco Bay. Mar Ecol Progress Series Oldendorf 44:167–173

Anderson TH, Taylor GT (2001) Nutrient pulses, plankton blooms, and seasonal hypoxia in western Long Island Sound. Estuaries 24:228–243

Behrenfeld MJ, Zbigniew K (1999) Widespread iron limitation of phytoplankton in the South Pacific ocean. Science 283:840–843

Blumberg AF, Pritchard DW (1997) Estimates of the transport through the East River, New York. J Geophys Res C: Oceans 102:5685–5703

Bowman MJ (1976) The tides of the East River, New York. J Geophys Res 81:1609–1616

Browning TJ, Bouman HA, Moore CM, Schlosser C, Tarran GA, Woodward EMS, Hender Son GM (2014) Nutrient regimes control phytoplankton ecophysiology in the South Atlantic. Biogeosciences 11:463–479

Capriulo GM, Smith G, Troy R, Wikfors GH, Pellet J, Yarish C (2002) The planktonic food web structure of a temperate zone estuary, and its alteration due to eutrophication. Hydrobiologia 475:263–333

Chekalyuk A, Hafez M (2008) Advanced laser fluorometry of natural aquatic environments. Limnol Oceanogr Methods 6:591–609

Cloern JE (1987) Turbidity as a control on phytoplankton biomass and productivity in estuaries. Cont Shelf Res 7:1367–1381

Cole BE, Cloern JE (1984) Significance of biomass and light availability to phytoplankton productivity in San-Francisco bay. Mar Ecol Prog Ser 17:15–24

Cole JJ, Caraco NF, Peierls BL (1992) Can phytoplankton maintain a positive carbon balance in a turbid, freshwater, tidal estuary? Limnol Oceanogr 37:1608–1617

Dandonneau Y, Neveux J (1997) Diel variations of in vivo fluorescence in the eastern equatorial Pacific: an unvarying pattern. Deep-Sea Research Part II-Topical Stud Oceanogr 44:1869–1880

Domingues RB, Anselmo TP, Barbosa AB, Sommer U, Galvao HM (2011a) Light as a driver of phytoplankton growth and production in the freshwater tidal zone of a turbid estuary. Estuar Coast Shelf Sci 91:526–535

Domingues RB, Anselmo TP, Barbosa AB, Sommer U, Galvao HM (2011b) Nutrient limitation of phytoplankton growth in the freshwater tidal zone of a turbid, Mediterranean estuary. Estuar Coast Shelf Sci 91:282–297

Galimany E, Rose JM, Dixon MS, Wikfors GH (2013) Quantifying feeding behavior of ribbed mussels (Geukensia demissa) in two urban sites (Long Island Sound, USA) with different seston characteristics. Estuaries Coasts 36:1265–1273

Gameiro C, Zwolinski J, Brotas V (2011) Light control on phytoplankton production in a shallow and turbid estuarine system. Hydrobiologia 669:249–263

Goeyens L, Kindermans N, Yusuf MA, Elskens M (1998) A room temperature procedure for the manual determination of urea in seawater. Estuar Coast Shelf Sci 47:415–418

Goto N, Kihira M, Ishida N (2008) Seasonal distribution of photosynthetically active phytoplankton using pulse amplitude modulated fluorometry in the large monomictic Lake Biwa, Japan. J Plankton Res 30:1169–1177

Grobbelaar JU (1985) Phytoplankton productivity in turbid waters. J Plankton Res 7:653–663

Hansen H, Koroleff F (1999) Determination of nutrients. In: Grasshoff K, Kremling K, Ehrhardt M (eds) Methods of seawater analysis. Wiley-VCH, Weinheim, pp 158–228

Harding LW, Meeson BW Jr, Fisher TR Jr (1986) Phytoplankton production in two East Coast estuaries: photosynthesis-light functions and patterns of carbon assimilation in Chesapeake and Delaware bays. Estuar Coast Shelf Sci 23:773–806
Hasle GR (1978) The inverted microscope method. In: Sournia A (ed) Phytoplankton manual. UNESCO, Paris, pp 88–96
Holmes RW (1970) The secchi disk in turbid coastal waters. Limnol Oceanogr 15:688–694
Irigoien X, Castel J (1997) Light limitation and distribution of chlorophyll pigments in a highly turbid estuary: the Gironde (SW France). Estuar Coast Shelf Sci 44:507–517
Johnson ZI, Shyam R, Ritchie AE, Mioni C, Lance VP, Murray JW, Zinser ER (2010) The effect of iron- and light-limitation on phytoplankton communities of deep chlorophyll maxima of the western Pacific Ocean. J Mar Res 68:283–308
Kim JK, Kraemer GP, Yarish C (2014) Field scale evaluation of seaweed aquaculture as a nutrient bioextraction strategy in Long Island Sound and the Bronx River Estuary. Aquaculture 433:148–156
Kocum E, Underwood GJC, Nedwell DB (2002) Simultaneous measurement of phytoplanktonic primary production, nutrient and light availability along a turbid, eutrophic UK east coast estuary (the Colne Estuary). Mar Ecol Prog Ser 231:1–12
Kromkamp J, Peene J, Van Rijswijk P, Sandee A, Goosen N (1995) Nutrients, light and primary production by phytoplankton and microphytobenthos in the eutrophic, turbid Westerschelde Estuary (The Netherlands). Hydrobiologia 311:9–19
Lewis WM Jr (1988) Primary production in the Orinoco River. Ecology 69:679–692
Li Y, Veilleux DJ, Wikfors GH (2009) Particle removal by Northern bay scallops *Argopecten irradians irradians* in a semi-natural setting: application of a flow-cytometric technique. Aquaculture 296:237–245
Li Y, Meseck SL, Dixon MS, Rivara K, Wikfors GH (2012) Temporal variability in phytoplankton removal by a commercial, suspended eastern oyster nursery and effects on local plankton dynamics. J Shellfish Res 31:1077–1089
Li Y, Meseck SL, Dixon MS, Rose JM, Smith BC, Wikfors GH (2013) Short term effects of a commercial eastern oyster nursery upon nutrient and plankton dynamics of a coastal embayment: observations from mesocosm experiments. Aquac Res 46:2049–2064
Liu S, Lin S (2008) Temporal and spatial variation of phytoplankton community in Long Island Sound Proceedings of the The 9th Biennial Long Island Sound Research Conference, 2008, 28–34
Lopez G, Carey D, Carlton JT, Cerrato R, Dam H, Digiovanni R, Elphick C, Michael F, Gobler C, Hice L, Howell P, Jordaan A, Lin S, Liu S, Lonsdale D, Mcenroe M, Mckown K, Mcmanus G, Orson R, Peterson B, Pickerell C, Rozsa R, Shumway SE, Siuda A, Streich K, Talmage S, Taylor G, Thomas E, Van Patten M, Vaudrey J, Yarish C, Wikfors GH, Zajac R (2014) Biology and ecology of Long Island sound. In: Latimer J, Tedesco MA, Swanson RL, Yarish C, Stacey PE, Garza C (eds) Long island sound prospects of the urban sea. Springer, New York, pp 285–479
Marmer HA (1935) Tides and currents in New York Harbor (Special Publication)
Mcminn A, Pankowskii A, Ashworth C, Bhagooli R, Ralph P, Ryan K (2010) In situ net primary productivity and photosynthesis of Antarctic sea ice algal, phytoplankton and benthic algal communities. Marine Biology 1345–1356
Meseck SL, Li Y, Dixon MS, Rivara K, Wikfors GH, Luther GI (2012) Effects of a commercial, suspended eastern oyster nursery upon nutrient and sediment chemistry in a temperate, coastal embayment. Aquac Environ Interact 3:65–79
Nixon SW (1995) Coastal marine eutrophication: a definition, social causes, and future concerns. Ophelia 41:199–219
O'donnel J, Wilson RE, Lwiza K, Whitney M, Bohlen WF, Codiga D, Fribance DB, Fake T, Bowman M, Varekamp JC (2014) The physical oceanography of Long Island Sound. In: Latimer J, Tedesco MA, Swanson RL, Yarish C, Stacey PE, Garza C (eds) Long Island sound prospects of the Urban Sea, New York, Heidelberg
Parker CA, O'reilly JE (1991) Oxygen depletion in Long Island sound: a historical perspective. Estuaries 14:248–264
Pourafrasyabi M, Ramezanpour Z (2014) Phytoplankton as bio-indicator of water quality in Sefid Rud River, Iran (South of Caspian Sea). Caspian J Environ Sci 12:31–40
Throndsen J (1978) Preservation and storage. In: Sournia A (ed) Phytoplankton manual. UNESCO, Paris, pp 69–74
Underwood GJC, Kromkamp J (1999) Primary production by phytoplankton and microphytobenthos in estuaries. Adv Ecol Res 29:93–153
Varekamp JC, Mcelroy AE, Mullaney JR, Breslin VT (2014) Metals, organic compounds, and nutrients in Long Island Sound: sources, magnitudes, trends, and impacts. In: Latimer J, Tedesco MA, Swanson RL, Yarish C, Stacey PE, Garza C (eds) Long Island sound prospects of the Urban sea. Springer, New York, pp 203–283
Wofsy SC (1983) A simple model to predict extinction coefficients and phytoplankton biomass in eutrophic waters. Limnol Oceanogr 28:1144–1155
Yamane Y, Utsunomiya T, Watanabe M, Sasaki K (2001) Biomass production in mixotrophic culture of Euglena gracilis under acidic condition and its growth energetics. Biotech Lett 23:1223–1228

8

Difference in benthic invertebrate communities of headwater streams can be detected using a short elevation gradient

Armin Namayandeh · David V. Beresford · Keith M. Somers · Peter J. Dillon

Abstract Communities of benthic invertebrates from a series of intermittent headwater streams spanning over a short elevation gradient (i.e., ~ 180–500 m) were investigated in April–July 2010 and May–July 2011. The main purpose of this study was to better understand whether the potential future effect of environmental change on biological communities of the Precambrian Shield's freshwaters can be detected using elevation as a substitute for time. Since obtaining long-term environmental data is a time-consuming process, substituting space for time could instead generate similar information in a shorter time. In this study, environmental differences associated with short elevation gradient were correlated with differences in benthic invertebrate communities. Therefore, elevation gradient provides a spatial proxy for anticipated future environmental change impacts over time. It was determined that water temperature accounts for the greatest variation in communities along the elevation. Many community metrics such as abundance, functional feeding groups, diversity, and evenness were significantly different based on the difference in elevation. Result indicates that even a short elevation gradient can potentially be used as a surrogate to look at the effect of environmental change.

Keywords Elevation · Benthic invertebrates · Communities · Headwater streams · Precambrian Shield

---

A. Namayandeh (✉)
Taxanama Corp., 25 Mabelle Ave., Toronto, ON, Canada
e-mail: a.namayan@taxanama.com

D. V. Beresford
Environmental and Life Sciences-Biology Department, Trent University, 2140 East Bank Drive, Peterborough, ON K9L 1Z8, Canada
e-mail: davidberesford@trentu.ca

K. M. Somers
Department of Ecology and Evolutionary Biology, University of Toronto, Toronto, Canada
e-mail: keith.somers@ontario.ca

P. J. Dillon
Environmental and Life Sciences-Chemistry Department, Trent University, 1600 West Bank Drive, Peterborough, ON K9J 7B8, Canada
e-mail: pdillon@trentu.ca

Introduction

Environmental factors such as water temperature, water velocity, food availability, and food type influence on the benthic invertebrate communities' structure along longitudinal stream profiles and with changes in altitude (Camargo and DeJalon 1995; Hawkins et al. 2000; Vannote et al. 1980). This is because a complex interaction exists between water temperature, flow regime, food quality, type and availability, and the substrate that determines environmental heterogeneity and contributes to species diversity (Minshall et al. 1985). This spatial heterogeneity has been investigated in terms of stream longitudinal profile from the headwater to the mouth within the context of the river continuum concept (e.g., Carter et al. 1996; Minshall et al. 1985; Scheibler et al. 2014; Scheibler and Debandi 2008; Vannote et al. 1980). However, few studies have compared streams at similar latitudes yet located at different elevations and inquired whether the spatial variation in altitude within the same geographic region can affect the community of benthos in streams (Jacobsen 2003, 2004, 2008; Jacobsen et al. 1997; Ramírez and Pringle 1998; Ramírez et al. 2008; Yule et al. 2009).

Jacobsen et al. (1997) have shown that when physical parameters such as size and current are accounted for, benthic assemblages from the tropical stream located at different altitude can be compared based on their thermal differences. In the same study, Jacobsen et al. (1997) have further illustrated that these streams are comparable with streams in different geographical regions (i.e., lowlands of Denmark). In addition, Jacobsen (2003, 2004) studies of Andean streams indicated that richness and diversity of benthic invertebrates of streams that are physico-chemically comparable changes along the altitudinal gradients. Therefore, altitude was not only useful to compare the benthic communities of several streams stretched along an elevation gradient, but was also a useful surrogate for detecting the effect of local environmental change on benthic communities. In areas with large elevation gradients, such as Andean regions, it would be easier to characterize and separate the streams spatially. This is because taxonomically different communities of benthic invertebrates are usually encountered. However, streams stretched over a short elevation gradient most probably have assemblage of similar invertebrates (i.e., taxonomically) that have differences only in abundance. The community difference based on the abundance can be detected if streams are in close geographical proximity, and therefore, species can migrate along the elevation gradient (Hughes et al. 2008).

Previous studies have proposed that freshwater systems of Ontario Precambrian Shield are vulnerable to effect of environmental change such as climatic change (Magnuson et al. 1997; Schindler et al. 1996). Obtaining long-term data on environmental variables is an expensive and time-consuming process. It should be possible to substitute space for time instead to generate similar information in a shorter period. The differences of environmental variables associated with an elevation gradient (e.g., temperature) corresponding to change in benthic invertebrate communities of comparable freshwaters can be potentially used for this purpose. In this study, we compared communities of benthic invertebrates from several hydrologically comparable intermittent headwater streams spanning an elevation gradient (i.e., 180–472 m) over a specific period (i.e., April–July 2010; May–July 2011) to determine whether their communities are different from one another based on community abundance and indices such as functional feeding groups (FFG), diversity, richness, and evenness. The objective was to determine whether the variance in biotic communities can be explained by local variable(s) that are associated with regional variable(s) as elevation changes. If variance is detected it would suggest that elevation can be a useful surrogate to look at the long-term effect of environmental change (e.g., climatic change) on freshwater biotic communities of the Precambrian Shield.

Methods

Study area

Streams were located from Georgian Bay- Massassauga Provincial Park to Algonquin Provincial Park with a rapid elevation increase extending from 180 to 472 m at 45°N on the Canadian Precambrian Shield (Fig. 1). Based on the region's topography, three headwater streams in Algonquin Provincial Park (i.e., D1, G1, and E1) were chosen at higher elevation. Two streams in Arrowhead Park (E2 and F1) and an additional stream near Harp Lake, Dorset (HP3A) were chosen at mid elevation. At the lower elevations, a stream in Massasauga Provincial Park (H1) and a stream in the District of Parry Sound (H2) were chosen. Period of the

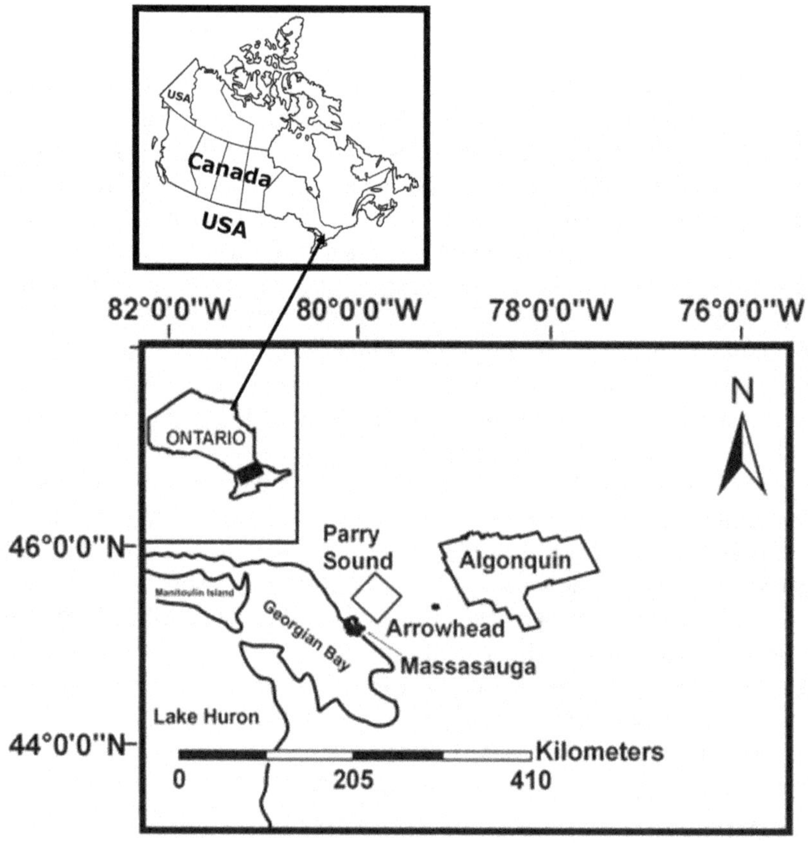

Fig. 1 Location of study streams in Precambrian shield, Ontario, Canada, 2010–2011

samplings was based on the start of the freshet each year, accessibility to the streams after the freshet, and also the time when streams dry up (i.e., presence of flow). The corridor from Massassauga to Algonquin sits on metamorphic bedrock with thin top soil layer. Streams were all fishless.

Sampling benthic invertebrates

Monthly samples of benthic invertebrates were obtained between April–July 2010 and May–July 2011 using a small net with a flexible frame and a mesh size of 250 μm. Samples were collected by walking along the width of the stream, vigorously kicking the substrate to disturb it to a depth of ~ 5 cm for about 3 min. Samples were collected from one reach (i.e., two riffles and one pool). In all streams, pools and riffles are near each other, so that with exception of sedentary benthos (e.g., bivalves), most benthic species occur in both habitats. The mesh size, sampling procedure including the effort time, and habitat types were determined and standardized based on a pilot study that we conducted in the previous year (April–June 2009). Samples of benthos transported in 3 L whirl-packs and preserved in ethyl alcohol anhydrous solution (85.5% ethanol and 13.7% methanol, and 0.85% ethyl acetate; ACP Chemical Inc). The benthos samples were sieved and sorted in the laboratory and all insects were identified to genera and species level. Individual species were subsequently placed in appropriate functional feeding groups based on Merrit et al. (2008) and Barbour et al. (1999). A detailed list and abundance of species collected are provided in Online Appendix 1.

Temperature and precipitation

Water and air temperatures were obtained using iBTag Alpha Mach Inc. temperature data loggers from April to July for 2010 and May to July at every 3 h interval for 2011, as no significant pattern of change were detected on hourly basis. For each study, stream two sets of loggers were installed: one at the designated reach and one monitoring the air temperature adjacent to the stream (i.e., attached to a tree adjacent to the designated reach). All the temperature loggers were synchronized, so that the air and the stream temperature recordings at all the elevations and streams were running simultaneously.

Degree day (DD) was calculated April–July 2010 and May–July 2011 based on accumulated temperature in each stream above 0 °C and below 16 °C. Poikilotherm development is best modelled using DD as the unit of temperature-time (Baskerville and Emin 1969; Beresford and Sutcliffe 2009). DD is defined as 1 °C per 24 h above some developmental threshold temperature. Based on the 2 years, our investigation and monitoring of the stream temperature we found that streams dry up as soon as water temperature exceeds 16 °C (i.e., being intermittent headwater streams), so we used the number of DD accumulated based on this measure to estimate the developmental threshold of benthos. In addition, since the streams dry up, there was a clear indication that ground water by means of water from the saturated zone had no influence on these streams or on their temperature (J.M. Buttle, School of Environment, Hydrology, Trent University—personal communication, November 05, 2010). We calculated mean monthly precipitation based on the data collected from the historical climate data of Environment Canada for Algonquin Provincial Park (streams D1, G1, and E1), Arrowhead Provincial Park (streams E2 and F1), Harp Lake, Dorset (stream HP3A), Massasauga Provincial Park (stream H1), and the District of Parry Sound (stream H2) (http://climate.weather.gc.ca/).

Chemical data

Stream pH and conductivity were measured monthly (April–July 2010, July 2011) using an Accumet AP85, Fisher Scientific Instrument. In addition, water samples were collected monthly for total phosphorus (TP) in 35 ml p-tubes and for total nitrogen (TN) dissolved organic and inorganic carbon (DOC and DIC) in 500 ml PET jars in 2010 and only in July for 2011 (Online Appendix 2). Chemical analyses were performed at the Dorset Environmental Science Centre, Dorset, Ontario or Trent University following the Ontario Ministry of Environment and Climate Change standard method (Sutey et al. 2015).

Data analysis

Ordinations of benthic community data were performed by non-metric multidimensional scaling (NMDS) based on the Bray-Curtis similarity index and were performed after fourth-root transformation to down weight the influence of dominant species and to normalize the data. The stress value function was used to assess the representations of NMDS solutions according to (Clark and Gorley 2015). The guideline indicates that a stress value of < 0.05 gives an excellent representation with no misinterpretation. The significance of any discrimination between sites was tested using analysis of similarity (ANOSIM) test with R ranging from -1 to 1 at $\alpha = 0.05$. Similarity percentage test (SIMPER) was used to calculate the contribution of individual species to similarity or dissimilarity between streams. NMDS analyses were performed using PRIMER v 7 (Clark and Gorley 2015).

A detrended correspondence analysis (DCA) of benthic communities was performed for 2010–2011. This was done to determine whether species have a linear or a unimodal distribution and to determine what ordination analysis is appropriate to correlate benthos abundance data with environmental variables. For each DCA analyses, we down-weighted the rare species. Based on the first axis lengths gradient of DCA in 2010 (i.e., 1.4), species had a linear distribution. Therefore, redundancy analyses (RDA) were used to correlate the communities of species in 2010 (May–July) to environmental variables (i.e., chemicals, water temperature, DD, precipitations, and catchment area). Result of first axis length gradient of DCA for July 2011 was inconclusive (i.e., 2.2); therefore, canonical correspondence analysis (CCA) and RDA were used to correlate July benthic data to July environmental data. For each set of RDAs and CCAs, we sequentially removed variables with high inflation rate (i.e., > 10) and a Monte-Carlo permutation test (999 unrestricted permutation) was ran to account for the variables that significantly (i.e., $P < 0.05$) contribute to the species variations. A correspondence analysis (CA) was performed on the benthos community 2010–2011. The first and second CA axis (CA1 and CA2) scores (i.e., for streams) obtained were correlated with environmental variable(s) that significantly accounted for variation in benthic communities along the elevation gradient. This will determine if the environmental variable correlates with the variation of benthic community in streams and whether this is elevational dependent. We fourth-root transformed the species relative abundance and normalized the environmental variables before these ordination analyses. The CANOCO for Windows 4.5 (ter Braak and Šmilauer 2002) was used for DCA, RDA, and CCA. CA analyses was conducted in PAST v 2.14 (Hammer et al. 2001). Species codes are provided in Online Appendix 3.

Following the ordination analyses of communities, a two-way ANOVA (i.e., difference in elevation and month) was performed in PAST v 2.14 (Hammer et al. 2001) on the communities' richness, Shannon-Weaver Diversity (H'), evenness, and FFGs. A two-way ANOVA was also used to compare the chemical variables between streams at different elevations in 2010. For 2011, only the July pH, TP, and TN data were available which we compared using a one-way ANOVA. In addition, linear regressions were performed to predict the effect of air temperature on stream temperature variation, 2010–2011. This analysis will determine whether change in air temperature along the elevation will influence the streams temperature variation. Analysis for each year was kept separate to determine if similar trend of variation (i.e., both biotic and abiotic) along the elevation can be detected.

Results

Environmental variables and variations in community data

The 2010–2011 DD accumulated was higher in higher elevation streams compared to those at lower (Table 1). The 2010 air temperature regression with streams temperature produced R^2 value of 0.86, which was significant ($F_{1,30df} = 186.4$, $P < 0.001$). The 2011 air temperature regression with stream temperature produced R^2 value of 0.75, which was significant ($F_{1,22df} = 64.6$, $P < 0.001$). This result indicates that change in air temperature along the elevation can account for variation in stream temperature.

Stepwise backwards elimination in RDA retained DD, catchment area, and water temperature as variables that were correlated with benthic communities in 2010 (April–July; Fig. 2a). Water temperature was the only factor in 2010 that significantly correlated with the communities ($P = 0.03$). CCA explained greater amounts of species variation in July 2011 compared to RDA (29.9 vs. 23.4%). Therefore, CCA results are presented here. Stepwise backward elimination in CCA retained catchment area, TP, TN, and water temperature as variables that were correlated with benthic communities in July 2011 (Fig. 2b). Water temperature was the only factor in July 2011 that significantly correlated with communities ($P = 0.002$). In 2010, CA1 did not significantly correlate with temperature ($R = -0.225$); however, CA2 significantly correlated with temperature ($R = 0.557$). For 2011, both CA1 and CA2 significantly correlated with stream temperature ($R = -0.67$ and $R = 0.78$). Results of the correspondence analysis are provided in Online Appendix 4.

Results of two-way ANOVA showed all FFG significantly differed between elevations in 2010 (Table 2A). Shannon-Weaver diversity index, evenness, and both CA1 and CA2 also differed significantly in 2010. However, richness did not vary significantly (Table 2A). Similar results were obtained for 2011 for FFG,

Table 1 Study streams' geographical location, elevation, location in Ontario Provincial Parks or district, mean DD accumulated ± SD, and mean stream temperature ± SD for April–July 2010 and May–July 2011

Streams	Latitude	Longitude	Elevation (m)	Provincial park/ district	DD April–July 2010	DD May–July 2011	Temperature April–July 2010	Temperature May–July 2011
D1	45°27′03″N	78°48′05″W	472	Algonquin	84.5 ± 11.2	23.6 ± 12.0	11.4 ± 3.2	12.9 ± 2.4
G1	45°28′16″N	78°48′17″W	445	Algonquin	113.8 ± 4.3	28.2 ± 4.1	11.3 ± 3.1	12.2 ± 2.0
E1	45°28′42″N	78°48′55″W	425	Algonquin	110.5 ± 3.9	28.0 ± 4.4	10.7 ± 3.4	11.3 ± 2.1
E2	45°25′51″N	79°11′17″W	373	Arrowhead	112.4 ± 4.1	27.7 ± 4.9	10.9 ± 3.3	11.6 ± 2.3
HP3A	45°22′24″N	79°08′56″W	362	Muskoka	75.5 ± 11.6	22.0 ± 12.9	11.8 ± 3.1	13.5 ± 1.7
F1	45°23′52″N	79°11′35″W	337	Arrowhead	75.4 ± 11.0	21.0 ± 12.3	12.0 ± 3.4	13.2 ± 2.1
H2	45°13′27″N	79°43′34″W	240	Parry sound	58.6 ± 8.2	22.4 ± 10.3	12.8 ± 2.6	13.7 ± 0.5
H1	45°07′29″N	79°59′52″W	185	Massasauga	58.6 ± 8.2	13.7 ± 9.3	12.8 ± 2.6	14.4 ± 1.6

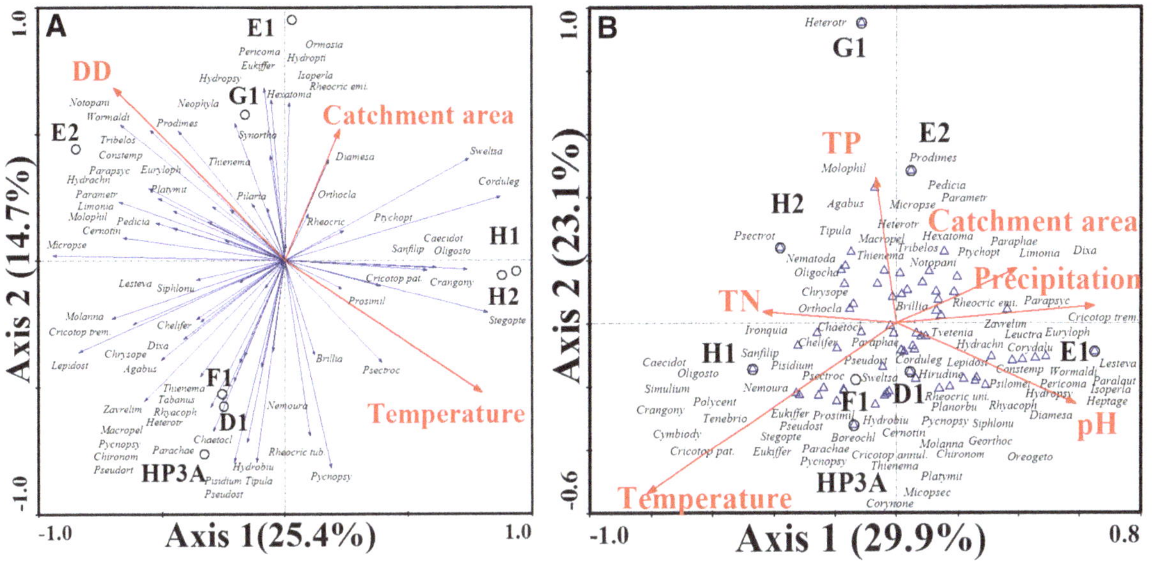

Fig. 2 Redundancy and canonical correspondence analyses of benthic communities and environmental variables, 2010–2011 **a** RDA April–July 2010 **b** CCA July 2011. Circles are study sites, and blue arrows (RDA) and triangles (CCA) are species, and red arrows are environmental variables

Table 2 Two-way ANOVA results of community variables, (A) April–July 2010 (B) May–July 2011

Variables	(A) 2010						(B) 2011					
	Elevation (8)		Month (4)		Interaction		Elevation (8)		Month (3)		Interaction	
	F	*P*	*F*	*P*	*F*	*P*	*F*	*P*	*F*	*P*	*F*	*P*
Shredders	14.3	< 0.001	271.6	< 0.001	14.3	< 0.001	17.9	< 0.001	332.2	< 0.001	17.6	< 0.001
Collector filterers	14.0	< 0.001	255.4	< 0.001	14.4	< 0.001	17.9	< 0.001	318.9	< 0.001	17.5	< 0.001
Collector gatherers	14.4	< 0.001	248.9	< 0.001	14.2	< 0.001	17.2	< 0.001	330.0	< 0.001	18.2	< 0.001
Scrapers	14.5	< 0.001	287.6	< 0.001	14.3	< 0.001	17.9	< 0.001	364.6	< 0.001	18.0	< 0.001
Predators	14.7	< 0.001	259.5	< 0.001	13.9	< 0.001	17.7	< 0.001	301.3	< 0.001	17.0	< 0.001
Parasites	14.4	< 0.001	290.5	< 0.001	14.4	< 0.001	18.0	< 0.001	364.7	< 0.001	18.0	< 0.001
Shannon–Weaver H'	12.5	< 0.001	88.8	< 0.001	14.1	< 0.001	10.5	< 0.001	92.1	< 0.001	14.6	< 0.001
Evenness	14.1	< 0.001	241.3	< 0.001	14.5	< 0.001	16.9	< 0.001	298.5	< 0.001	18.2	< 0.001
Richness	1.6	0.14	240.1	< 0.001	2.4	0.01	1.5	0.2	183.4	< 0.001	1.9	0.05
CA1	17.2	< 0.001	262.1	< 0.001	11.8	< 0.001	15.1	< 0.001	160.6	< 0.001	7.4	< 0.001
CA2	12.8	< 0.001	237.9	< 0.001	11.7	< 0.001	16.1	< 0.001	283.4	< 0.001	13.5	< 0.001

The number of months and elevation sites (i.e., streams) is in the brackets

Shannon-Weaver diversity index (H'), evenness, and CA1 and CA2, and richness did not significantly vary again (Table 2B).

Based on the results of the two-way ANOVA water temperature significantly differed between streams at different elevations (i.e., lower to higher elevations) in 2010. However, precipitation did not vary significantly in 2010 (Table 3A). Result of the two-way ANOVA indicates that for 2010, all water chemistry variables were significantly different between streams at different elevations (Table 3A). Results of the 2011 two-way ANOVA water temperature showed significant difference between streams at different elevations. However, precipitation did not vary significantly in 2011 (Table 3B). Result of the one-way ANOVA indicates that July 2011 water chemistry variables did not vary significantly between streams at different elevations (Table 3B).

Table 3 (A) Two-way ANOVA results of environmental variables, April–July 2010. (B) Two-way ANOVA results of temperature and precipitation May–July 2011 and one-way ANOVA of July 2011 chemical variables

Variables	(A) 2010						(B) 2011					
	Elevation (8)		Month (4)		Interaction		Elevation (8)		Month (3)		Interaction	
	F	*P*	*F*	*P*	*F*	*P*	*F*	*P*	*F*	*P*	*F*	*P*
pH	12.9	< 0.001	215.2	< 0.001	15.4	< 0.001	0.91	0.55	–	–	–	–
DOC (mg/L)	2.5	0.02	7.2	< 0.001	1.2	0.29	–	–	–	–	–	–
DIC (mg/L)	7.3	< 0.001	11.3	< 0.001	2.6	< 0.001	–	–	–	–	–	–
TP (μg/L)	1.6	0.16	6.0	< 0.001	1.6	0.1	0.96	0.52	–	–	–	–
TN (μg/L)	13.7	< 0.001	227.5	< 0.001	13.4	< 0.001	1.0	0.49	–	–	–	–
Conductivity (μS/cm)	30.6	< 0.001	351.9	< 0.001	28.3	< 0.001	–	–	–	–	–	–
Temperature (°C)	3.3	0.005	164.7	< 0.001	1.4	0.16	3.9	0.002	261.6	< 0.001	1.9	0.05
Precipitation (mm)	0.3	0.94	123.2	< 0.001	0.34	0.99	0.1	0.99	127.6	< 0.001	0.03	0.99

The number of months and elevation sites (i.e., streams) is in the brackets

DOC dissolved organic carbon, *DIC* dissolved inorganic carbon, *TP* total phosphorus, *TN* total nitrogen

Community data 2010–2011

NMDS ordination of 2010 abundance data (April–July) in two-dimensional space yielded a clear separation between streams (stress value 0.04). Streams E2, F1, and HP3A and D1 sit in center of biplot; E1 at bottom-left and G1 at middle-top of biplot; H1 and H2 at right of the biplot (Fig. 3a). One-way nested ANOSIM yields a significant difference between benthic communities at different elevations (global test $R = 0.380$, $P = 0.001$). Pairwise ANOSIM test indicates that D1 and E1 are significantly different from H1 and H2 (global test $R_{D1–H1} = 0.854$, $P_{D1–H1} = 0.03$; $R_{D1–H2} = 0.635$, $P_{D1–H2} = 0.03$; $R_{E1–H1} = 0.896$, $P_{D1–H1} = 0.03$; $R_{E1–H2} = 0.833$, $P_{E1–H2} = 0.03$). E2 and F1 significantly differed from H1 and H2 (global test $R_{E2–H1} = 0.729$, $P_{E2–H1} = 0.03$; $R_{E2–H2} = 0.635$, $P_{E2–H2} = 0.03$; $R_{F1–H1} = 0.688$, $P_{F1–H1} = 0.03$; $R_{F1–H2} = 0.479$, $P_{F1–H2-} = 0.03$). HP3A significantly differed from H1 (global test $R = 0.573$, $P = 0.03$). NMDS ordination of 2011 abundance data (stress value 0.04) in two-dimensional space yielded a clear separation between streams that are located at high and mid elevations (i.e., D1, G1, E1, E2, HP3A, and F1) from those at the lower elevation (i.e., H1 and H2). The 2011 result was similar to 2010; however, stream F1 sit closer to H2 (Fig. 3b). One-way nested ANOSIM yield a significant difference between benthic communities at different elevations (global test $R = 0.496$, $P = 0.001$). Pairwise ANOSIM test for 2011 was slightly different. E1 and G1 were significantly different from H1 (global test $R_{E1–H1} = 0.926$, $P_{D1–H1} = 0.03$; $R_{G1–Ha} = 0.778$, $P_{G1–H1} = 0.03$).

Based on the 2010 results of SIMPER test on benthic communities, the highest dissimilarity was between streams E1 and H1 (SIMPER average dissimilarity = 72.0%) with top five contributor species *Crangonyx* sp. (5.2%), *Caecidotea* sp. (4.4%), *Prosimulium* sp. (4.1%), *Parapsyche apicalis* (4.0%), and *Notopanisus* sp. (3.8%). Streams G1 and H1 showed the second highest dissimilarity (SIMPER average dissimilarity = 71.6%)

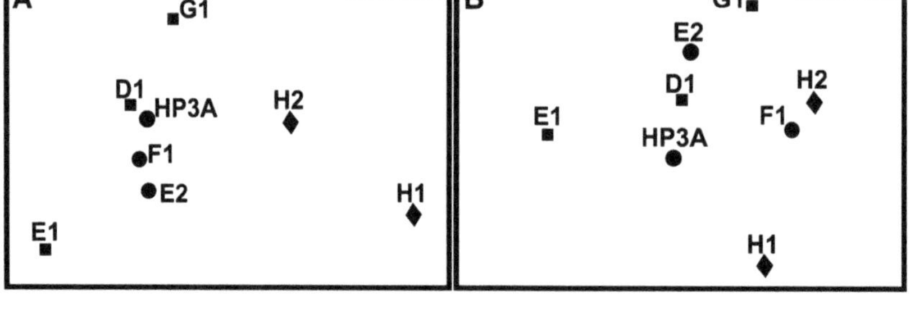

Fig. 3 Non-metric multidimensional scaling analyses of benthic communities, 2010–2011 **a** April–July 2010 **b** May–July 2011. Diamonds are lower elevation streams (185–240 m), circles are mid elevation streams (337–373 m), and squares are higher elevation streams (425–472 m)

with top five contributor species being *Crangonyx* sp. (7.3%), *Caecidotea* sp. (6.2%), *Micropsectra polita* (5.5%), *Prosimulium* sp. (5.5%), and *Sweltsa naica* (3.9%). Results of SIMPER test for 2011 were slightly different. The highest dissimilarity was between streams E1 and F1 (SIMPER average dissimilarity = 73.0%) with top five contributor species *Siphlonurus marshalli* (3.4%), *Isoperla richardsoni* (3.1%), *Parametriocnemus lundbeckii* (2.7%), *Micropsectra polita* (2.5%), and *Sweltsa naica* (2.4%). Streams E1 and H1 had the second highest dissimilarity (SIMPER average dissimilarity = 67.5%) with top five contributor species *Simulium* sp. (4.1%), *Nemoura trispinosa* (3.7%), *Crangonyx* sp. (3.1%), *Thienemannimyia* group (3.1%), and *Siphlonurus marshalli* (3.0%).

Discussion

Delineating the effect of temperature

In this study, streams of similar size and same order were chosen, and in addition, we eliminated or minimized the effect of the predatory fish, bedrock geology and ground water influence to limit the local environmental variables that may obscure the effect of elevation. Therefore, due to close geographical proximity (i.e., < 100 km) and similarity in habitats (i.e., first-order streams in mixed-deciduous forest), much of the community differences can be inferred to be caused by elevation, with any differences in the hydrological properties of the streams (e.g., order: substratum, size) having minimal effect on benthic communities' composition. This argument is further supported by the fact that the variation in the different stream communities was mainly attributed to differences in temperature between streams. Parameters such as water chemistry and catchment area did not explain the observed community variations. We also found that changes in air temperature along the elevation gradient accounted for variations in stream temperature. Given that precipitation (i.e., another climatic factor) did not vary with elevation, we argue that any changes in air temperature due to elevation directly influence stream temperature.

From this, we infer that temperature associated with elevation can account for differences in abundance, diversity and FFG of communities. This is supported by the results of the 2010–2011 pairwise ANOSIM and SIMPER analyses, which indicate that the main differences in benthic communities were between high and low elevation streams, with the greatest difference accounted by cold-stenothermic species, e.g., insects such as Plecoptera species *Sweltsa naica*, *Nemoura trispinosa* and *Isoperla richardsoni*, Ephemeroptera species *Siphlonurus marshalli*, Trichoptera species *Parapsyche apicalis*, and Diptera species *Prosimulium* sp., *Simulium* sp., *Parametriocnemus lundbeckii* and *Micropsectra polita*. Haidekker and Hering (2008) have previously shown that in small streams with differences of up to 3 °C, cold-stenothermic insects can account for majority of the variation in benthic communities. Given that mean April–July 2010 and May–July 2011, temperature difference between high and low elevation streams (i.e., E1, D1, G1, and H1 and H2) was 2.1 and 3.1 °C, respectively, it appears that only the higher elevations had streams close with optimal temperature condition (i.e., cooler) for high benthic insect abundance and diversity.

Distinguishing the effect of water temperature on communities of streams benthos from other environmental variables (i.e., physical and chemical) is a challenging task. This is especially true if streams are of different size, depth, flow, and substrate types (Arai et al. 2015; Duggan et al. 2007; Phillips et al. 2015; Quinn and Hickey 1990; Statzner and Higler 1986). Headwater streams naturally control for many confounding factors as they have significantly lower or no anthropogenic impact and lower variation in flow regimes and water quality year around (Arai et al. 2015; Cummins 1974; Karr and Dudley 1981; Moore and Wondzell 2005; Richardson and Danehy 2007). Therefore, temperature plays a significant role for benthic invertebrates of headwater streams (Arai et al. 2015; Chang et al. 2012). For example, few eurythermic species are found in headwater streams in general (Haidekker and Hering 2008). Increasing a stream's temperature can have a significant impact on physiology, ecology, and life history of organisms living in streams (Angilletta 2009; Sheldon and Tewksbury 2014a, b). We are still at the preliminary stage of understanding of how temperature change shifts the communities in freshwaters as a whole and in long term. Moreover, consensus on how this change occurs varies perhaps due to differences in geographical regions, scale of the study and type of the freshwaters. In this study, data are only suggestive that temperature accounts for variation among community of benthos (i.e., based on the CA analysis species scores correlation with stream temperature, RDA and CCA analyses, DD, and temperature data).

Intermittent nature of the streams, flow regime, and drawbacks

The intermittent natures of the streams in addition to their inaccessibility were major drawbacks in our study. This means that many routine or daily measurements such as water chemistry, DO, and flow regime were not obtainable throughout the sampling season. The year around flow of the streams has influence on abundance and function of benthic communities (Boulton and Lake 1988; Hose et al. 2005), yet cannot be determined with certainty in this study. In the 2010, freshet started in April and stream flow lasted until late July with all streams at higher elevation (i.e., D1, G1, and E1) and those at mid elevation (i.e., E2, HP3A, and F1) having flow during this period. However, in 2011, freshet started in mid-May and stream flow lasted until mid-July with many streams experiencing (i.e., at all elevations) drought after this period. These combined with the 2 year accumulated DD indicate that streams at mid-to-high elevation dry up much later in summer compared to those of lower elevation which further could indicate a prolonged period of drought for the latter. Given that precipitation did not significantly varied between streams, we can suggest that difference of air temperature driven by difference in elevation may accounts for considerable variation in flow regimes as well. That is if we exclude the influence of the ground water or subsurface flow which were minimized in this study. Therefore, at least for the summer months when the community data are obtained we can expect a lower community measures such as abundance, diversity, and functional differences in low elevation streams compared to those at mid–higher elevation. Results of DD and temperature data suggest that streams water at lower elevations evaporate faster than those in mid–higher elevation. Furthermore, similarity of benthic community of stream F1 located at mid elevation (at lower range for this elevation) to stream at low elevation in 2011 suggests that during prolonged low flow regimes, streams at mid-higher elevation can eventually resemble those of the low elevation. This is because shorter growth periods for benthos can limits their resources such as food, timing of emergence, and synchrony of emergence among competing species (Hynes 1958; Boulton 2003; Cowx et al. 1984) and hence limit space and time for many benthic invertebrates. Therefore, based on the 2010–2011 results of abundance, diversity, and FFG, we can only suggest that higher elevation streams provide a better condition for insect's growth and colonization during the months that optimal conditions are not obtainable in lower elevations. This also suggests that flow regime change in this region will be more detectable for streams of lower elevation. However, further increase in temperature and drought can eventually impact streams at higher elevations as well.

Elevation and community of benthos

Past studies in their accounting the community differences of invertebrate communities along elevation gradient (e.g., Huey 1978, Jacobsen et al. 1997; Jacobsen 2003, 2004; Lawton et al. 1987; MacArthur 1972; McCoy 1990; Stevens 1992; Scheibler et al. 2014; Vinson and Hawkins 1998) have used large elevations (i.e., $\gg$ 100 m of difference). These studies in general encountered communities with different taxonomic groups stretched along large elevation. However, in this study, communities are stretched over a short elevation gradient (i.e., 287 m of difference). Therefore, the difference in communities is mainly driven by the change in abundance of benthos along the elevation, and not necessarily driven by different taxonomically structured communities (Hughes et al. 2008). Therefore, for geographical areas such as the Precambrian Shield of Canada which large elevation gradient is not obtainable, task of discriminating communities of benthic invertebrates should rely on this abundance difference and probably lowest possible taxonomic identification of benthos (i.e., genus and species) is also required.

The 2010–2011 result obtained from richness did not produce a support for our objectives of showing difference in communities of benthos along a short elevation gradient. The fact that richness did not significantly vary among the streams is due to the fact that the difference in communities is not based on difference in the number of taxonomically different species but on differences in abundance. An effect on richness per se might be found across a larger elevation gradient (e.g., Jacobsen 2003; McCoy 1990; Monoghan et al. 2000; Ramírez and Pringle 1998; Stout and Vandermeer 1975). We found that only the abundance of different FFG varies between streams of different elevation and not the type of FFG which further indicates that communities are not taxonomically different. Therefore, result of the richness obtained in this study is not comparable with other studies having large elevation gradients (e.g., Jacobsen 2003; McCoy 1990; Monoghan et al. 2000; Ramírez and Pringle 1998; Stout and Vandermeer 1975), so that variation in abundance (i.e., species and FFG) and diversity is better measures for community difference compared to richness using low elevation gradients. Richness, however, varied monthly within and between streams. Therefore, we can assume that certain taxa could locally go extinct during the months that streams dry up or harsh stream conditions (e.g. high temperature) sets in. This also means that populations of these species can thrive in other elevations with suitable environmental conditions (i.e., via migration). Many benthic invertebrates in this study streams (i.e., insects) have high natural dispersal capability and, therefore, capable of rapid colonization of other waters.

In this study, we showed that difference between benthic invertebrate communities of several physico-chemically comparable head water streams stretched along a short elevation gradient can be detected. The variation among benthic communities was mainly based on their abundance, FFG and diversity and not richness. Furthermore, among several obtainable and measured physico-chemical variables, stream temperature associated with altitudinal difference accounted significantly for the variation among benthos. Therefore, effects of change in environmental variables (e.g., temperature) on freshwater communities, associated with short elevational difference, can be inferred and used as a proxy for anticipated environmental change (e.g., climate) over time.

Acknowledgements We thank the personnel at Ontario Provincial Parks (MNRF), all the personnel in Wildlife Research Station, Algonquin Provincial Park, the summer students in Dr. Dillon's lab, and the staff of Ontario Ministry of the Environment, Dorset Environmental Science Centre. And finally, we thank Dr. D. Currie of Royal Ontario Museum and University of Toronto. Financial support for the project was provided by Ontario Ministry of Environment through a grant to Dr. Peter Dillon, Trent University.

References

Angilletta MJ (2009) Thermal adaptation: a theoretical and empirical synthesis. Oxford University Press, Oxford

Arai R, Nukazawa K, Kazama S, Takemon Y (2015) Variation in benthic invertebrate abundance along thermal gradients within headwater streams of a temperate basin in Japan. Hydrobiologia 762:55–63. https://doi.org/10.1007/s1075

Barbour MT, Gerristen J, Snyder BD, Stribling JB (1999) Rapid bioassessment protocols for use in streams and wadeable rivers: periphyton, benthic macroinvertebrates, and fish. US Environmental Agency Office of Water, Washington DC

Baskerville GL, Emin P (1969) Rapid estimation of heat accumulation from maximum and minimum temperatures. Ecology 50:514–517. https://doi.org/10.2307/1933912

Beresford DV, Sutcliffe JF (2009) Sampling designs of insect time series data: are they all irregularly spaced? Oikos 118:115–121. https://doi.org/10.1111/j.1600-0706.2008.16883.x

Boulton AJ (2003) Parallels and contrasts in the effects of drought on stream macroinvertebrate assemblages. Freshw Biol 48:1173–1185. https://doi.org/10.1046/j.1365-2427.2003.01084.x

Boulton AJ, Lake PS (1988) Australian temporary streams—some ecological characteristics. Verh Int Ver Theor Angew Limnol 23:1380–1383. https://doi.org/10.1080/03680770.1987.11898025

Camargo JA, DeJalon DG (1995) Assessing the influence of altitude and temperature on biological monitoring of freshwater quality: a preliminary investigation. Environ Monit Assess 35:227–238. https://doi.org/10.1007/BF00547634

Carter JL, Fend SV, Kennelly SS (1996) The relationships among three habitat scales and stream benthic invertebrate community structure. Freshw Biol 35:109–124. https://doi.org/10.1046/j.1365-2427.1996.d01-450.x

Chang H, Jung IW, Steele M, Gannett M (2012) Spatial patterns of March and September streamflow trends in Pacific Northwest streams, 1958–2008. Geogr Anal 44: 177–201. https://doi.org/10.1111/j.1538-4632.2012.00847.x

Clark KR, Gorley RN (2015) PRIMER v7: user manual/tutorial. PRIMER–E, Plymouth, p 296

Cowx IG, Young WO, Hellawell JM (1984) The influence of drought on fish and invertebrate populations of an upland stream in Wales. Freshw Biol 14:165–177. https://doi.org/10.1111/j.1365-2427.1984.tb00030.x

Cummins KW (1974) Structure and function of stream ecosystems. Bioscience 24:631–641. https://doi.org/10.2307/1296676

Duggan I, Boothroyd I, Speirs D (2007) Factors affecting the distribution of stream macroinvertebrates in geothermal areas: Taupo Volcanic Zone, New Zealand. Hydrobiologia 592:235–247. https://doi.org/10.1007/s10750-007-0748-9

Haidekker A, Hering D (2008) Relationship between benthic insects (Ephemeroptera, Plecoptera, Coleoptera, Trichoptera) and temperature in small and medium-sized streams in Germany: a multivariate study. Aqua Ecol 42:463–481. https://doi.org/10.1007/s10452-007-9097-z

Hammer, Ø, Harper, DAT, Ryan PD (2001) PAST: paleontological statistics software package for education and data analysis. Palaeontologia Electronica 4(1):9

Hawkins CP, Norris RH, Gerritsen J, Hughes RM, Jackson SK, Johnson RK, Stevenson RJ (2000) Evaluation of the use of landscape classification for the prediction of freshwater biota: synthesis and recommendation. J N Am Benthol Soc 19:541–556. https://doi.org/10.2307/1468113

Hose GC, Jones P, Lim RP (2005) Hyporheic macroinvertebrates in riffle and pool areas of temporary streams in south eastern Australia. Hydrobiologia 532:81–90. https://doi.org/10.1007/s10750-004-9016-4

Huey RB (1978) Latitudinal pattern of between-elevation faunal similarity: mountains might be higher in tropics. Am Nat 112:225–229. https://doi.org/10.1086/283262

Hughes JM, Schmidt DJ, McLean A, Wheatley A (2008) Population genetic structure in stream insects: what have we learned. In: Lancaster J, Briers RA (eds) Aquatic insects, Proceedings of the Royal Entomological Society's 24th Symposium, pp 268–288

Hynes HBN (1958) The effect of drought on the fauna of a small mountain stream in Wales. Verh Int Ver Limnol 46:823–833. https://doi.org/10.1080/03680770.1956.11895473

Jacobsen D (2003) Altitudinal changes in diversity of macroinvertebrates from small streams in the Ecuadorian Andes. Arch Hydrobiol 158(2):145–167. https://doi.org/10.1127/0003-9136/2003/0158-0145

Jacobsen D (2004) Contrasting patterns in local and zonal family richness of stream invertebrates along an Andean altitudinal gradient. Freshw Biol 49:1293–1305. https://doi.org/10.1111/j.1365-2427.2004.01274.x

Jacobsen D (2008) Tropical high-altitude streams. In: Dudgeon D (ed) Aquatic ecosystems: tropical stream ecology. Elsevier Science, London, pp 219–256

Jacobsen D, Schultz R, Encalada A (1997) Structure and diversity of stream invertebrate assemblages: the influence of temperature with altitude and latitude. Freshw Biol 38:247–261. https://doi.org/10.1046/j.1365-2427.1997.00210.x

Karr J, Dudley D (1981) Ecological perspective on water quality goals. Environ Manag 5:55–68. https://doi.org/10.1007/BF01866609

Lawton JH, MacGarvin M, Heads PA (1987) Effects of altitude on the abundance and species richness of insect herbivores on bracken. J Anim Ecol 56:147–160. Stable URL: http://www.jstor.org/stable/4805

MacArthur RH (1972) Geographical ecology: patterns in the distributions of species. Harper and Row, New York

Magnuson JJ, Webster KE, Assel RA, Bowser CJ, Dillon PJ, Eaton JG, Evans HE, Fee EJ, Hall RI, Mortsch LR, Schindler DW, Quinn FH (1997) Potential effects of climate changes on aquatic systems: Laurentian Great Lakes and Precambrian Shield Region. Hydrol Process 11:825–871. https://doi.org/10.1002/(sici)1099–1085(19970630)11:8<825:aid–hyp509>3.0.co;2–g

McCoy ED (1990) The distribution of insects along elevational gradients. Oikos 58:313–322

Merrit RW, Cummins KW, Berg MB (2008) An introduction to aquatic insects of North America, 4th edn. Kendall/Hunt Publishing Company, Iowa

Minshall GW, Cummins KW, Petersen RC, Cushing CE, Bruns DA, Sedell JR, Vannote RL (1985) Developments in stream ecosystem theory. Can J Fish Aqua Sci 42:1045–1055. https://doi.org/10.1139/f85-130

Monoghan KA, Peck MR, Brewin PA, Masiero M, Zarate E, Turcotte P, Ormerod SJ (2000) Macroinvertebrate distribution in Ecuadorian hill streams: the effect of altitude and land use. Arch Hydrobiol 149:421–440

Moore RD, Wondzell SM (2005) Physical hydrology and the effects of forest harvesting in the Pacific Northwest: a review. J Am Water Res Assoc 41:753–784. https://doi.org/10.1111/j.1752-1688.2005.tb03770.x

Phillips ID, Pollock MS, Bowman MF, McMaster DG, Chivers DP (2015) Thermal alteration and macroinvertebrate response below a large Northern Great Plains Reservoir. J Gt Lakes Res 41:155–163. https://doi.org/10.1016/j.jglr.2015.07.005

Quinn J, Hickey C (1990) Characterization and classification of benthic invertebrate communities in 88 New Zealand rivers in relation to environmental factors New Zealand. J Mar Freshw Res 24:387–409. https://doi.org/10.1080/00288330.1990.9516432

Ramírez A, Pringle CM (1998) Structure and production of a benthic insect assemblage in a neotropical stream. J N Am Benthol Soc 17:443–463. https://doi.org/10.2307/1468365

Ramírez A, Pringle CM, Wantzen KM (2008) Tropical stream conservation. In: Dudgeon D (ed) Aquatic ecosystems: tropical stream ecology. Elsevier Science, London, pp 285–304

Richardson JS, Danehy RJ (2007) A synthesis of the ecology of headwater streams and their riparian zones in temperate forests. For Sci 53:131–147

Scheibler EE, Debandi GO (2008) Spatial and temporal patterns in the aquatic insect community of a high-altitude Andean stream (Mendoza, Argentina). Aqu Insects 30:145–161. https://doi.org/10.1080/01650420701880974

Scheibler EE, Claps MC, Roig-Juñent SA (2014) Temporal and altitudinal variations in benthic macroinvertebrate assemblages in an Andean river basin of Argentina. J Limnol 73:92–108. https://doi.org/10.4081/jlimnol.2014.789

Schindler DW, Bayley SE, Parker BR, Beaty KG, Cruikshank DR, Fee EJ, Schindler EU, Stainton MP (1996) The effect of climate warming on properties of boreal lakes and streams at the Experimental Lake Area, northwestern Ontario. Limnol Ocean 41:1004–1017. https://doi.org/10.4319/lo.1996.41.5.1004

Sheldon KS, Tewksbury JJ (2014a) The impact of seasonality in temperature on thermal tolerance and elevational range size. Ecology 95:2134–2143. https://doi.org/10.1890/13-1703.1

Sheldon KS, Tewksbury JJ (2014b) The impact of seasonality in temperature on thermal tolerance and elevational range size. Ecology 95:2134–2143. https://doi.org/10.1890/13-1703.1

Statzner B, Higler B (1986) Stream hydraulics as a major determinant of benthic invertebrate zonation patterns. Freshw Biol 16:127–139. https://doi.org/10.1111/j.1365-2427.1986.tb00954.x

Stevens GC (1992) The elevational gradient in altitudinal range: and extension of Rapoport's latitudinal rule to altitude. Am Nat 140:893–911. https://doi.org/10.1086/285447

Stout J, Vandermeer J (1975) Comparison of species richness for stream-inhabiting insects in tropical and mid-latitude streams. Am Nat 109:263–280. https://doi.org/10.1086/282996

Sutey P, Anderson J, Xu R, Rusak JA, Pereira C, Masters C, Evans D, Thomson C (2015) User Guide for the collection and submission of water samples. Ministry of Environment and Climate Change Chemistry Laboratory, Dorset, Ontario version 2016 2.0. https://desc.ca/water_chemistry. Accessed 30 Apr 2018 ter Braak CJF, Šmilauer P (2002). Canoco Reference Manual and CanoDraw for Windows User´s Guide. Software for Canonical Community Ordination (version 4.5) Microcomputer Power. Ithaca, New York

Vannote RL, Minshall GW, Cummins KW, Sedel JR, Cushing CE (1980) The river continuum concept. Can J Fish Aqua Sci 37:130–137. https://doi.org/10.1139/f80-017

Vinson MR, Hawkins CP (1998) Biodiversity of stream insects: variation at local, basin, and regional scales. Annu Rev Entomol 43:271–293. https://doi.org/10.1146/annurev.ento.43

Yule CM, Leong MY, Liew KC, Ratnarajah L, Schmidt K, Wong HM, Pearson RG, Boyero L (2009) Shredders in Malaysia: abundance and richness are higher in cool upland tropical streams. J N Am Benthol Soc 28:404–415. https://doi.org/10.1899/07-161.1

9

Characterizing a novel predator–prey relationship between native *Diplonychus esakii* (Heteroptera: Belostomatidae) and invasive *Gambusia affinis* (Teleostei: Poeciliidae) in central China

Xu Ouyang · Jiancao Gao · Bojian Chen · Zaizhao Wang · Hong Ji · Martin Plath

Abstract A considerable body of the literature considers the potential impact of exotic predators on native prey organisms, while comparatively, few studies have asked whether and how native predators include novel prey types into their diet spectrum. Here, we asked whether the native aquatic heteropteran *Diplonychus esakii* preys on the highly invasive western mosquitofish (*Gambusia affinis*), which has been introduced to southern China and threatens native fish species through competition and predation on their fry. We conducted 48-h prey choice experiments under semi-natural conditions. In a 'no-choice' experiment (one predator and one potential prey; $n = 200$), we found the heteropterans to prey more on large-bodied fish, a pattern that was also described for other belostomatids, while prey sex had no effect on capture rates. Moreover, large-bodied heteropterans caught more fish than small-bodied individuals. However, overall capture rates in our study were low (11.5–30%) compared to studies on other belostomatids, which explains why subsequent binary prey choice experiments using one predator and two prey—either large and small females or male and female (with smaller sample sizes of $n = 20$ and 30, respectively)—did not confirm the results of our first experiment. Our study exemplifies how a pattern of body size-dependent predation can arise in a novel (not coevolved) predator–prey interaction. We tentatively argue that the observed pattern could be driven by intrinsic features of the predator, namely, altered prey preferences with increasing age coupled with a general preference for large-bodied prey, or changing nutritional needs at different developmental stages.

Keywords Aquatic invasion · Mosquitofish · Neozoan · Predator–prey interaction

Introduction

Biological invasions challenge the conservation of native biodiversity and can have profound impacts on ecological processes and evolutionary dynamics (Gurevitch and Padilla 2004; Kistner and Dybdahl 2013; Stuart et al. 2014). Several studies focused on the question of whether and how invasive predators affect native prey species (Gerard et al. 2014; Nunes et al. 2014). For example, predation experiments by Błońska et al. (2015) demonstrated that the Ponto–Caspian racer goby *Babka gymnotrachelus* preys more on the native

X. Ouyang · J. Gao · B. Chen · Z. Wang · H. Ji · M. Plath
College of Animal Science and Technology, Northwest A&F University, Yangling 712100, Shaanxi, People's Republic of China

M. Plath (✉)
Department of Ecology and Evolution, Goethe University Frankfurt, Max-von-Laue-Straße 13, 60438 Frankfurt am Main, Germany
e-mail: mplath-zoology@gmx.de

amphipod *Gammarus fossarum* than invasive *Dikerogammarus villosus* and *Pontogammarus robustoides* in its invasive range in Poland. Fewer studies have asked how invasive prey species integrate into existing food webs (e.g., Weerman et al. 2014; Cabrera-Guzmán et al. 2015; Raspi et al. 2015). One important step towards understanding trophic relationships between native predators and invasive prey species is to examine predators' feeding preferences, for instance, via standardized food choice experiments (see Horstkotte and Plath 2008; Plath et al. 2011).

Our present study provides insights into a novel predator–prey interaction between the predatory aquatic heteropteran *Diplonychus* cf. *esakii* and invasive western mosquitofish (*Gambusia affinis*) in central China. The members of the family Belostomatidae (Heteroptera: Hemiptera; giant water bugs) are voracious sit-and-wait predators with extra-oral digestion that prey on a variety of invertebrates and vertebrates (Menke 1979). They are typically found sitting along the water's edge posing with their front legs below and their abdominal spiracles above the water surface, from where they catch bypassing prey with their raptorial forelegs that are strongly incrassate, with the femora often grooved to accept the tibiae (Menke 1979). Upon capture, belostomatids inject toxins causing prey paralysis and digestive enzymes causing tissue necrosis (Swart and Felgenhauer 2003; Swart et al. 2006).

While the large-bodied members of the subfamily Lethocerinae feed on small snakes, anurans, and fish, the much smaller members of the subfamily Belostomatinae mainly feed on various invertebrates and anuran larvae, but do not regularly prey on fish (Cullen 1969; Babbitt and Jordan 1996; Swart and Felgenhauer 2003; Mori and Ohba 2004; Ohba and Nakasuji 2006; Swart et al. 2006; Ohba 2011). Prey selection in belostomatines has been thoroughly investigated in the predator–prey relationship between *Belostoma* cf. *bakeri* and a co-occurring fish, the cave molly (*Poecilia mexicana*) in a South-Mexican sulfur cave. Due to the toxic effects of hydrogen sulfide (H_2S) and the associated extreme hypoxia in the water, cave mollies must spent considerable time performing aquatic surface respiration (Plath et al. 2007), rendering them vulnerable to attacks by giant water bugs (Tobler et al. 2009). Prey choice experiments found the heteropterans to preferentially prey upon large-bodied fish (Plath et al. 2003; Tobler et al. 2007), and to prefer gestating over non-gestating females (Plath et al. 2011), and males over females (Tobler et al. 2008). While higher predation risk of large-bodied fish could be explained by better (non-visual) detectability of surfacing large-bodied individuals, higher predation on males and gestating females is likely a function of increased energy demands of males and gestating females, leading to increased aquatic surface respiration and, thus, increase likelihood of falling victim to heteropteran predators (Tobler et al. 2008; Plath et al. 2011).

In contrast to the aforementioned system involving coevolved predatory and prey species, our current study focused on a novel predator–prey relationship between the belostomatine *D. esakii* and their potential exotic prey, Western mosquitofish, *Gambusia affinis* (like *P. mexicana*, a member of the neotropical family Poeciliidae). Since its introduction to China in 1927 for malaria prophylaxis, *G. affinis* has spread widely in the Yangtze River drainage (Li and Xie 2002). Studies on interspecific ecological relationships between invasive *G. affinis* and native fish species, such as *Tanichthys albonubes* and *Oryzias latipes*, suggest that *G. affinis* threatens the survival of these species (Chen et al. 1989; Chen 2010), highlighting the need for investigations into predation on invasive *G. affinis* by native faunal elements in China, and our present study represents a first step in this direction.

Our study was motivated by the observation that, while belostomatines do not usually prey on fish (see above), they occasionally seem to function even as top predators, e.g., in the aforementioned Mexican sulfur cave with unusually high densities of fish (mean $\pm$ SE: 162.3 $\pm$ 16.1 individuals m^{-2}; Jourdan et al. 2014) and *B. bakeri* (1.12 $\pm$ 0.43 individuals m^{-2}; Tobler et al. 2007). Moreover, a previous study found congeneric *Diplonychus rusticus* to prey on small fish under natural conditions (Hazarika and Goswami 2012). During field work, we repeatedly noted high densities of *D. esakii* (and other aquatic heteropterans) in several stagnant water bodies in China, in which the species regularly co-occurs with the abundant invasive *G. affinis*—including the study site we chose for our present study (Fig. 1a). Therefore, we tested whether some common patterns of prey selection in the well-researched *B. bakeri–P. mexicana* predator–prey relationship (Plath and Tobler 2010) would also be apparent in the novel predator–prey interaction between *D. esakii* and *G. affinis*.

Using prey choice experiments under semi-natural conditions, we first experimentally confirmed the assumption that *D. esakii* prey on *G. affinis*, as previous studies found other belostomatines (including members of the genus *Diplonychus*) to include fish prey (Plath et al. 2003, 2011; Tobler et al. 2007, 2008, 2013; Hazarika and Goswami 2012). Having confirmed this assumption, (1) we then asked if a

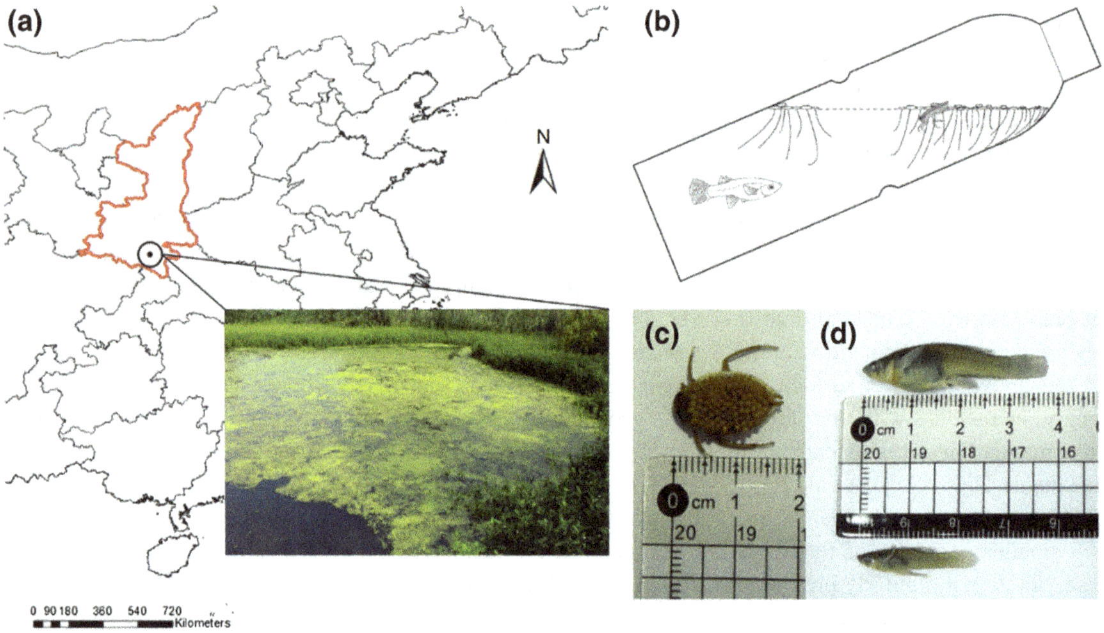

Fig. 1 Study site and exemplary photos of experimental test subjects. **a** Study site near Ankang in Shaanxi Province (*highlighted*) in central China. *Insert* shows the abandoned fish pond in which predation experiments were conducted. **b** Schematic view on an experimental unit containing one predator and one prey fish. **c** Adult male *Diplonychus* cf. *esakii* carrying a clutch on its abdomen; **d** adult female (*above*) and male *Gambusia affinis* (*below*) collected at our study site (photographs were taken from ethanol-preserved specimens)

pattern of size-specific predation exists. We predicted that *D. esakii* would show a predilection for larger over small-bodied mosquitofish, which would be congruent with optimal foraging theory (Pyke 1984). (2) As previous studies found male-biased predation by *B.* cf. *bakeri* on another poeciliid (Tobler et al. 2008), we predicted a similar pattern in the predatory interaction between *D. esakii* and *G. affinis*, as poeciliid males expose themselves more to predators during their continuous mate searching and mating behavior (Godin 1995). (3) Furthermore, we asked if the predators' body size would affect their prey selection (e.g., in the form of large-bodied predators preferring large prey items and small-bodied predators preferring smaller prey). This prediction was based on previous studies showing that different instar stages in belostomatids differ in metabolic rates and foraging tactics (Biesmeijer and Tóth 1998; Cloarec 1990).

Materials and methods

We conducted two types of experiments: (a) a 'no-choice' predation experiment (i.e., one predator and one prey fish per test run) used randomly assigned combinations of water bugs and mosquitofish to establish general patterns of body size- and sex-dependent predation. (b) In the latter experiment, we found large individuals to prey more on mosquitofish than small-bodied individuals (see "Results"); hence, we conducted binary prey choice experiments (i.e., one predator and two prey fish per test run) with only large-bodied (adult) *D. esakii* to verify the effect of prey body size (higher predation on large- compared to small-bodied fish), and we tested if an effect of prey sex (higher predation on male vs. female mosquitofish) might become apparent in this experimental situation.

Study site

We conducted our experiments between May 20th and August 25th 2015 in an abandoned fish pond located near the Aquaculture Demonstration Center of Northwest A&F University, close to the national road G316 in

Ankang, Shaanxi Province, China (E 108.807, N 32.73409; 222 m altitude). The pond was surrounded by dense riparian vegetation, and a large portion of the pond's surface was covered with floating filamentous green algae and duckweed (*Lemna* sp.; Fig. 1a).

We collected adult and juvenile *G. affinis* and *D. esakii* using dip nets (1 mm mesh size) and kept them separated by species in aerated buckets in the shade (for less than 1 h) until they were used in one of the following experiments. We first checked all collected fish for signs of attacks from heteropteran predators (puncture wounds) and made sure that only fish without such wounds entered the experiments. Heteropterans collected for the prey choice experiments included males carrying clutches on their abdomen and individuals without clutches; however, reliably determining the sex and the exact developmental stages of all test subjects was not feasible under field conditions. We collected several heteropterans and stored them in 96% ethanol for later species identification in the laboratory. Following the descriptions and keys provided by Miyamoto and Lee (1966) and Ding (2004), the collected heteropterans were identified as *Diplonychus* cf. *esakii*, originally described from Chejudo, Korea.

We initially collected a larger number of *G. affinis* than needed in our prey choice experiments. We used the total sample of all collected fish to calculate the adult sex ratio (numbers of males/numbers of females) in our study population; this information was important for the discussion of our results on potential sex-specific predation. Fishes not used in the prey choice experiments were measured for the standard length (SL, from the tip of the snout to the posterior end of the caudal peduncle) so as to provide information on body size distribution patterns. Furthermore, to provide empirical data on population densities of *D. esakii* as mentioned in the "Introduction", we determined densities once (May 22nd, 2015) by counting larvae and adult water bugs (imagines) in six randomly selected 1 $\times$ 1 m sampling squares along the edge of the pond through total counts.

Total counts of *D. esakii* in six randomly selected quadrants resulted in mean ($\pm$SD) estimated densities of 2.5 $\pm$ 1.9 (1–6) adult individuals and 18.8 $\pm$ 5.8 (11–28) nymphs of different instar stages per square meter. We collected $N = 260$ *G. affinis* in preparation of our no-choice experiment, which covered a wide range of body size (6.2–39.6 mm). Fifty-eight adult individuals were males (SL, mean $\pm$ SD = 16.41 $\pm$ 1.40 mm, range = 11.2–18.8 mm), and 119 were females (23.27 $\pm$ 4.39 mm, 13.2–39.6 mm). This resulted in an adult sex ratio of 0.49. Amongst the inspected fish ($N = 260$), no cases of puncture wounds resulting from heteropteran attacks were observed.

We caught $N = 232$ adult *G. affinis* in preparation of the binary prey choice experiments (this time leaving out juvenile fish). Ninety-six individuals were males (SL = 19.90 $\pm$ 1.98 mm, range = 15.8–30.1 mm) and 136 females (32.49 $\pm$ 3.46 mm, 22.9–39.9), and the adult sex ratio was 0.71. Four female *G. affinis* had puncture wounds on their caudal peduncle (one per fish), presumably stemming from attacks by heteropterans or other predatory aquatic arthropods.

Experimental setup

Control experiment

Before we conducted our main experiments, we ran controls ($N = 10$ replicates for the no-choice experiment with four cases of juveniles and equal numbers of males and females; $N = 5$ replicates for the binary choice experiment with one male and one female per replicate) to assess baseline mortality and feasibility of our approach. We used experimental containers as described for our main experiments (see below) but did not introduce *D. esakii*, but only *G. affinis* and some floating vegetation. Standard lengths were 12.23 $\pm$ 0.67 mm (11.5–12.8 mm; males), 35.50 $\pm$ 3.86 mm (31.8–39.5 mm; females), and 8.90 $\pm$ 1.35 mm (7.1–10.2 mm; juveniles) in the control runs of the no-choice experiment, and 20.60 $\pm$ 3.43 mm (16.7–24.1 mm; males) and 33.84 $\pm$ 5.93 mm (23.7–38.4 mm; females) in the control runs of the binary choice experiments. No mortality was detected in any of the control runs after 48 h.

No-choice predation experiment

Our general experimental approach followed well-established protocols (Tobler et al. 2007; Plath et al. 2011). In the no-choice predation experiment ($N = 200$ trials), a randomly selected individual of *G. affinis* (SL = 16.67 ± 6.12 mm, size range = 6.2–31–0.5 mm) as well as one individual of *D. esakii* (total length, TL = 12.44 ± 3.32 mm, size range = 3.9–17.3 mm) were introduced into each experimental container, a transparent plastic bottle (600 ml). Some pieces of floating vegetation were added to provide opportunities for the water bugs to sit and prevent them from drowning (Fig. 1b). The bottles were perforated by 12 small holes to maintain water and oxygen exchange throughout the entire experiment. 20 bottles per batch were fixed in a PVC rack and then partially submerged near the edge of the pond and covered with riparian vegetation. A piece of green plastic net was placed between adjacent bottles to prevent visual contact between test subjects in different bottles.

Binary prey choice experiments

Our binary prey choice experiments largely followed the experimental approach described above, but we used only large-bodied (adult) heteropterans (TL = 16.24 ± 0.72 mm, size range = 13.8–17.5 mm) and offered two different mosquitofish in each experimental container. In the first experiment, we introduced one large (SL = 37.32 ± 1.13 mm) and one small female mosquito fish (29.92 ± 2.39 mm; paired *t* test: $t_{19} = 14.01$, $p < 0.0001$). The second experiment tested for sex-specific predation. We were unable to collect sufficient numbers of size-matched pairs of male and female *G. affinis*, as mosquitofish males are, on average, much smaller than females (Reznick 1981). Thus, we introduced one male (20.27 ± 1.07 mm) and one small-bodied adult female mosquitofish (32.17 ± 2.08 mm; paired *t* test: $t_{29} = 26.82$, $p < 0.0001$), while making an attempt to keep the size difference at a minimum.

The acclimation phase was kept short (30 min), because the total duration of our experiments (48 h) provided a sufficient time for all test subjects to settle and for the water bugs to exert predation. We terminated all tests after 48 h and screened the fish for signs of predation, as evidenced by puncture wounds (resulting from attempted attacks) or cases of mortality (i.e., successful attacks). In some cases, a partially lysed carcass was retrieved. All fish and heteropterans were measured for body size upon termination of a test (standard length, see above, in case of fish and total length, from the anterior tip of the head to the posterior tip of the abdomen, in case of the heteropterans).

Statistical analysis

Data from the no-choice experiment included juvenile mosquitofish (whose sex could not be determined) and adult fish. In a first analysis using only the subset of data involving adult fish ($N = 41$ males, $N = 79$ females), we tested for an effect of prey sex on predation by *D. esakii*. We used 'predation success' (yes/no) as the dependent variable in a binary logistic regression with a stepwise backward elimination procedure (based on likelihood ratios) and coded 'predator body size' and 'prey body size' as covariates and 'prey sex' as a factor, including all two-way and three-way interactions. No sex effect was detected (see "Results"), and so we proceeded with an analysis that included all test runs (i.e., also juvenile mosquitofish) while excluding the factor 'sex' from the analysis. Finally, to corroborate both statistically significant effects of this analysis (see "Results"), we applied independent-samples *t* tests to compare the body sizes of fish that were or were not preyed upon and *D. esakii* that did or did not catch fish.

When analyzing the results from the two binary prey choice experiments, we first asked what factors influence overall predation by *D. esakii* (i.e., independent of whether large/small or male/female fish were more preyed upon). We used 'predation success' (yes/no) as the dependent variable in binary logistic regressions while including the water bugs' body size as a covariate and 'presence of eggs' as a factor to test if males carrying a clutch would show lower capture rates (see Crowl and Alexander 1989 for predation by *Belostoma flumineum* on *G. affinis*). Note that 'presence of eggs' was not included in the previous analysis as a low number of $n = 2$ males carrying eggs precluded inclusion of this factor. We also included the interaction term 'presence of eggs × body size' and excluded non-significant terms in a stepwise backward elimination procedure as mentioned above. Low capture rates (three out of 20 bottles, i.e., 15%) in the experiment on size-dependent predation precluded further analysis of effects of prey body size, while sex-specific predation was evaluated for the subset of bottles in which predation had occurred by comparing observed frequencies of female and male predation against the distribution expected by chance using a χ^2 test.

All data were analyzed using the SPSS statistical package (version 19, IBM) and are presented as mean ± SD.

Results

No-choice predation experiment

Throughout our predation experiments, we observed only cases of predation by *D. esakii* on *G. affinis*, but not vice versa. In the no-choice experiment, predation by *D. esakii* occurred in 11.5% of cases (i.e., in 23 out of the $N = 200$ independent replicates). Seven cases of predation involved male *G. affinis*, 14 female fish, and two cases juveniles. Our first binary logistic regression model using only the subset of sexually mature fish (−2 log likelihood = 105.64, Nagelkerke $R^2 = 0.08$) found neither 'sex' nor any interaction involving 'sex' to have statistically significant effects, and so they were excluded during the stepwise elimination procedure (Wald < 0.61, $df = 1$, $p > 0.30$).

In the model including both adult and immature mosquitofish (−2 log likelihood = 126.41, Nagelkerke $R^2 = 0.15$), both covariates (SL of the mosquitofish: $B \pm$ SE $= 0.11 \pm 0.04$, Wald = 7.81, $df = 1$, $p = 0.01$; TL of the heteropterans: $B \pm$ SE $= 0.22 \pm 0.09$, Wald = 6.32, $df = 1$, $p = 0.01$) had significant effects and were retained during the stepwise elimination procedure, while all other factors and interactions were excluded (Wald < 0.01, $df = 1$, $p > 0.97$). To visualize both effects, we calculated residuals (differences between observed and predicted probabilities) for each covariate and plotted the resulting data against the respective other covariate (Fig. 2a, b).

Post hoc independent samples *t* tests corroborated that fishes that were preyed upon (20.24 ± 4.91 mm, range: 10.8–29.5 mm, $N = 23$) were significantly larger than those that were not preyed upon (16.20 ± 6.12 mm, 6.2–31.5 mm, $N = 177$; $t_{198} = 3.04$, $p = 0.003$). Likewise, individuals of *D. esakii* that preyed on mosquito fish (14.16 ± 2.69 mm, 7.7–17.1 mm) were significantly larger than their counterparts that did not (12.21 ± 3.34 mm, 3.9–17.3 mm, $t_{198} = 2.69$, $p = 0.008$).

Binary prey choice experiments

In our first binary prey choice experiment, we tested for size-specific predation on *G. affinis* females by *D. esakii*. Predation occurred in only three out of 20 bottles (15%), whereby two of the three captured fish were small and one was large. This low overall predation rate precluded further statistical analysis of size-specific predation. In the final analytical model on overall capture success, only the covariate 'body size' of the water bugs was retained ($B \pm$ SE $= -0.11 \pm 0.04$, Wald = 7.75, $df = 1$, $p = 0.005$). *Diplonychus esakii* that caught fish (16.03 ± 0.61 mm, $N = 3$) were slightly smaller than those that did not catch any fish (16.46 ± 0.63 mm, $N = 17$).

Our second binary prey choice experiment tested for sex-specific fish predation by *D. esakii*. Predation was twice as high as in our first binary prey choice experiment (i.e., nine out of 30 bottles, 30%). Among the nine fish caught by *D. esakii*, a non-significant tendency was seen for higher predation on males (seven cases) than females (two cases; Chi-square test: $\chi^2 = 2.78$, $df = 1$, $p = 0.096$). Again, only the covariate 'body size' had a statistically significant effect in the final model on overall capture success ($B \pm$ SE $= -0.06 \pm 0.03$, Wald = 4.85, $df = 1$, $p = 0.028$), and water bugs that captured *G. affinis* (15.68 ± 0.94 mm, $N = 9$) were again slightly smaller than those that did not (16.29 ± 0.51 mm, $N = 21$).

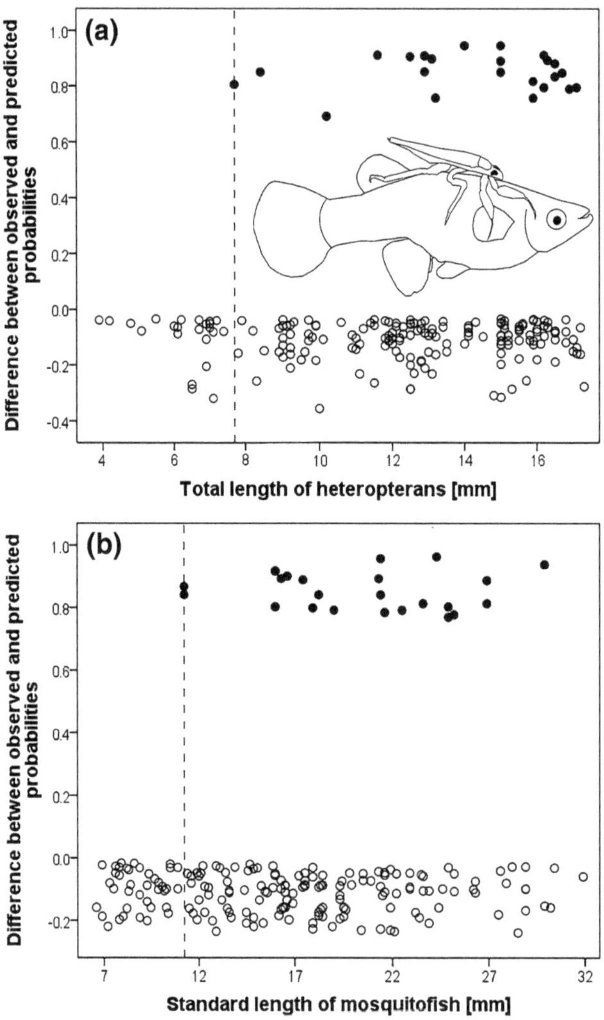

Fig. 2 Graphic illustration of the main effects in the multiple regression analysis on predation by the heteropteran *D.* cf. *esakii* on mosquitofish (*G. affinis*); **a** shows the effect of the heteropterans' body size (total length), while **b** shows the effect of prey body size (standard length). Depicted are residuals (difference between observed and predicted probabilities), whereby positive values and *black circles* indicate cases of successful predation, while negative values and *open circles* indicate no predation. *Dotted lines* indicate the minimum body size at which predation was observed (**a**: 7.70 mm; **b**: 10.80 mm)

Discussion

In the no-choice experiment, we cohabitated one individual of the aquatic heteropteran *D. esakii* with one individual of their potential exotic prey, the western mosquitofish (*G. affinis*). *Gambusia affinis* that were attacked were significantly larger than those that were not attacked, and *D. esakii* that preyed on *G. affinis* were themselves larger than those that did not. However, in this and in subsequent binary prey choice experiments, we found predation rates to be low (11.5–30%). For comparison, when offering two different-sized *Poecilia mexicana* females as potential prey in a similar experimental design, predation rates by *Belostoma* cf. *bakeri* were as high as 44% (Tobler et al. 2007). Low predation rates in our present study likely explain why our subsequent binary choice experiments (with smaller sample sizes) did not corroborate an effect of prey body size. Only a non-significant tendency was seen for higher predation on males than on females. In contrast to our no-choice experiment, smaller *D. esakii* showed slightly higher overall predation than larger individuals in the binary choice experiments. Note, however, that we used only comparatively large-bodied (adult) predators in the binary choice experiments, such that our results are not necessarily conflicting (see below). Whether or not the relatively small dimensions of our experimental containers

(600 ml) played a role in affecting the outcome of our second experiment will need to be determined in future studies using larger containers (compare 1.5 L as used in Plath et al. 2011).

Our study corroborates predation by *D. esakii* on *G. affinis* and demonstrates that this exotic prey species is readily included into the prey spectrum of *D. esakii*. The overall low predation rate could be a result of both species not sharing a coevolutionary history, assuming that the predator does not adequately recognize its exotic prey (David et al. 2017), and *D. esakii* could evolve to be more efficient in the future (Carlsson et al. 2009). Another possible explanation for the low predation rate could be the short acclimation period. Stress resulting from mechanical handling affects individuals for different periods of time (Harak et al. 1998); however, we believe that this explanation is unlikely as the total duration of our tests was rather long (48 h). Some of the fish that we collected at our study site showed puncture wounds. However, some co-occurring aquatic insects (such as *Ranatra* cf. *chinensis* and *Nepa* cf. *cinerea*, both Nepidae) also prey on fish (Rao 1976; Bailey 1986), and so it remains to be determined whether or not *D. esakii* prey on mosquito fish not only in our experimental containers, but also in their natural environment. We are aware that fish prey certainly constitutes only a part of the total diet spectrum of *D. esakii*—as indicated by comparatively low overall predation rates—and so several points in our discussion will benefit from future experiments offering different types of invertebrate prey as well. For example, mosquito larvae (Culicidae) are a common prey type of giant water bugs and other aquatic heteropterans, and recent studies even recommended using belostomatids as new biological control agents of mosquitoes, i.e., for malaria control (Mukherjee and Datta 1999; Saha et al. 2010, 2014).

Using prey choice experiments involving individuals of the comparatively large-sized *Lethocerus americanus* (Lethocerinae), as well as three juvenile (small) and three adult (large) *G. affinis*, Schumann et al. (2012) found giant water bugs to preferentially prey on adult mosquitofish, characterizing the species as a specialized predator of adult fish. In our present study, we also detected a pattern of body size-dependent predation (question 1). Our finding of *D. esakii* preying more on larger *G. affinis* is congruent with optimal foraging theory (Stephens et al. 2007), whereby individuals accept opportunity and energetic costs of (potentially unsuccessful) attacks only when the expected benefits (in terms of net energy uptake) are high. Indeed, both male and female *G. affinis* showed pronounced variation in body size, which is congruent with studies from the species' natural distribution range (Campton and Gall 1988). Alternatively, higher predation on larger mosquitofish may be due to a greater detectability by *D. esakii*, as larger fishes produce stronger vibration, e.g., when surfacing (Plath et al. 2011).

Only a non-significant tendency towards male-biased predation was uncovered but only in our binary choice tests (question 2). By contrast, a strong signal of male-biased predation was uncovered in a study on predation by *B.* cf. *bakeri* on *P. mexicana* in a hydrogen sulfide-rich cave (Tobler et al. 2008). The authors reported on higher oxygen demands of male fish leading to more aquatic surface respiration under hypoxic conditions, thus exposing males to elevated levels of predation risk. Moreover, male poeciliid fishes are more active than females, frequently switching between shoals in search of females (Andreev 1994; Griffiths and Magurran 1998; Magurran 1998), again increasing the likelihood of detection by predators. Even though Cech et al. (1985) noted that sex had little effect on metabolic rates when determining oxygen consumption rates of *G. affinis* in flow-through respirometry chambers while controlling for body mass differences, mate searching, and mating behavior of *G. affinis* males in our binary choice experiment could still result in slightly higher oxygen consumption and occasional aquatic surface respiration, or it simply renders males more detectible and thus vulnerable to predation by *D. esakii*. Slightly male-biased predation could be responsible for the observed female-biased sex ratio of our study population of *G. affinis* (see also Hildebrand 1927).

In our no-choice experiment, we used the full range of body sizes of *D. esakii* encountered at our study site (12.44 ± 3.32 mm), and large-bodied heteropterans were more likely to prey upon *G. affinis* (question 3). By contrast, only large-bodied, adult heteropterans (16.19 ± 0.76 mm) were used in the binary choice experiment, and we did not find the same effect. We argue that predatory strategies and/or energetic demands of *D. esakii* may change during ontogeny. Cloarec (1990) studied predatory tactics adopted by *Diplonychus indicus* during different developmental stages (see also Ohba and Nakasuji 2006 for *Lethocerus deyrolli*). Overall, predation success increased with age, and one type of predatory behavior ('lunge', the sudden acceleration from an ambush spot with one stroke of the swimming legs) decreased with age, while another hunting strategy became more frequent ('strike', involving the closing of the claws with or without projection of the forelegs). It seems reasonable to assume that small-bodied *D. esakii* also adopted the active (lunge) strategy, which may render them less efficient fish predators. Venkatesan and Muthukrishnan (1987) reported that earlier instar stages of *Diplonychus indicus* showed lower assimilation rates when feeding on fish fingerlings compared to tests in which *Culex* and *Aedes* larvae were offered. Moreover, life spans of nymphs were lowest when specimens were only provided with fish fingerlings. We argue that large-bodied *D. esakii* likely face a higher risk of being detected by their prey; hence, employing the 'lunge' strategy would lead to considerable energy losses at a comparably low net energy uptake (i.e., successful capture rate). This ought to favor the use of the 'strike' strategy, which is more likely to be successful when preying on fish. Note that *D. esakii* that were used in the binary choice experiment were uniformly large (adult) individuals, and a "reversed" body size effect within the much smaller body size range was detected. Thus, the possibility remains that aging results in lower metabolic rates and translates into reduced predatory behavior (Biesmeijer and Tóth 1998). Moreover, senescence can be associated with loss of muscle mass and deterioration of neural functioning (Rose 1991; Boggs 2009), possibly impeding predatory behavior of older *D. esakii.*

Suares and Ramadoss (2003) reported that both nymphs and females of the congeneric *Diplonychus rusticus* preferred (largely sessile) *Chironomus* spp. larvae as prey, while male *D. rusticus* mainly preyed on (actively moving) *Culex quinquefasciatus* larvae, likely reflecting higher encounter rates with *C. quinquefasciatus* and a more elaborate predatory behavior in males. A pattern of sex-specific prey choice did not become apparent in our present study (i.e., at least those individuals we could unequivocally identify as males, as they were carrying clutches on their backs, did not show different patterns of prey selection). Several studies on *Belostoma flumineum* showed that brooding an egg raft increased drag during swimming (Crowl and Alexander 1989; Kight et al. 1995). This cost of paternal care not only impaired the predation efficiency of *B. flumineum*, but also their agility during escape behavior. In our present study, however, we found no effect of 'presence of eggs' (see "Results"). Based on the argumentation outlined above, *D. esakii* that showed predation in our experiments may adopt a 'strike' strategy, and so their swimming ability likely had a negligible effect on capture success.

Species become invasive if they establish self-sustaining populations in a new range; however, native faunal elements can slow down this process or even prevent it altogether (Babbitt and Jordan 1996; Locke et al. 2014). For example, Cabrera-Guzmán et al. (2015) found that although invasive cane toads (*Rhinella marina*) contain potent defensive chemicals, native rodents (*Melomys burtoni*, *Rattus colletti,* and *R. tunneyi*) readily killed and consumed cane toads. Weerman et al. (2014) discovered that spat of invasive Pacific oyster (*Crassostrea gigas*) is more susceptible to predation by indigenous brown shrimps (*Crangon crangon*) compared to the spat of native bivalves. In the latter examples, native predators potentially constrain the invasion success or exotic prey species, and we argue that our present study—in conjunction with future studies on other potential predators of *G. affinis*—may help understand how local predator communities affect the further spread of the species.

Acknowledgements We are grateful to Ke Changsen for accommodation at the Aquaculture Demonstration Station in Ankang. Financial support came from Northwest Agriculture and Forestry University (Z111021403).

References

Andreev OA (1994) The behavior of male and female guppies in an unfamiliar environment. J Ichthyol 34:139–143

Babbitt K, Jordan F (1996) Predation on *Bufo terrestris* tadpoles: effects of cover and predator identity. Copeia 2:485–488

Bailey PCE (1986) The feeding behaviour of a sit-and-wait predator, *Ranatra dispar* (Heteroptera: Nepidae): description of behavioural components of prey capture, and the effect of food deprivation on predator arousal and capture dynamics. Behaviour 97(1):66–92

Biesmeijer JC, Tóth E (1998) Individual foraging, activity level and longevity in the stingless bee *Melipona beecheii* in Costa Rica (Hymenoptera, Apidae, Aeliponinae). Insectes Soc 45(4):427–443

Błońska D, Grabowska J, Kobak J, Jermacz Ł, Bącela-Spychalska K (2015) Feeding preferences of an invasive Ponto–Caspian goby for native and non-native gammarid prey. Freshw Biol 7:1–9. doi:10.1111/fwb.12647

Boggs CL (2009) Understanding insect life histories and senescence through a resource allocation lens. Funct Ecol 23(1):27–37

Cabrera-Guzmán E, Crossland MR, Pearson D, Webb JK, Shine R (2015) Predation on invasive cane toads (*Rhinella marina*) by native Australian rodents. J Pest Sci 88:143–153. doi:10.1007/s10340-014-0586-2

Campton DE, Gall GAE (1988) Responses to selection for body size and age at sexual maturity in the mosquitofish, Gambusia affinis. Aquaculture 68(3):221–241

Carlsson NOL, Sarnelle O, Strayer DL (2009) Native predators and exotic prey—an acquired taste? Front Ecol Eviron 7(10):525–532

Cech JJJr, Massingill MJ, Vondracek B, Linden AL (1985) Respiratory metabolism of mosquitofish, *Gambusia affinis*: effects of temperature, dissolved oxygen, and sex difference. Environ Biol Fish 13(4):297–307

Chen GZ (2010) Interspecific relationship between the invasive species *Gambusia affinis* and the native endangered species *Tanichthys albonubes*. Dissertation, Jinan University

Chen YR, Yu HH, Chu XL (1989) Taxonomy and distribution of the genus *Oryzias* in Yunnan, China. Acta Zootax Sin 2:239–246

Cloarec A (1990) Variations of predatory tactics of a water bug during development. Ethology 86:33–46

Crowl TA, Alexander E Jr (1989) Parental care and foraging ability in male water bugs (*Belostoma flumineum*). Can J Zool 67:513–515

Cullen MJ (1969) The biology of giant water bugs (Hemiptera: Belostomatidae) in Trinidad. Proc R Ent Soc Lond 44:123–136

David P, Thébault E, Anneville O, Duyck PF, Chapuis E, Loeuille N (2017) Impacts of invasive species on food webs: a review of empirical data. Adv Ecol Res 56:1–60

Ding JH (2004) The taxonomic study on some families of aquatic Heteroptera from China. Dissertation, Nankai University

Gerard A, Jourdan H, Cugniere C, Millon A, Vidal E (2014) Is naïveté forever? Alien predator and aggressor recognition by two endemic island reptiles. Naturwiss 101(11):921–927. doi:10.1007/s00114-014-1233-8

Godin JGJ (1995) Predation risk and alternative mating tactics in male Trinidadian guppies (*Poecilia reticulata*). Oecologia 103(2):224–229

Griffiths SW, Magurran AE (1998) Sex and schooling behaviour in Trinidadian guppies. Anim Behav 56:689–693

Gurevitch J, Padilla DK (2004) Are invasive species a major cause of extinctions? Trends Ecol Evol 19:470–474

Harak M, Kuusik A, Hiiesaar K, Metspalu L, Luik A, Tartes U (1998) Calorimetric investigations on physiological stress in *Tenebrio molitor* (Coleoptera, Tenebrionidae) pupae. Thermochim Acta 309(1–2):57–61

Hazarika R, Goswami MM (2012) Feeding behavior of *Diplonychus rusticus* Fabricius (Hemiptera, Belostomatidae) on fish and fish food. Int Proc Chem Biol Environ 40:136–140

Hildebrand SF (1927) Sex ratio in *Gambusia*. Biol Bull 53(5):390–404

Horstkotte J, Plath M (2008) Divergent evolution of feeding substrate preferences in a phylogenetically young species flock of pupfish (*Cyprinodon* spp.). Naturwiss 95(12):1175–1180. doi:10.1007/s00114-008-0439-z

Jourdan J, Bierbach D, Riesch R, Schießl A, Wigh A, Arias-Rodriguez L, Indy JR, Klaus S, Zimmer C, Plath M (2014) Microhabitat use, population densities, and size distributions of sulfur cave-dwelling *Poecilia mexicana*. PeerJ 2:e490. doi:10.7717/peerj.490

Kight SL, Sprague J, Kruse KC, Johnson L (1995) Are egg-bearing male water bugs, *Belostoma flumineum* Say (Hemiptera: Belostomatidae), impaired swimmers? J Kans Entomol Soc 68(4):468–470

Kistner EJ, Dybdahl MF (2013) Adaptive responses and invasion: the role of plasticity and evolution in snail shell morphology. Ecol Evol 3(2):424–436. doi:10.1002/ece3.471

Li ZY, Xie Y (2002) Invasive species in China. China Forestry Publishing House, Beijing, p 88

Locke SA, Bulté G, Marcogliese DJ, Forbes MR (2014) Altered trophic pathway and parasitism in a native predator (*Lepomis gibbosus*) feeding on introduced prey (*Dreissena polymorpha*). Oecologia 175:315–324. doi:10.1007/s00442-014-2898-6

Magurran AE (1998) Population differentiation without speciation. Philos Trans Roy Soc Lond B 353:275–286

Menke A (1979) Family Belostomatidae—giant water bugs. The semiaquatic and aquatic hemiptera of California (Heteroptera: Hemiptera). University of California Press, Berkeley, pp 76–86

Miyamoto S, Lee CE (1966) Heteroptera of Quelpart Island (Chejudo). Sieboldia 3:313–426

Mori A, Ohba S (2004) Field observations of predation on snakes by the giant water bug. Bull Herpetol Soc Japan 2004:78–81

Mukherjee B, Datta NC (1999) New biological control agents of mosquito. Environ Ecol 17(3):622–625

Nunes AL, Orizaola G, Laurila A, Rebelo R (2014) Rapid evolution of constitutive and inducible defenses against an invasive predator. Ecology 95(6):1520–1530

Ohba S (2011) Field observation of predation on a turtle by a giant water bug. Entomol Sci 14(3):364–365. doi:10.1111/j.1479-8298.2011.00450.x

Ohba S, Nakasuji F (2006) Dietary items of predacious aquatic bugs (Nepoidea: Heteroptera) in Japanese wetlands. Limnology 7:41–43. doi:10.1007/s10201-006-0161-5

Plath M, Tobler M (2010) Subterranean fishes of Mexico (*Poecilia mexicana*, Poeciliidae). In: Trajano E, Bichuette ME, Kapoor BG (eds) Biology of subterranean fishes. Science Publishers, Taylor & Francis Inc, Enfield, pp 281–330

Plath M, Parzefall J, Schlupp I (2003) The role of sexual harassment in cave and surface dwelling populations of the Atlantic molly, *Poecilia mexicana* (Poeciliidae, Teleostei). Behav Ecol Sociobiol 54(3):303–309. doi:10.1007/s00265-003-0625-0

Plath M, Tobler M, Riesch R, de Leon FJG, Giere O, Schlupp I (2007) Survival in an extreme habitat: the roles of behaviour and energy limitation. Naturwiss 94(12):991–996. doi:10.1007/s00114-007-0279-2

Plath M, Riesch R, Culumber ZW, Streit B, Tobler CM (2011) Giant water bug (*Belostoma* sp.) predation on a cavefish (*Poecilia mexicana*): effects of female body size and gestational state. Evol Ecol Res 13:133–144

Pyke GH (1984) Optimal foraging theory—a critical review. Annu Rev Ecol Syst 15:523–575

Rao TKR (1976) Bioecological studies on some aquatic Hemiptera-Nepidae. Entomon 1:123–132

Raspi A, Abdimaleki R, Fallahzadeh M, Saghaei N, Benelli G (2015) The Oriental drosophilid *Cacoxenus* (*Gitonides*) *campsiphallus*, a predator of invasive mealybugs: first record for Palearctic region and female's description. J Asia-Pac Entomol 18(3):525–528

Reznick D (1981) "Grandfather effects": the genetics of interpopulation differences in offspring size in the mosquito fish. Evolution 35(5):941–953

Rose MR (1991) Evolutionary biology of aging. Oxford University Press, New York

Saha N, Aditya G, Saha GK, Hampton SE (2010) Opportunistic foraging by heteropteran mosquito predators. Aquat Ecol 44(1):167–176

Saha N, Aditya G, Saha GK (2014) Prey preferences of aquatic insects: potential implications for the regulation of wetland mosquitoes. Med Vet Entomol 1(28):1–9

Schumann DA, Cavallaro MC, Hoback WW (2012) Size selective predation by *Hydrophilis triangularis* (Coleoptera: Hydrophilidae) and *Lethocerus americanus* (Hemiptera: Belostomatidae). J Kansas Entomol Soc 85(2):155–159

Stephens DW, Brown JS, Ydenberg RC (2007) Foraging: behavior and ecology. University of Chicago Press, Chicago

Stuart YE, Campbell TS, Hohenlohe PA, Reynolds RG, Revell LJ, Losos JB (2014) Rapid evolution of a native species following invasion by a congener. Science 346:463–466. doi:10.1126/science.1257008

Suares MC, Ramadoss R (2003) Predatory strategy of the water bug, *Diplonychus rusticus* Fabr. (Heteroptera: Belostomatidae) through their immature stages and adulthood. J Entomol Res 27(3):189–195

Swart CC, Felgenhauer BE (2003) Structure and function of the mouthparts and salivary gland complex of the giant waterbug, *Belostoma lutarium* (Stål) (Hemiptera: Belostomatidae). Ann Entomol Soc Am 96(6):870–882. doi:10.1603/0013-8746(2003)096[0870:SAFOTM]2.0.CO;2

Swart CC, Deaton LE, Felgenhauer BE (2006) The salivary gland and salivary enzymes of the giant waterbugs (Heteroptera; Belostomatidae). Comp Biochem Phys 145(1):114–122. doi:10.1016/j.cbpa.2006.05.010

Tobler M, Schlupp I, Plath M (2007) Predation of a cave fish (*Poecilia mexicana*, Poeciliidae) by a giant water-bug (*Belostoma*, Belostomatidae) in a Mexican sulphur cave. Ecol Entomol 32(5):492–495. doi:10.1111/j.1365-2311.2007.00892.x

Tobler M, Franssen C, Plath M (2008) Male-biased predation on a cave fish by a giant water bug. Naturwiss 95:775–779

Tobler M, Riesch RW, Tobler CM, Plath M (2009) Compensatory behaviour in response to sulphide-induced hypoxia affects time budgets, feeding efficiency, and predation risk. Evol Ecol Res 11:935–948

Tobler M, Roach K, Winemiller KO, Morehouse RL, Plath M (2013) Population structure, habitat use, and diet of giant waterbugs in a sulfidic cave. Southw Nat 58:420–426

Venkatesan P, Muthukrishnan S (1987) Impact of predation and food utilization on reproduction of *Diplonychus indicus*, and *Danatra filiformis*. Proc Anim Sci 96(3):293–304

Weerman EJ, Eriksson BK, Olff H, van der Heide T (2014) Predation by native brown shrimp on invasive Pacific oyster spat. J Sea Res 85:126–130. doi:10.1016/j.seares.2013.04.010

10

Surface water quality and deforestation of the Purus river basin, Brazilian Amazon

Eduardo Antonio Ríos-Villamizar · Maria T. F. Piedade · Wolfgang J. Junk · Andréa Viviana Waichman

Abstract In the last years, deforestation constitutes a threat for the aquatic ecosystems. This paper aims to characterize the water quality of the Purus river in the Brazilian Amazon, and investigate the relations between water quality and deforestation of the Purus river basin over a 9-year period, as well as to quantify the Purus river basin's land cover changes (%) in a 5-year period. Sampling data from upstream to downstream show a decrease in pH-value, dissolved oxygen, electrical conductivity, and total suspended solids. Correlation analysis revealed a significant negative correlation of the accumulated total deforestation values (km^2) with the pH-value (in all the study sites), and a significant positive correlation with temperature (only in two sites). However, the deforestation rates (km^2/year) did not present, in none of the study stations, any significant correlation with water quality parameters. It seems that the effects of deforestation on water quality are related not with the rate but with the total area deforested. It was estimated that the basin's forested area decreased by 5.17%. Since similar attributes are common in other basins of the whitewater systems of the Brazilian Amazon, this results may be seen as a warning on the effects of deforestation on water quality (reduction in pH and increment in temperature values), in larger areas than those of our study sites. To maintain the conservation and preservation status of the Purus river basin, it is necessary, the implementation of a transboundary watershed management program that could serve as a conservation model for Brazil and other countries of the Amazonian region.

Keywords Amazon basin · Deforestation · Purus · Water quality

Introduction

The Amazon, that presents different characteristics among others Brazilian hydrographic areas, is one of the world's unique regions containing most of the usable water resources in Brazil. This water potential displays

---

E. A. Ríos-Villamizar (✉) · M. T. F. Piedade
Instituto Nacional de Pesquisas da Amazônia/Ecologia, Monitoramento e Uso Sustentável de Áreas Úmidas Amazônicas (Grupo MAUA/CDAM/INPA), Campus I, Av. André Araújo, 2.936, Petrópolis, Caixa Postal 478, CEP: 69067-375 Manaus, Brazil

e-mail: eduardorios17@hotmail.com

W. J. Junk
Instituto Nacional de Ciência e Tecnologia em Áreas Úmidas (INCT-INAU-UFMT), Cuiabá, Brazil

A. V. Waichman
Universidade Federal do Amazonas - UFAM/ICB, Manaus, Brazil

privileged place into the speeches on Amazonian sustainable development. Water chemistry provides important parameters for quantifying biogeochemical cycles and determines management options in river systems and wetlands. The first scientific classification of Amazonian rivers was elaborated by Sioli (1956) who used water color, transparency, pH and electrical conductivity to explain limnological characteristics of the large Amazonian rivers and correlated these characteristics to the geological properties of the river catchments, a landscape ecology approach. Whitewater rivers (such as the Amazon, Juruá and Madeira) are turbid and have their origins in the Andes, from which they transport large amounts of nutrient-rich sediments. Their waters have near neutral pH and relatively high concentrations of dissolved solids indicated by the electric conductivity that varies between 40 and 140 $\mu S\ cm^{-1}$. Blackwater rivers (such as the Negro River) drain the Precambrian Guiana shield, which is characterized by large areas of white sands (podzols), their waters show low quantities of suspended matter but high amounts of humic acids that give the water a brownish-reddish color. The pH values of such rivers are in the range of 4–5 and their electrical conductivity is $<20\ \mu S\ cm^{-1}$. Clearwater rivers (such as the Tapajós and Xingu rivers) have their upper catchments in the Cerrado region of the Central Brazilian archaic shield. The transparency of their greenish waters is above 1.5 m, with low amounts of sediments and dissolved solids, electrical conductivity is in the range of 10–20 $\mu S\ cm^{-1}$, and pH that varies between 6 and 7 in large rivers.

Ríos-Villamizar et al. (2014) classified a number of rivers using the combination of several parameters such as alkali (Na, K) and alkali-earth (Ca, Mg) metals, major anions, electrical conductivity, pH, N_{total}, P_{total}, water color and suspended sediment load to distinguish the three classical water types (white, black and clear) and to separate other water bodies with intermediate position, concluding that many rivers and streams have to be considered as mixed waters resulting from the influence of lower order tributaries with different physico-chemical properties of their waters. According to this classification system, the Purus river is a whitewater river but many of its tributaries are of intermediate type (Ríos-Villamizar et al. in preparation) if considering that the geology of the pre-Andean zone is rather heterogeneous with large old sedimentation areas (paleo-varzeas), which sediments are strongly weathered, but still have a higher bioelement content than the tertiary sediments of the central Amazon basin and the soils on the archaic shields.

The Purus river is one of the chief tributaries of the Solimões/Amazon River system, being one of the longest rivers in South America, rising in the eastern lowlands of Peru and flowing about 3380 kilometres (km) before entering Solimões river at the northwestern Brazil. The estimated water volume that is generated within the Purus river basin is about 8500 $m^3\ s^{-1}$, and the total area of this basin is about 375,458 km^2. The Purus basin is classified into the basin group that still is in a high conservation status in the Brazilian Amazon (Fig. 1). However, in recent years, the cattle ranching, the advance of the agricultural frontier mainly for soybean production and the associated deforestation constitute a threat for the aquatic ecosystems.

Materials and methods

The aims of this paper are to diagnose water quality changes at the main channel of the Purus river, and to relate these with deforestation rates (km^2/year) and accumulated total deforestation values (km^2) of the Purus

Fig. 1 A section of the middle/lower Purus river basin

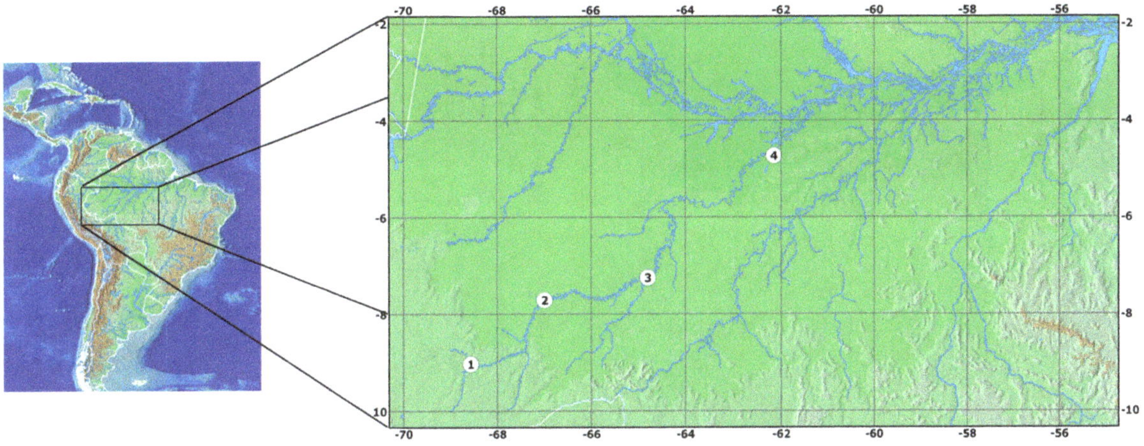

Fig. 2 Location of the ANA's water quality monitoring stations on the Purus river: *1* Seringal Caridade (Municipal District of Boca do Acre); *2* Seringal Fortaleza (Municipal District of Pauini); *3* Lábrea (Municipal District of Lábrea); *4* Arumã (Municipal District of Beruri). Classes: *green color* forest; *blue color* hydrography

basin, over a 9-year period, to detect spatial and temporal relationships, using Spearman's rank and Pearson correlation analysis, which were performed using Open Stat 4.0. The time series of data were constituted by the results of, at least, three samplings per year in each monitoring station. The information on deforestation (PRODES 2007) was reported by the Brazilian National Institute of Spatial Research (INPE), and the data regarding to water quality were recorded by the Brazilian National Agency of Water (ANA) (HIDROWEB 2007), see Fig. 2. Water samples were collected in the center of the river channel using acid-washed polyethylene bottles, which were rinsed with the water being collected and the samples were manually collected beneath the surface and kept cool until the time of analysis. Water samples were filtered through Whatman GF/F fiberglass filters (0.45 μm). The values of electrical conductivity (μS cm^{-1} at 25 °C), dissolved oxygen (mg O_2 l^{-1}), temperature (°C) and pH were measured in the field (on site). In the laboratory, the values of turbidity (NTU) and total suspended solids (mg l^{-1}) were analyzed. All the analyses were carried out by standard methods (APHA 2005). The Purus river basin's land cover changes (%), in a 5-year period, also were identified (diagnosed) by comparing two maps which used remote sensing and geographical information systems (GIS) techniques.

Results and discussion

Water quality trends

The waters of the Purus river basin showed characteristics of Amazonian natural waters (Furch and Junk 1997). The electrical conductivity and the total suspended solids decrease from upstream to downstream of the Purus river, probably explained because of the increase of activities associated with the use of land, such as deforestation, cattle breeding and agriculture, concentrated in the municipalities of Boca do Acre and Lábrea, which are located in the Amazonas State in Brazil. It could be related with four deforestation fronts, corresponding with migratory processes, proceeding from the neighboring Brazilian States (Acre and Rondônia), which are stimulated by the agriculture expansion and wood predatory exploration (Sanches et al. 2007). Nevertheless, it was observed an increase in turbidity from upstream to downstream that would be explained by the soil loss and sediments transport from all the contribution areas along the basin (Leonardo 2003). This phenomenon also could be responsible for the slight decrease in dissolved oxygen in the same direction. Since it would be expected a general decreasing trend for total suspended solids and turbidity, the observed opposite trend for these two parameters, as well as the increasing trend for turbidity could be explained by possible measurement inaccuracies of the turbidity meter used. The pH-value decreased from upstream to downstream and the temperature shows a slight increase in the same direction (Fig. 3). This last trend could be due to the entrance of acidic blackwater tributaries which may contribute to pH and temperature changes.

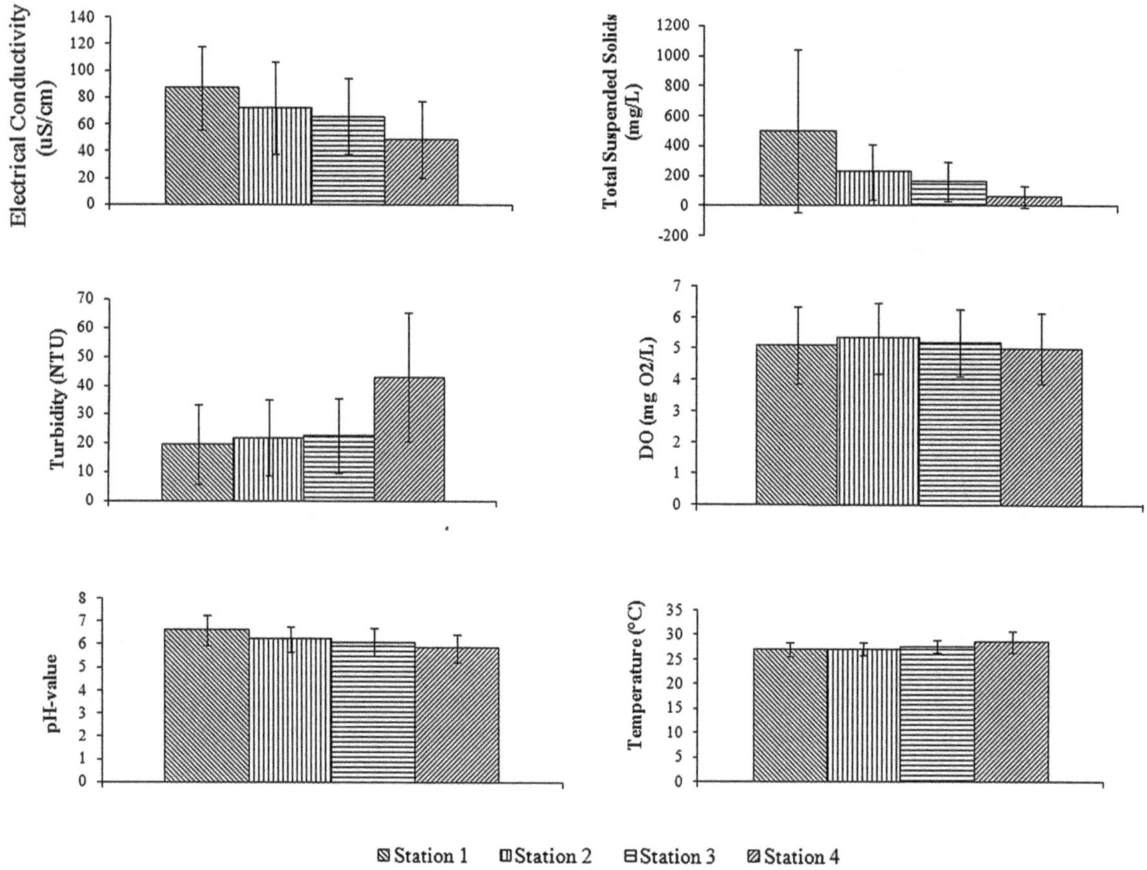

Fig. 3 Average values and standard deviation (SD) of the water quality variables for each monitoring station of the ANA in the Purus river (1998–2006)

Relations of the water quality with deforestation variables

Table 1 presents the accumulated total deforestation data for the years 2000–2006 and the deforestation rates for the years 2001–2006, respectively.

The higher deforestation rates were observed in Boca do Acre and Lábrea, with larger total deforested areas probably related with the access by road to the floodplain areas. The lower rates were observed in Pauini and Beruri (Arumã) where the access to the floodplain forest by road is almost inexistent. No relation was found among deforestation rates and the water quality parameters at the four sites. On the other hand, ATD showed

Table 1 Deforestation levels for the municipal districts of the study (PRODES 2007)

Year	Boca do Acre		Pauini		Lábrea		Beruri	
	ATD	DR	ATD	DR	ATD	DR	ATD	DR
2000	1165.9	–	146.8	–	1245.1	–	181.4	–
2001	1216.2	50.3	154.7	7.9	1419.8	174.7	189.2	7.8
2002	1319.8	103.6	164.8	10.1	1625.9	206.1	191.7	2.5
2003	1570.6	250.8	177.5	12.7	2083.5	457.6	197.2	5.5
2004	1693.7	123.1	184.5	7.0	2449.5	366.0	200.2	3.0
2005	1747.3	53.6	193.5	9.0	2631.3	181.8	200.9	0.7
2006	1831.9	84.6	201.1	7.6	2910.5	279.2	203.3	2.4

– Not reported

ATD accumulated total deforestation (km^2), *DR* deforestation rate (km^2/year)

Table 2 Correlation coefficient analysis among DR, ATD and water quality variables in the Purus river basin (2000–2006)

	ATD 1	DR 1	ATD 2	DR 2	ATD 3	DR 3	ATD 4	DR 4
Temp1	0.403**	0.216**						
pH 1	−0.883*	0.132**						
E.C 1	−0.584**	−0.326**						
Turb 1	−0.417**	−0.042**						
D.O 1	0.441**	0.356**						
TSS 1	−0.011**	−0.364**						
Temp 2			0.863*	0.035**				
pH 2			−0.765*	−0.382**				
E.C 2			−0.107**	−0.543**				
Turb 2			−0.631**	0.071**				
D.O 2			0.671**	−0.219**				
TSS 2			0.572**	0.322**				
Temp 3					0.793*	0.174**		
pH 3					−0.971*	−0.430**		
E.C 3					−0.433**	−0.558**		
Turb 3					−0.760*	−0.124**		
D.O 3					0.941*	0.388**		
TSS 3					−0.159**	0.193**		
Temp 4							0.511**	−0.455**
pH 4							−0.828*	0.598**
E.C 4							−0.284**	0.680**
Turb 4							0.154**	0.399**
D.O 4							−0.259**	0.305**
TSS 4							0.848*	−0.506**

1 Station 1, *2* Station 2; *3* Station 3, *4* Station 4, *Temp* temperature (°C), *E.C* electrical conductivity (μS cm^{-1}), *Turb* turbidity (NTU), *D.O* dissolved oxygen (mg l^{-1} O_2), *TSS* total suspended solids (mg l^{-1}), *ATD* accumulated total deforestation (km^2), *DR* deforestation rate (km^2 $year^{-1}$)

* Significant ($P < 0.05$); ** non-significant ($P > 0.05$)

relationships with water quality mainly in station 3 (Lábrea) where larger ATD values were observed. The ATD values correlates with changes in temperature in stations 2 and 3, with total suspended solids in station 4, with dissolved oxygen in station 3, with pH-value in all the four stations, and with turbidity in station 3 (Table 2). It is remarkable, the significant negative relationship between the ATD and pH-value, indicating that increase in the ATD will contribute to decrease the water pH-value. Comparable pattern was already obtained in other basins in USA, Brazil, Indonesia and India (Kunkle 1974; Neal et al. 1992; Arceivala and Asolekar 2006; Hirano et al. 2007). Deforestation likely contributes to increase the concentration of nitrates (nitrification) in the river water, and the increased hydrogen ions, released through increased nitrification, cause the water pH-value decrease (Lampert and Sommer 1997; Arceivala and Asolekar 2006). Since deforestation often causes the loss of soil carbon and a net release of CO_2, the increasing deforestation processes in some parts of the basin may also contribute to the intensification of carbon (CO_2) release from soils, with associated changes in content and composition of this gas on river water, which may result in alterations on water pH-value (Mizuno and Mori 1970; Wissmar et al. 1981; Goudie 2000; Soulsby et al. 2002; Hirano et al. 2007; Iyobe and Haraguchi 2008; Lal et al. 2015).

Land use changes

In the Purus river basin, as a whole, the highest percentage of the area is represented by forested land since it covered 85.8% of the total area in 2007. However, this basin suffered a relative intense process of land use and changes on the cover composition in the period 2003–2007. Forests were cleared to give way to expansion of

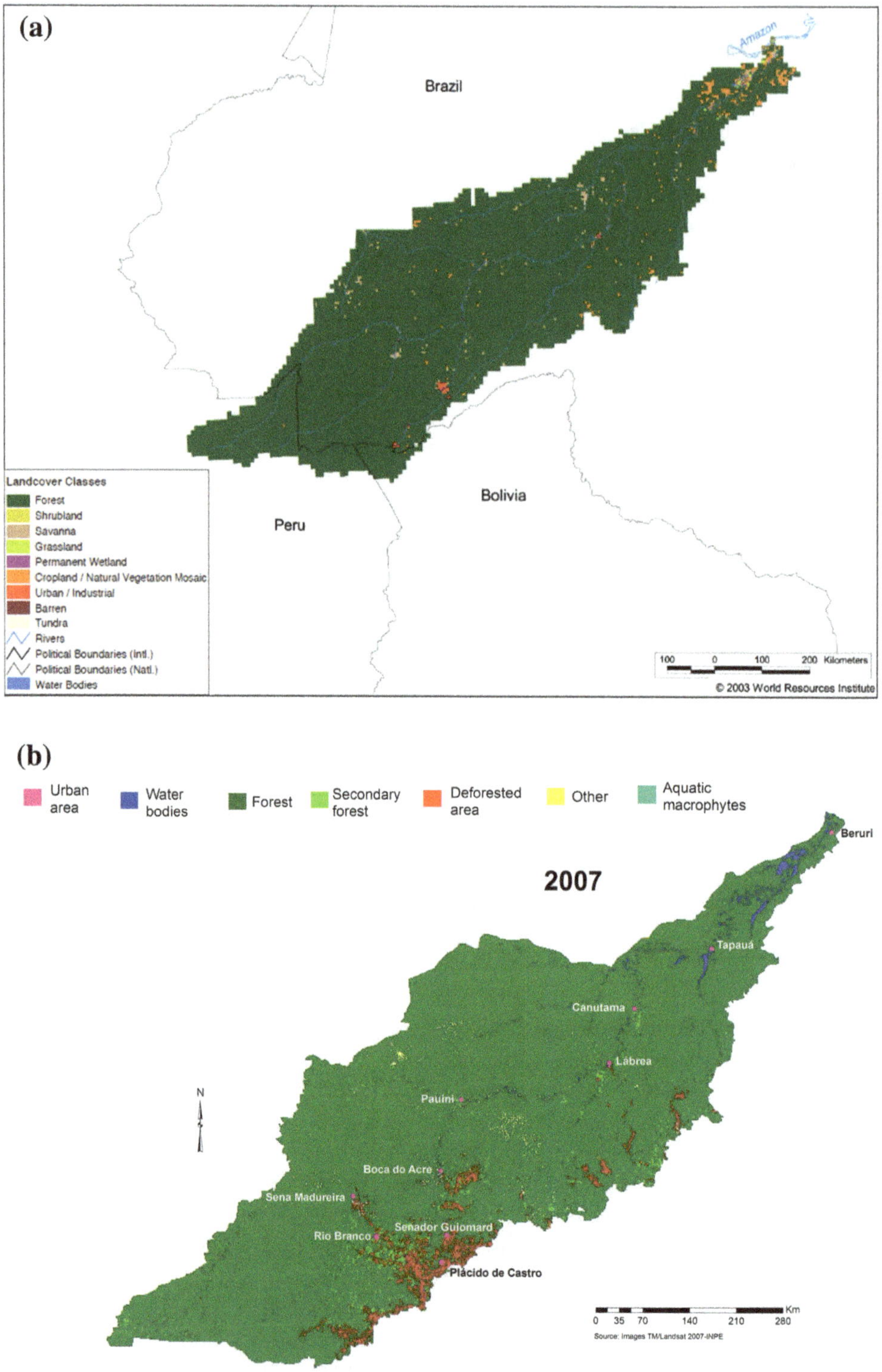

Fig. 4 Comparison of land cover and use maps in the Purus river basin. **a** 2003 (Water Resources eAtlas 2003); **b** 2007 (modified from Ríos-Villamizar et al. 2011)

cattle ranching and farming lands on this basin. According to the spatial sequence displayed by the INPE data, the deforestation rates are further increasing at the municipal districts located at south and southeast of the Amazon region. In this sense, the larger deforested areas along the basin are concentrated close to urban areas of the municipal districts of Rio Branco, Sena Madureira, Plácido de Castro and Senador Guiomard, in the

Table 3 Land cover variables percents in the Purus river basin on 2003 and 2007

Land cover variables	2003 (%)[a]	2007 (%)[b]
Urban area	0.1	0.13
Deforested area	6.7[c]	5.48
Forest cover	91.0	85.83
Hydrography	–	1.6
Wetlands (*Várzea*)	7.0	5.9
Aquatic macrophytes	–	0.01
Secondary forest	–	0.89
Other	[d]	0.23[e]

– Not reported

[a] Water Resources eAtlas (2003) (Basin area: 371,042.0 km^2)

[b] Ríos-Villamizar et al. (2011) (Basin area: 375,458.5 km^2)

[c] Percent loss of original forest cover

[d] This category includes percent grassland, savanna and shrubland (3.9%), and percent cropland (3.9%)

[e] This category includes cloudy area

Acre State; and Lábrea and Boca do Acre, in the Amazonas State (Fig. 4b). It is evident that the reduction of forest cover was roughly 5.17% in the period 2003–2007 (Fig. 4; Table 3). Another important aspect of the identified changes is that wetlands area lost approximately 1.1% in the same period, but this variation would also be due to the effects of the flood pulse, during natural seasonal periods, on the rivers and lakes' water level (Junk et al. 1989; Junk and Wantzen 2004).

Conclusions

The analyzed waters of the Purus river basin showed typical characteristics of Amazonian natural waters. We observed clear trends in changes of physico-chemical properties of water from upstream to downstream of the Purus river such as decrease in pH-value, dissolved oxygen, electrical conductivity, total suspended solids, and increase in turbidity. Only the pH-value presented significant negative relationships with the ATD in all the study sites, indicating that increase in the ATD will contribute to decrease the water pH-value. On the other hand, the ATD values correlated positively and significantly with changes of temperature only in stations 2 and 3, indicating that increase in the ATD will contribute to increase the water temperature. The DR did not present, in none of the study stations, any significant relation with the water quality parameters. Then, the deforestation levels have not yet caused a large effect on the water quality, which is probably most influenced by hydrological and climatic factors such as river discharge, river level, and pluvial precipitations, among others; but this will be addressed in forthcoming papers.

It seems that the effects of deforestation on water quality are related not with the rate but with the total area deforested, and they become apparent after a minimal size of deforested area is achieved.

Therefore, despite the good conservation status for most parts of the Purus river basin, impacts on water quality caused by human activities are evident, especially close to urban areas, at a local scale. Since similar attributes are common in other basins of the whitewater systems of the Brazilian Amazon, this may be seen as a warning on the effects of deforestation on water quality in larger areas than those of our study sites. Actions to control the deforestation in this basin would need to be taken to maintain its conservation status and, in view of that, it would be necessary to implement a transboundary watershed management program for conservation and preservation purposes, and this program could serve as a conservation model for Brazil and other Amazonian countries.

Acknowledgements This work was funded by Brazilian National Scientific Council (CNPq), Tropical Forest Protection Program (PPG-7), Grant Number 556899/2005-9. We thank the postgraduate Program of Environmental Sciences and Sustainability in the Amazon (PPG/CASA/UFAM), the postgraduate Program in Climate and Environment at the National Institute of Amazonian Research (INPA/UEA), the Ecology, Monitoring and Sustainable Use of Wetlands Group (MAUA/CDAM/INPA), CAPES/

CNPq—IEL Nacional—Brasil, Programa de Apoio à Fixação de Doutores no Amazonas (FIXAM/AM) and Programa de Apoio à Participação em Eventos Científicos e Tecnológicos (PAPE), from the Fundação de Amparo à Pesquisa do Estado do Amazonas (FAPEAM/SECTI/AM) for financial support. We also thank Kyara Martins Formiga by suggestions and elaboration of the study area map.

References

APHA, AWWA and WEF (2005) Standard methods for the examination of water and wastewater, 21st ed. American Public Health Association, Washington, D.C

Arceivala SJ, Asolekar SR (2006) Wastewater treatment for pollution control and reuse. Tata McGraw-Hill Education. New Delhi. https://books.google.com.br/books?isbn=0070620997. Accessed 17 Nov 2016

Furch K, Junk WJ (1997) Physicochemical conditions in the floodplains. In: Junk WJ (ed) The central Amazon floodplain: ecology of a pulsing system. Springer-Verlag, Berlin, pp 69–108

Goudie A (2000) The human impact on the natural environment. MIT Press, Cambridge

HIDROWEB (2007) Hydrologic information system of the ANA. http://hidroweb.ana.gov.br/. Accessed 20 Jun 2007

Hirano T, Segah H, Harada T et al (2007) Carbon dioxide balance of a tropical peat swamp forest in Kalimantan, Indonesia. Glob Change Biol 13:412–425

Iyobe T, Haraguchi A (2008) Ion flux from precipitation to peat soil in spruce forest–Sphagnum bog communities in the Ochiishi district, eastern Hokkaido, Japan. Limnology 9:89–99

Junk WJ, Wantzen KM (2004) The flood pulse concept: new aspects, approaches, and applications—an update. In: Proceedings of the 2nd international symposium on the management of large rivers for fisheries, Bangkok, Thailand, vol 2, pp 117–149

Junk WJ, Bayley PB, Sparks RE (1989) The flood pulse concept in river-floodplain-systems. Can Special Publ Fish Aquat Sci 106:110–127

Kunkle SH (1974) Water—its quality often depends the forester. Technical document Unasylva 105 FAO. http://www.fao.org/docrep/e7730e/e7730e02.html. Accessed 27 Apr 2008

Lal R, Singh BR, Mwaseba DL, Kraybill D, Hansen D O, Eik LO (2015) Sustainable intensification to advance food security and enhance climate resilience in Africa. Springer. Switzerland. https://books.google.com.br/books?isbn=9783319093604. Accessed 18 Nov 2016

Lampert W, Sommer U (1997) Limnoecology: the ecology of lakes and streams. Oxford University Press, New York

Leonardo HCL (2003) Soil and water quality indicators to evaluate the sustainable use of the rio Passo Cue watershed in western Paraná. Master Dissertation, University of São Paulo, Brazil

Mizuno T, Mori S (1970) Preliminary hydrobiological survey of some southeast asian inland waters. Biol J Linn Soc 2:77–117

Neal C, Forti MC, Jenkins A (1992) Towards modelling the impact of climate change and deforestation on stream water quality in Amazonia: a perspective based on the MAGIC model. Sci Total Environ 127:225–241

PRODES (2007) Via satellite Brazilian Amazonian forest monitoring project—INPE deforestation database. http://www.dpi.inpe.br/prodesdigital/prodesmunicipal.php. Accessed 25 Apr 2007

Ríos-Villamizar EA, Junior AFM, Waichman AV (2011) Water physico-chemical characterization and soil use in the Purus river basin, western Brazilian Amazon. Rev Geogr Acadêmica 5(2):54–65

Ríos-Villamizar EA, Piedade MTF, Costa JG et al (2014) Chemistry of different Amazonian water types for river classification: a preliminary review. WIT Trans Ecol Environ 178:17–28

Sanches MV, Assis FP, Bueno CR et al (2007) Environmental and sustainability analysis of the Amazonas State. United Nations Publication, Santiago

Sioli H (1956) Über Natur und Mensch im brasilianischen Amazonasgebiet. Erdkunde 10(2):89–109

Soulsby C, Gibbins C, Wade AJ et al (2002) Water quality in the Scottish uplands: a hydrological perspective on catchment hydrochemistry. Sci Total Environ 294:73–94

Water Resources eAtlas (2003) Watersheds of south America (SA01 Amazon, Purus). The world conservation union (IUCN), IWMI, RAMSAR, WRI, pp 1–2

Wissmar RC, Richey JE, Stallard RF et al (1981) Plankton metabolism and carbon processes in the Amazon river, its tributaries and floodplain waters, Peru-Brazil, May–June 1977. Ecology 62:1622–1633

11

Baseline survey of the fish fauna of a highly eutrophic estuary and evidence for its colonisation by Goldfish (*Carassius auratus*)

James R. Tweedley · Chris S. Hallett · Stephen J. Beatty

Abstract This study represents the first quantitative survey of the fish fauna of the highly eutrophic Vasse and Wonnerup estuaries, part of the Ramsar-listed Vasse-Wonnerup Wetland System in south-western Australia. Sampling at five sites in each of these estuaries occurred in January 2012 (austral summer) to provide a species inventory and determine whether the number of species, total density and fish community composition differed between the two water bodies. A total of 18,148 fish were recorded, representing six species across four families. Three species that can complete their life cycle within estuaries, i.e. the atherinids *Lepthatherina wallacei* and *Atherinosoma elongata* and the gobiid *Pseudogobius olorum,* dominated the fish fauna, accounting for >99% of all fish collected. No significant inter-estuary differences were observed in the mean number of species, mean total density or fish community composition. Although the fish community was depauperate in terms of the number of species, total density was high, reflecting the presence of permanent and seasonal barriers to the immigration of marine species into these estuaries and the highly productive nature of this system, respectively. Two introduced freshwater species, i.e. the Eastern Gambusia *Gambusia holbrooki* and the Goldfish, *Carassius auratus*, were recorded in the Vasse Estuary. As *C. auratus* was found in mesohaline conditions, individuals may be able to use the estuary as a 'saltbridge' to gain access to other tributaries and/or the Wonnerup Estuary, and thus expand their distribution. These findings are of concern given the potential deleterious biological and ecological effects of these alien species.

Keywords Introduced species · Ramsar · South-western Australia · Vasse-Wonnerup · Wetland

Introduction

It is widely recognised that, despite their many valuable ecosystem services (Costanza et al. 1997; Creighton et al. 2015; Sheaves et al. 2015), temperate estuaries are the most degraded of all aquatic ecosystems (Jackson et al. 2001). Microtidal systems, i.e. those with a tidal range <2 m, located in Mediterranean climate regions are particularly susceptible to degradation, due to their limited tidal water movement, highly seasonal rainfall and, in some cases, ephemeral connection to the ocean (Potter et al. 2010; Tweedley et al. 2016). These characteristics increase the residency time of water in the estuary and facilitate the trapping of fine sediments and contaminants (Patchineelam et al. 1999; Tweedley et al. 2016). Given the 'natural susceptibility' of microtidal estuaries, such as those in south-western Australia, to environmental degradation and the effects on

J. R. Tweedley (✉) · C. S. Hallett · S. J. Beatty
School of Veterinary and Life Sciences, Centre for Fish and Fisheries Research, Murdoch University, 90 South Street, Murdoch, WA 6150, Australia
e-mail: j.tweedley@murdoch.edu.au

these systems of increased urbanisation and eutrophication, it is unsurprising that the environmental health of some of these systems has been shown to have declined since the 1980s (Wildsmith et al. 2009; Tweedley et al. 2012, 2014).

One such system in south-western Australia that has experienced a suite of deleterious anthropogenic impacts is the Vasse-Wonnerup; a shallow, intermittently-open, nutrient-enriched system located near the town of Busselton, Western Australia (Brearley 2005; Commonwealth of Australia 2002). The system is comprised of the Vasse and Wonnerup estuaries and provides habitat, at any one time of the year, for over 37,500 water birds comprising ~90 species, a function that is recognised by its designation as a Wetland of International Importance under the Ramsar Convention in 1990 (Lane et al. 2007). Although the importance of the Vasse-Wonnerup is well recognised, the system and surrounding land have been subjected to extensive anthropogenic modification. For example, much of the catchment has been cleared, primarily for cattle grazing; extensive drainage networks have been constructed; several rivers that used to flow into the system have been diverted to the sea and permanent surge barriers have been installed at the junction between each of the estuaries and Wonnerup Inlet to prevent seawater intrusion (Lane et al. 1997; Fig. 1). Moreover, the large amounts of fertilizer applied to agricultural land, combined with animal waste discharged from pastures into the estuaries, have resulted in the Vasse-Wonnerup becoming "the most grossly enriched major wetland system known in Western Australia" (McAlpine et al. 1989). Without management intervention nutrient loads are expected to increase further over the next 20 years due to increased urbanisation and more intensive agriculture (Department of Water 2010).

These anthropogenic pressures have had numerous impacts on the ecological health of the system. Fish kills, in particular, occur regularly in the Vasse-Wonnerup, with reports of such events dating back to 1905 and at least nine kills have occurred between 1984 and 2013 (Hart 2014). Anoxia, high water temperatures and

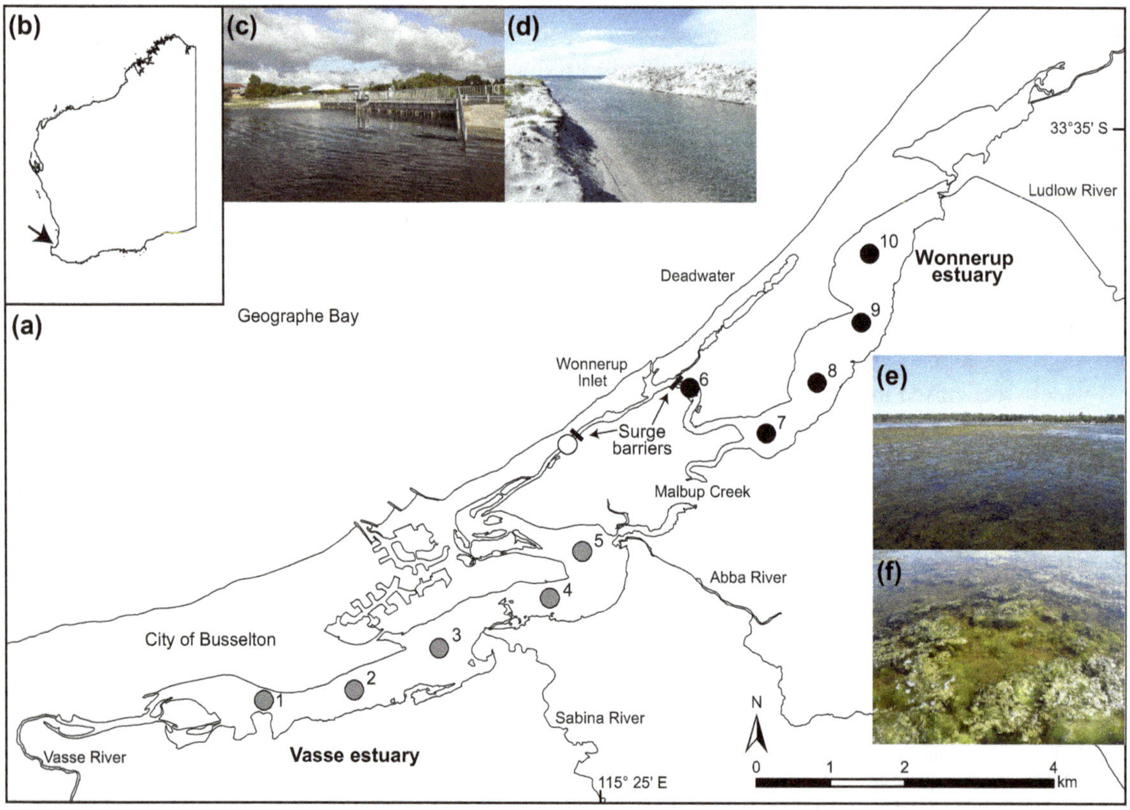

Fig. 1 **a** Map of the Vasse-Wonnerup Wetland System, showing the location of sites sampled in the Vasse (*dark gray filled circle*) and Wonnerup (*filled circle*) estuaries in January 2012. **b** denotes the location of Vasse-Wonnerup in Western Australia. **c** shows the Vasse surge barrier, which prevents the upstream flow of water, **d** the mouth of the estuary in Wonnerup Inlet and **e**, **f** an algal bloom in the Vasse Estuary at the time of sampling. *Open circle* denotes the location where additional physico-chemical measurements were recorded

hypersalinity, macroalgal blooms and/or toxic phytoplankton have all been considered the causes of these events (Lane et al. 1997; Hart 2014). However, despite the frequency of fish kills and the importance of this system to recreational and commercial fishers (Elscot 2000), no quantitative survey of the fish fauna of this estuary has been conducted. Such a study is fundamental to understanding the ecological functioning of this system, particularly as fish constitute a substantial component of the diet of at least 24 bird species found in the Vasse-Wonnerup (Wetland Research and Management 2007). Moreover, a population of the highly invasive Goldfish (*Carassius auratus*) is known to reside in the Lower Vasse River, which flows into the Vasse Estuary (Morgan and Beatty 2007; Beatty et al. 2017), thus posing a further potential threat to the ecology of this system.

Given the range of natural and anthropogenic pressures that threaten this internationally important system, the collection of the first quantitative data on the fish fauna would also allow the development of ecologically relevant tools for monitoring the future health of the system (Harrison and Whitfield 2004; Hallett et al. 2012; Tweedley et al. 2017). Such tools could help to provide a rigorous assessment of the effectiveness of management actions (Hallett et al. 2016), including recent efforts to reduce nutrient inputs (Department of Water 2010). The collection of these quantitative data would also help determine the level of threat posed by the population of invasive *C. auratus* in the Lower Vasse River.

The overall aim of this study was to conduct a baseline survey of the fish fauna of the Vasse and Wonnerup estuaries, the areas of the wetland system that are most important for the avian fauna (Lane et al. 1997). Specifically, the objectives were to (1) provide a quantitative description of the fish communities of the Vasse and Wonnerup estuaries, (2) determine whether the number of species, density and composition of the fish faunas differed between the two estuaries and (3) determine whether *C. auratus* were present in either of the estuaries.

Materials and methods

Sampling regime

Five sites in the shallow waters of each of the Vasse and Wonnerup estuaries were sampled in January 2012 (Fig. 1). At that time, the bar at the mouth of Wonnerup Inlet, which is intermittently-open to the ocean, was open having been artificially breached on December 14, 2011 and two narrow chutes (40 cm in width) in the Vasse and Wonnerup surge barriers were opened periodically to allow the movement of fish into and out of the estuaries (Fig. 1c). At each site, two samples of the fish fauna were collected using a seine net that was 21.5 m long and consisted of two 10 m long wings (6 m of 9 mm mesh and 4 m of 3 mm mesh) and a 1.5 m long bunt made of 3 mm mesh, fished to a depth of 1.5 m and swept an area of ~ 116 m^2. Upon capture, all fish were immediately euthanized in an ice slurry (Murdoch University Animal Ethics Permit R2450_11). The total number of individuals of each fish species in each sample was then recorded and the total length of each individual measured to the nearest 1 mm, except when a large number of any one species was caught, in which case the lengths of a representative subsample of 100 fish were measured. Each species was assigned to an estuarine usage guild, based on the results of numerous studies of the biology of fish species in south-western Australian estuaries (e.g. Potter and Hyndes 1999). The ways in which fish use estuaries were considered to comprise four categories, i.e. marine, estuarine, diadromous and freshwater, with each containing multiple guilds (Potter et al. 2015a, b).

Salinity, water temperature and dissolved oxygen concentration and saturation were measured in the middle of the water column at each site at the time of fish collection and on the same day immediately upstream of the Vasse surge barriers using a Yellow Springs Instrument 556 water quality meter.

Statistical analyses

Environmental conditions and univariate measures of diversity

One-way analysis of variance (ANOVA) was employed to determine whether any of a suite of physico-chemical variables, i.e. salinity, water temperature and dissolved oxygen concentration and saturation, and biotic variables, i.e. number of species and total density, differed significantly between the Vasse and Wonnerup estuaries. Prior to undertaking ANOVA, the data for each of the five dependent variables were subjected to the Anderson–Darling and *F* test routines in Minitab 15 to ascertain the type of transformation, if any, required to meet the test assumptions of normality and equal variance among *a priori* groups. This showed that the data for each of the physico-chemical variables and the number of species required no transformation, but that total density required a square-root transformation. In all ANOVA tests the term Estuary (2 levels; Vasse and Wonnerup) was considered fixed and the null hypothesis of no significant differences between *a priori* groups was rejected if the significance level (p) was <0.05.

Multivariate analyses of fish faunal composition

The abundances of the various fish species recorded were dispersion-weighted to down-weight the effects of those species whose numbers exhibited erratic differences among replicate samples due to schooling (Clarke et al. 2006), followed by a square-root transformation to down-weight the contributions of species with consistently high values (across replicates within a group) in relation to those with consistently low values (Veale et al. 2014). The two replicate samples from each site were averaged and the resultant pre-treated data used to construct a Bray–Curtis resemblance matrix which was, in turn, subjected to a one-way analysis of similarities (ANOSIM) test (Clarke and Green 1988) in PRIMER v7. This test was employed to determine whether the composition of the fish faunas of the two estuaries differed. The null hypothesis that there were no significant differences in fish faunal composition between *a priori* groups (i.e. the two estuaries) was rejected if the significance level (p) was <0.05. The extent of any such significant difference was determined by the magnitude of the test statistic (R), which typically ranges between ~0, i.e. no group differences, to 1, i.e. the similarities between samples from different groups are always less than those between samples belonging to the same group. Non-metric Multidimensional Scaling (nMDS) ordination was then employed to display visually the ways in which the samples from each *a priori* group were distributed in low dimensional space according to their faunal compositions (Clarke 1993).

Similarity percentages (SIMPER; Clarke et al. 2014a) were used to identify those species that typified the fish fauna in each estuary and, if appropriate, those that were responsible for distinguishing between them. A shade plot was constructed from the dispersion-weighted, square-root transformed and averaged data and used to display the trends exhibited by the abundance of all fish species recorded. This shade plot is a simple visualisation of the frequency matrix, where a white space for a species demonstrates that that species was never collected, while the depth of shading from grey to black is linearly proportional to its abundance (Clarke et al. 2014b). Species (y axis) were clustered based on their Bray–Curtis similarities and placed in optimum serial order, constrained by the cluster dendrogram (Clarke et al. 2014a). Sites (x axis) were ordered based on their location in the two estuaries from most upstream to downstream, i.e. 1–5 in the Vasse Estuary and 10–6 in the Wonnerup Estuary (see Fig. 1).

Results

Environmental conditions

Salinities in January 2012 ranged from 17.1 to 19.4 in the Vasse Estuary and 14.2 to 17.3 in the Wonnerup Estuary. One-way ANOVA demonstrated that salinity differed significantly between estuaries ($p = 0.005$), with, on average, higher values being recorded in the Vasse (18.3 ± 0.44 SE) than the Wonnerup (15.7 ± 0.51 SE) Estuary (Table 1). Water temperature ranged from 24 to 31 °C across the ten sites, but did not differ significantly between the two estuaries ($p = 0.678$), being, on average, ~28 °C (Table 1). Dissolved oxygen saturation was similar between estuaries ($p = 0.520$), exceeding 100% at nine of the ten sites, reaching >200% in both estuaries and as high as 257%, likely due to the large quantities of macroalgae present (Table 1; Fig. 1e, f). However, concentrations were, on average, 11.92 (±0.66 SE) and 13.80 (±2.18 SE) mgL^{-1} in the Vasse and Wonnerup estuaries, respectively, and were found by one-way ANOVA not to differ significantly ($p = 0.434$).

Table 1 Salinity, water temperature (°C), dissolved oxygen (DO) concentration (mgL^{-1}) and saturation (%) in the water column at each of the ten sites in the Vasse and Wonnerup estuaries in January 2012

Site	Salinity	Water temp.	DO conc.	DO sat.
(a) Vasse Estuary				
1	19.00	29.10	10.90	155.00
2	18.60	30.40	14.40	211.00
3	17.50	29.70	12.20	177.00
4	17.10	29.00	11.10	159.00
5	19.40	25.90	11.00	158.00
Mean	18.32	28.82	11.92	172.00
SE	0.44	0.77	0.66	10.53
(b) Wonnerup Estuary				
6	17.30	29.60	17.60	254.00
7	15.60	31.00	11.90	175.00
8	14.20	29.20	18.10	257.00
9	15.40	27.20	6.28	88.00
10	16.20	24.00	15.10	197.00
Mean	15.74	28.20	13.80	194.20
SE	0.51	1.21	2.18	30.96

Mean and standard errors for each parameter in each estuary are also provided

Water physico-chemical conditions were also measured immediately upstream of the Vasse surge barrier (Fig. 1). While water temperature there was similar to that recorded in the main body of the two estuaries (27.4 °C), the salinity was substantially greater (35.7) and the dissolved oxygen concentration was far lower and hypoxic (1.5 mgL^{-1}; 23% saturation).

Fish species ranking by abundance and life cycle contribution

A total of 18,148 fish were caught in the nearshore waters of the Vasse and Wonnerup estuaries in January 2012. These fish comprised six species, representing four families. The Atherinidae and Gobiidae were the most speciose, each represented by two species, and together accounted for >99% of all fish collected (Table 2).

Table 2 Mean density (fish 100 m^{-2}; $\bar{X}$), standard deviation (SD), percentage contribution to the overall catch (%C), rank by density (R), mean total length (mm; L) and length range (Lr) of each fish species recorded in the Vasse and Wonnerup estuaries in January 2012

Species	Vasse Estuary							Wonnerup Estuary					
	LCG	$\bar{X}$	SD	%C	*R*	*L*	Lr	$\bar{X}$	SD	%C	*R*	*L*	Lr
Lepthatherina wallacei	EF	**810.0**	**1319.6**	**72.58**	**1**	**34.17**	**20–86**	**29.4**	**66.8**	**6.55**	**3**	**32.7**	**21–46**
Atherinosoma elongata	EF	**198.0**	**324.8**	**17.74**	**2**	**39.35**	**22–68**	**236.3**	**324.0**	**52.68**	**1**	**37.6**	**22–60**
Pseudogobius olorum	EF	**106.0**	**147.1**	**9.50**	**3**	**29.62**	**18–53**	**182.4**	**378.1**	**40.67**	**2**	**25.0**	**16–43**
*Gambusia holbrooki**	FEO	1.1	3.3	0.10	4	24.92	18–43						
*Carassius auratus**	FEO	0.8	2.2	0.07	5	43.50	43–44						
Favonigobius lateralis	EM							0.43	1.09	0.10	4	52.00	47–57
Total number of species		5						4					
Mean total density		1116.0						448.5					

Abundant species in each estuary, i.e. those that contributed ≥5% to the catch, are highlighted in bold. The life cycle guild (LCG; see Potter et al. [2015a] for full details) of each species is also provided (*EF* estuarine & freshwater, *EM* estuarine & marine and *FEO* freshwater estuarine-opportunist). *Asterisk* denotes that a species has been introduced

The atherinids *Lepthatherina wallacei* and *Atherinosoma elongata,* together with the gobiid *Pseudogobius olorum,* dominated the fish fauna of the Vasse Estuary, ranking first, second and third, respectively. All three species are capable of completing their lifecycle in estuaries. The same three species were also the most abundant in the Wonnerup Estuary, but differed in their rank order (Table 2). Another goby species, *Favonigobius lateralis*, was found only in the Wonnerup Estuary, while the reverse was true for two introduced freshwater species, namely the Eastern Mosquitofish *Gambusia holbrooki* and the Goldfish *Carassius auratus.*

Mean number of species, density and faunal composition

There was no significant difference ($p = 0.614$) in the mean number of species between the Vasse (3.2) and Wonnerup estuaries (2.8; Fig. 2a). Although the mean total density appeared to differ among estuaries, i.e. 1116 and 449 individuals 100 m^{-2} in the Vasse and Wonnerup estuaries, respectively, ANOVA did not detect a significant difference in density between the two estuaries ($p = 0.482$). Such a result reflects the large inter-site variability in density (Fig. 2b), which ranged between 13 and 3607 individuals 100 m^{-2} in the Vasse Estuary and between 9 and 1144 individuals 100 m^{-2} in the Wonnerup Estuary.

The composition of the fish community was shown by one-way ANOSIM not to differ significantly between the Vasse and Wonnerup estuaries ($p = 0.413$; $R = 0.008$). This is shown on the associated nMDS ordination plot, where the points representing the sites sampled in the two estuaries do not form discrete groups (Fig. 3). This reflects the fact that SIMPER identified the same three species as typifying the fish fauna of both the Vasse and Wonnerup estuaries, namely *L. wallacei, A. elongata* and *P. olorum.* This is illustrated on the shade plot, where at least one of these three species was found at each site and, at seven of the ten sites, all three species were recorded in moderate abundances (Fig. 4). The shade plot also showed that the introduced species only occurred in the Vasse Estuary, with *C. auratus* being found at the two most upstream sites, while *G. holbrooki* was present at the most upstream site and also further downstream (Fig. 4). The estuarine & marine gobiid *F. lateralis* was found only at the most downstream site in the Wonnerup Estuary.

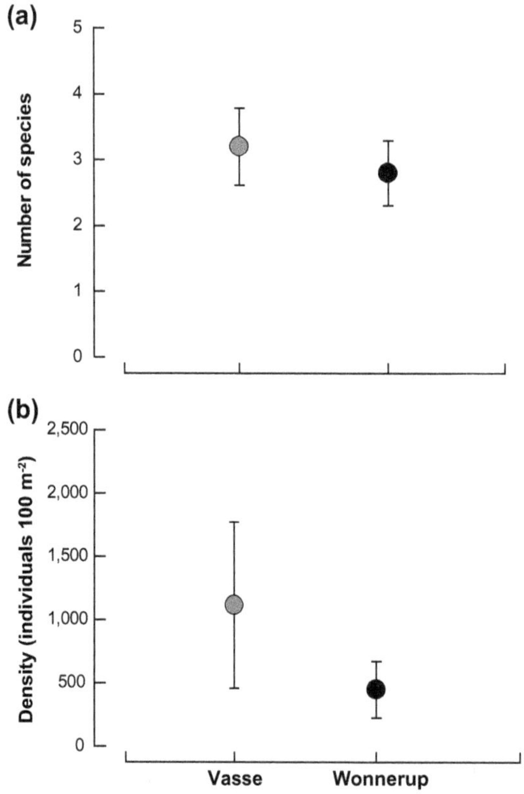

Fig. 2 Mean values ± 1 standard error for the **a** number of species and **b** density of fishes recorded in the Vasse (*dark gray filled circle*) and Wonnerup (*filled circle*) estuaries in January 2012

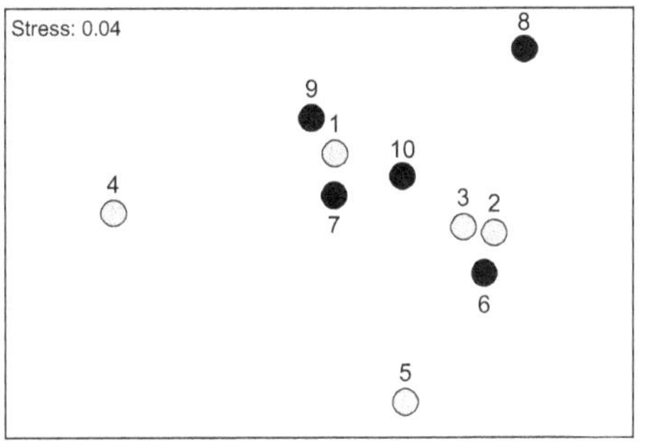

Fig. 3 Non-metric Multidimensional Scaling (nMDS) ordination plot derived from a Bray–Curtis resemblance matrix constructed from the dispersion-weighted, square-root transformed and averaged densities of the fish species recorded at each of the ten sites in the Vasse (*dark gray filled circle*) and Wonnerup (*filled circle*) estuaries in January 2012. The *numbers* above or below the *points* denote the site that the point represents (see Fig. 1)

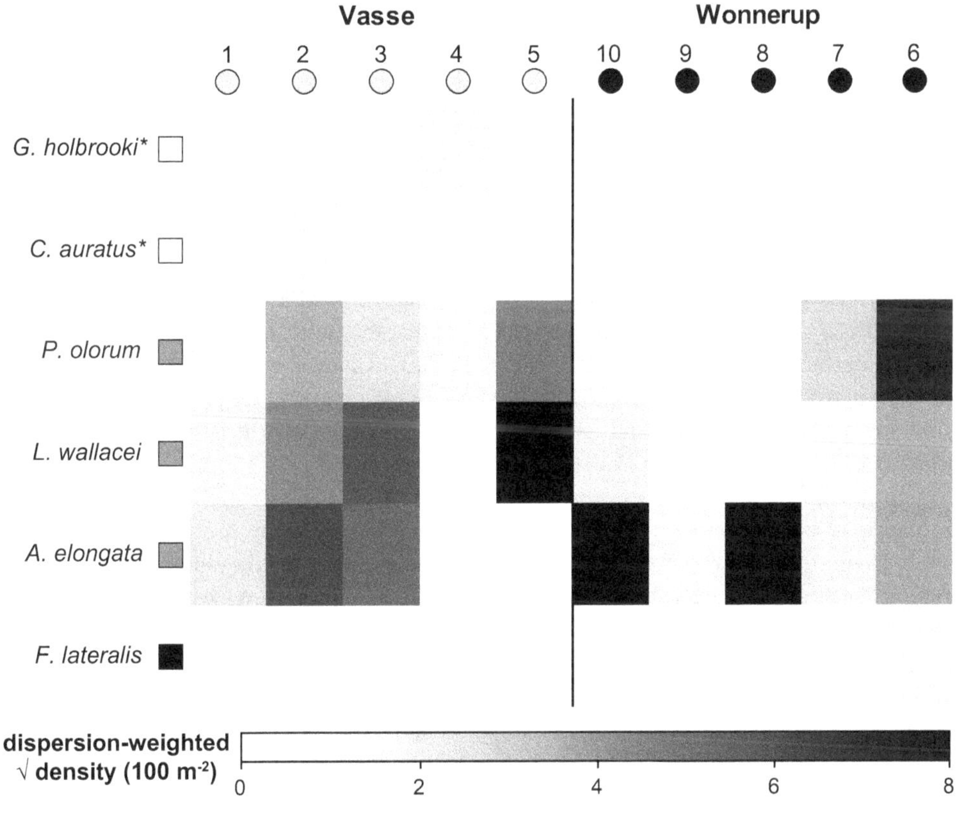

Fig. 4 Shade plot of the dispersion-weighted, square-root transformed and averaged densities ($100\ m^{-2}$) of each fish species recorded at each of the ten sites in the Vasse and Wonnerup estuaries. *White* areas denote the absence and *grey scale* the abundance of a species. Vasse Estuary (*dark gray filled circle*); Wonnerup Estuary (*filled circle*). Freshwater estuarine-opportunist (*open square*); Estuarine & freshwater (*gray filled square*); Estuarine & marine (*filled square*). *Asterisk* denotes that a species has been introduced

Discussion

Species abundances and faunal composition

The fish fauna of the Vasse and Wonnerup estuaries in January 2012 was dominated by the atherinids *Lepthatherina wallacei* and *Atherinosoma elongata* and the gobiid *Pseudogobius olorum*. These species, all of which complete their entire lifecycle within estuaries, comprised over 99.8 and 99.9% of the total catches in the Vasse and Wonnerup estuaries, respectively. The prevalence of these species in estuaries, particularly those with a seasonal connection to the ocean, has been reported elsewhere in south-western Australia (Hoeksema et al. 2009; Valesini et al. 2009). The large abundances of these species in these systems, many of which undergo marked changes in salinity (Chuwen et al. 2009), reflect the fact that these species are highly euryhaline (Hoeksema et al. 2006; Veale et al. 2014). The ability to osmoregulate in a broad range of salinities is advantageous in the Vasse and Wonnerup estuaries as salinities can change from >90 in April to <1 in June (Lane et al. 2011).

The total number of six fish species recorded during the present study is far lower than that recorded in the shallow, nearshore waters of the nearby permanently open Swan-Canning (61), Peel-Harvey (53) and Leschenault (43) estuaries (Valesini et al. 2014; Veale et al. 2014; Potter et al. 2016). It is also less than the number of species found in estuaries on the south coast of Western Australia (i.e. Broke, Wilson and Irwin inlets; 11–20 species), which like the Vasse-Wonnerup, all maintain a seasonal connection to the ocean (Hoeksema et al. 2009). The relatively depauperate nature of the fish fauna of the Vasse and Wonnerup estuaries reflects the fact that the bar at the mouth of Wonnerup Inlet is only open to the ocean for part of the year, thus reducing the opportunity for marine species to penetrate the system (Strydom 2015; Tweedley et al. 2016), and also the fact that, although this study was undertaken during a period when the bar at the mouth of Wonnerup Inlet was open, the movements of marine species into the estuaries are inhibited by the surge barriers (Fig. 1c).

In contrast to the number of species, the total densities of fish were far higher in the Vasse and Wonnerup estuaries than in many of those systems mentioned above (Tweedley 2011). It is thus relevant that, when comparing the fish faunas of five estuaries on the south coast of Western Australia, Hoeksema et al. (2009) noted that *L. wallacei, A. elongata* and *P. olorum,* the three most abundant species in the Vasse and Wonnerup estuaries, were far more abundant in the eutrophic Wellstead Estuary than in the oligotrophic Broke Inlet. These authors attributed the large differences in the abundances of these species to the dense stands of macrophytes that occur in the former system. It stands to reason, therefore, that the macrophytes and algae that proliferate in the Vasse and Wonnerup estuaries (Brearley 2005; Chambers et al. 2013; Fig. 1e, f) may play a similar role in supporting the large fish populations reported in the current study. For example, such areas are known to provide a preferred habitat and an abundance of prey for these fish species (Prince et al. 1982; Humphries et al. 1992; Humphries and Potter 1993).

There was no significant difference between the fish faunas of the two estuaries. This result reflects the fact that (1) the faunas of both estuaries were dominated to a similar extent by the same three species, (2) the marked variability in the densities of fish recorded at the various sites within either estuary and (3) that the environments of both estuaries were broadly similar at the time of sampling. Although salinities were significantly higher in the Vasse (18.3) than Wonnerup (15.7) Estuary, these differences were not great enough to inhibit colonisation by the three dominant species, all of which are euryhaline (Hoeksema et al. 2006; Veale et al. 2014). Furthermore, although three of the six species recorded were unique to a particular estuary, none of these species occurred in appreciable densities; for example, *F. lateralis* was only caught at a single site in the Wonnerup Estuary and constituted only five individuals.

Records of introduced species and their possible impacts

To date, 13 introduced species have been recorded in natural freshwater systems of south-western Australia (Morgan et al. 2004; Beatty and Morgan 2013) and two of these species, *G. holbrooki* and *C. auratus,* have been recorded in the lower reaches of the Vasse River (Morgan and Beatty 2007; Beatty et al. 2017). This study has confirmed the presence of these species in the Vasse Estuary, and thus within the boundaries of the Ramsar listing. Such a finding is important as introduced fish have been identified by the Australian Federal Government as a key threat to the ecological character of the Vasse-Wonnerup Wetland System (Australian Government 2011). The poeciliid *G. holbrooki*, for example, is known to behave in a highly antagonistic manner to other small fishes, causing damage to their fins and even death (Schoenherr 1981; Rowe et al. 2007), and to predate on the eggs and larvae of fish and amphibians (Ivantsoff and Aarn 1999; Pyke 2005). As this species is highly tolerant of changes in salinity, water temperature, dissolved oxygen and turbidity (Cherry et al. 1976; Hubbs 2000) and is extremely abundant in disturbed habitats near urban areas (Lloyd et al. 1986), it is likely to be able to survive in the Vasse Estuary despite the large seasonal changes in hydrology. Furthermore, by feeding on zooplankton such as *Daphnia,* which feed on phytoplankton, *G. holbrooki* may facilitate, if present in great enough densities, the development of algal blooms, which regularly occur in the system (Hurlbert et al. 1972; Ho et al. 2011).

Of greater concern, however, was the collection of *C. auratus* from the Vasse Estuary. Individuals of this species have been caught frequently and in large numbers from the fresh waters of the lower Vasse River since 2003 (Morgan and Beatty 2007), and an acoustic tracking study revealed *C. auratus* entered the uppermost (i.e. fresh) section of the Vasse Estuary, raising concern over the potential for further downstream colonisation (Beatty et al. 2017). This cyprinid is a benthic herbivore and through its vigorous feeding activities uproots and consumes vegetation, increases turbidity by resuspending sediments and releases nutrients into the water column (Richardson et al. 1995). Furthermore, *C. auratus* may exacerbate the regular algal blooms that occur in this system, as the growth of cyanobacteria is stimulated by passage through the intestine of this species (Kolmakov and Gladyshev 2003). From a biological perspective, *C. auratus* is a known vector for the introduction of parasites and diseases. For example, it is the likely source of the parasitic copepod *Lernaea cyprinacea* in Western Australia and responsible for the introduction, establishment and spread of Goldfish Ulcer Disease in Australia, which is caused by variants of the pathogenic bacterium *Aeromonas salmonicida* (Humphrey and Ashburner 1993; Marina et al. 2008). Therefore, its establishment in other estuarine or riverine habitats of the Vasse-Wonnerup system is of great concern.

Although *C. auratus* have previously been recorded in estuaries elsewhere around the world (Maes et al. 1998; Martinho et al. 2007; Henderson and Bird 2010), such records have been typically limited to the upper, oligohaline reaches of these systems and/or have coincided with periods of higher freshwater flows and flushing. The presence of *C. auratus* in the Vasse Estuary during the austral summer, albeit before the system becomes markedly hypersaline (Lane et al. 2011), indicates that this species is capable of surviving in mesohaline environments. *Carassius auratus,* in the current study, were found in waters with a salinity of 19, which suggests that this population may have a greater tolerance to salinity than that found in populations elsewhere in the world. For example, Lawson and Alake (2011) found that, although individuals had a high level of acute tolerance to salinities between 1 and 5, none of the fish exposed to higher salinities (i.e. 6–10) remained alive after seven days and, in another study, juvenile *C. auratus* exposed to salinities of 10 died within 24 h (Jasim 1988). Schofield et al. (2006) found that *C. auratus* was able to persist in salinities ≤ 10, but exhibited significant mortality at 15 and 20. The ability of *C. auratus*, in the current study, to survive for a protracted period of time in mesohaline salinities raises the concerning possibility that this species might use the estuary as a 'saltbridge' (Bringolf et al. 2005; Schofield et al. 2006; Brown et al. 2007) to access and expand its distribution into new tributaries, i.e. the Sabina and Abba rivers, and to the Wonnerup Estuary and Ludlow River through Malbup Creek, which connects the two estuaries during wet winters (Fig. 1). It is relevant that such movements have been recorded in other freshwater fish species, such as Zander *Stizostedion lucioperca*, that have previously been considered stenohaline (Brown et al. 2001; 2007).

Future research

This study represents the first quantitative survey of the fish fauna of the Vasse and Wonnerup estuaries, part of the internationally important Ramsar-listed Vasse-Wonnerup Wetland System, and will provide useful baseline information against which to assess future changes in faunal composition. However, further work is needed to build upon the preliminary findings of this study and to provide a more complete understanding of the ecology of the fish communities of this system. As environmental conditions, such as salinity, water temperature, dissolved oxygen concentration and the composition and abundance of macrophytes change markedly throughout the year (Lane et al. 2011; Tweedley et al. 2013), future work should be undertaken on a seasonal basis, noting that not all areas of the estuaries will contain water at all times of the year. Second, this study focused only on the Vasse and Wonnerup estuaries and no attempt was made to survey other areas of the system, such as Wonnerup Inlet, the Deadwater and the Lower Vasse River Wetlands (Fig. 1). These areas are of significant ecological interest. For example, the Deadwater is known to local fisherman as an important area for juveniles of marine species (Elscot 2000) and has, like the Vasse and Wonnerup estuaries, experienced high nutrient levels, poor water quality, toxic algal blooms and fish kills (Lane et al. 1997; Hart 2014). Given the presence of *C. auratus* in the Vasse Estuary and their potential to use the estuary as a 'saltbridge', a laboratory study to determine the chronic and acute salinity tolerance of this population of *C. auratus* would be valuable, together with sampling of the other tributaries to determine if this invasive species has expanded its range into other water bodies connected to the Vasse Estuary.

Acknowledgements The authors gratefully acknowledge the support given to them during the fieldwork by Alan Clarke (Department of Biodiversity, Conservation and Attractions), Kath Lynch (Department of Water and Environmental Regulation), and Jane Chambers (Murdoch University). Jane Chambers is also thanked for providing the water physico-chemical data. This project was supported by the South West Catchments Council, through funding from the Australian Government, and separately from Murdoch University.

References

Australian Government (2011) Caring for our country site investment guide: Vasse Wonnerup and Geographe Bay. Australian Government, Canberra

Beatty SJ, Morgan DL (2013) Introduced freshwater fishes in a global endemic hotspot and implications of habitat and climatic change. Bioinvasions Rec 2:1–9

Beatty SJ, Allen MG, Whitty JM, Lymbery AJ, Keleher JJ, Tweedley JR, Ebner BC, Morgan DL (2017) First evidence of spawning migration by goldfish *Carassius auratus* and the implications for control of a globally invasive species. Ecol Freshw Fishes 26:444–455

Brearley A (2005) Ernest Hodgkin's Swanland, 1st edn. University of Western Australia Press, Crawley

Bringolf RB, Kwak TJ, Cope WG, Larimore MS (2005) Salinity tolerance of flathead catfish: implications for dispersal of introduced populations. Trans Am Fish Soc 134:927–936

Brown JA, Moore WM, Quabius ES (2001) Physiological effects of saline waters on zander. J Fish Biol 59:1544–1555

Brown J, Scott D, Wilson R, Gherardi F (2007) Do estuaries act as saline bridges to allow invasion of new freshwater systems by non-indigenous fish species? Biological invaders in inland waters: profiles, distribution, and threats, vol 2. Invading Nature—Springer Series in Invasion Ecology. Springer, The Netherlands, pp 401–414

Chambers JM, Tweedley JR, Clarke A, Paice R (2013) A snapshot survey of the distribution and abundance of macrophytes, macroalgae, phytoplankton and macroinvertebrates of the Vasse-Wonnerup lagoons January 2012. Murdoch University, Perth

Cherry DS, Guthrie RK, Rodgers JH, Cairns J, Dickson KL (1976) Responses of mosquitofish (*G. affinis*) to ash effluent and thermal stress. Trans Am Fish Soc 105:686–694

Chuwen BM, Hoeksema SD, Potter IC (2009) The divergent environmental characteristics of permanently-open, seasonally-open and normally-closed estuaries of south-western Australia. Estuar Coast Shelf Sci 85:12–21

Clarke KR (1993) Non-parametric multivariate analyses of changes in community structure Australian. J Ecol 18:117–143

Clarke KR, Green RH (1988) Statistical design and analysis for a 'biological effects' study. Mar Ecol Prog Ser 46:213–226

Clarke KR, Chapman MG, Somerfield PJ, Needham HR (2006) Dispersion-based weighting of species counts in assemblage analyses. Mar Ecol Prog Ser 320:11–27

Clarke KR, Gorley RN, Somerfield PJ, Warwick RM (2014a) Change in marine communities: an approach to statistical analysis and interpretation, 3rd edn. PRIMER-E Ltd, Plymouth

Clarke KR, Tweedley JR, Valesini FJ (2014b) Simple shade plots aid better long-term choices of data pre-treatment in multivariate assemblage studies. J Mar Biol Assoc UK 94:1–16

Commonwealth of Australia (2002) Australian catchment, river and estuary assessment 2002, vol 1. National Land and Water Resources Audit, Canberra

Costanza R, d'Arge R, de Groot R, Farber S, Grasso M, Hannon B, Limburg K, Naeem S, O'Neil RV, Paruelo J, Raskin RG, Sutton P, van den Belt M (1997) The value of the world's ecosystems services and natural capital. Nature 387:253–260

Creighton C, Boon PI, Brookes JD, Sheaves M (2015) Repairing Australia's estuaries for improved fisheries production—what benefits, at what cost? Mar Freshw Res 66:493–507

Department of Water (2010) Vasse Wonnerup Wetlands and Geographe Bay water quality improvement plan. Perth, Australia

Elscot SV (2000) Monitoring of fish behaviour in the lower reaches of the Vasse-Wonnerup Wetland System during the summer of 1999/2000. Department of Conservation and Land Management and the Geographe Catchment Council, Busselton

Hallett CS, Valesini FJ, Clarke KR, Hesp SA, Hoeksema SD (2012) Development and validation of fish-based, multimetric indices for assessing the ecological health of Western Australian estuaries. Estuar Coast Shelf Sci 104–105:102–113

Hallett CS, Valesini F, Elliott M (2016) A review of Australian approaches for monitoring, assessing and reporting estuarine condition: III. Evaluation against international best practice and recommendations for the future. Environ Sci Policy 66:282–291

Harrison TD, Whitfield AK (2004) A multi-metric fish index to assess the environmental condition of estuaries. J Fish Biol 65:683–710

Hart BT (2014) Independent review of the current and future management of water assests in the Geographe catchment. Water Science, Echuca, Australia

Henderson PA, Bird DJ (2010) Fish and macro-crustacean communities and their dynamics in the Severn Estuary. Mar Pollut Bull 61:100–114

Ho SS, Bond NR, Lake S (2011) Comparing food-web impacts of a native invertebrate and an invasive fish as predators in small floodplain wetlands. Mar Freshw Res 62:372–382

Ho SS, Bond NR, Lake S (2011) Comparing food-web impacts of a native invertebrate and an invasive fish as predators in small floodplain wetlands. Mar Freshw Res 62:372–382

Hoeksema SD, Chuwen BM, Hesp SA, Hall NG, Potter IC (2006) Impact of environmental changes on the fish faunas of Western Australian south-coast estuaries. Centre for Fish and Fisheries Research Murdoch University, Perth

Hoeksema SD, Chuwen BM, Potter IC (2009) Comparisons between the characteristics of ichthyofaunas in nearshore waters of five estuaries with varying degrees of connectivity with the ocean. Estuar Coast Shelf Sci 85:22–35

Hubbs C (2000) Survival of *Gambusia affinis* in a hostile environment. Southwest Nat 45:521–522

Humphrey JD, Ashburner LD (1993) Spread of the bacterial fish pathogen *Aeromonas salmonicida* after importation of infected goldfish, *Carassius auratus*, into Australia. Aust Vet J 70(453):454

Humphries P, Potter IC (1993) Relationship between the habitat and diet of three species of atherinids and three species of gobies in a temperate Australian estuary. Mar Biol 116:193–204

Humphries P, Potter IC, Loneragan NR (1992) The fish community in the shallows of a temperate Australian estuary: relationships with the aquatic macrophyte *Ruppia megacarpa* and environmental variables. Estuar Coast Shelf Sci 34:325–346

Hurlbert SH, Zedler J, Fairbanks D (1972) Ecosystem alteration by mosquitofish (*Gambusia affinis*) predation. Science 175:639–641

Ivantsoff W, Aarn (1999) Detection of predation on Australian native fishes by *Gambusia holbrooki*. Mar Freshw Res 50:467–468

Jackson JBC, Kirby MX, Berger WH, Bjorndal KA, Botsford LW, Bourque BJ, Bradbury RH, Cooke R, Jon E, Estes JA, Hughes TP, Kidwell S, Lange CB, Lenihan HS, Pandolfi JM, Peterson CH, Steneck RS, Tegner MJ, Warner RR (2001) Historical overfishing and the recent collapse of coastal ecosystems. Science 293:629–638

Jasim BM (1988) Tolerance and adaptation of goldfish *Carassius auratus* (L.) to salinity. J Biol Sci Res 19:149–153

Kolmakov VI, Gladyshev MI (2003) Growth and potential photosynthesis of cyanobacteria are stimulated by viable gut passage in crucian carp. Aquat Ecol 37:237–242

Lane JA, Hardcastle KA, Tregonning RJ, Holtfreter S (1997) Management of the Vasse-Wonnerup wetland system in relation to sudden, mass fish deaths. Vasse Estuary Tech Work Gr, Busselton

Lane JAK, Clarke AG, Pearson GB (2007) Waterbirds of the Vasse-Wonnerup Wetlands in 1998–2000 and some comparisons with earlier data. Western Australian Department of Environment and Conservation, Busselton

Lane JAK, Clarke AG, Winchcombe YC (2011) Depth, salinity and temperature profiling of Vasse-Wonnerup wetlands in 1998-2000. Western Australian Department of Environment and Conservation, Busselton

Lawson EO, Alake SA (2011) Salinity adaptability and tolerance of hatchery reared comet goldfish *Carassius auratus* (Linnaeus 1758). Int J Zool Res 7:68–76

Lloyd LN, Arthington AH, Milton DA (1986) The mosquitofish – a valuable mosquito-control agent or a pest? In: Kitching RL (ed) The Ecology of Exotic Animals and Plants Some Australian Case Histories. Wiley, Brisbane, pp 6–25

Maes J, Taillieu A, Van Damme PA, Cottenie K, Ollevier F (1998) Seasonal patterns in the fish and crustacean community of a turbid temperate estuary (Zeeschelde Estuary, Belgium). Estuar Coast Shelf Sci 47:143–151

Marina H, Beatty SJ, Morgan DL, Doupé RG, Lymbery AJ (2008) An introduced parasite, *Lernaea cyprinacea* L., found on native freshwater fishes in the south west of Western Australia. J R Soc West Aust 91:149–153

Martinho F, Leitão R, Viegas I, Dolbeth M, Neto JM, Cabral HN, Pardal MA (2007) The influence of an extreme drought event in the fish community of a southern Europe temperate estuary. Estuar Coast Shelf Sci 75:537–546

McAlpine KW, Spice JF, Humphries R (1989) The environmental condition of the Vasse-Wonnerup wetland system and a discussion of management options. Western Australian Environmental Protection Authority Technical Series 31:1–35

Morgan DL, Beatty SJ (2007) Feral Goldfish (*Carassius auratus*) in Western Australia: a case study from the Vasse River. J R Soc West Aust 90:151–156

Morgan DL, Gill HS, Maddern MG, Beatty SJ (2004) Distribution and impacts of introduced freshwater fishes in Western Australia. NZ J Mar Freshw Res 38:511–523

Patchineelam SM, Kjerfve B, Gardner LR (1999) A preliminary sediment budget for the Winyah Bay estuary, South Carolina, USA. Mar Geol 162:133–144

Potter IC, Hyndes GA (1999) Characteristics of the ichthyofaunas of southwestern Australian estuaries, including comparisons with holarctic estuaries and estuaries elsewhere in temperate Australia: a review. Aust Ecol 24:395–421

Potter IC, Chuwen BM, Hoeksema SD, Elliott M (2010) The concept of an estuary: a definition that incorporates systems which can become closed to the ocean and hypersaline Estuarine. Coast Shelf Sci 87:497–500

Potter IC, Tweedley JR, Elliott M, Whitfield AK (2015a) The ways in which fish use estuaries: a refinement and expansion of the guild approach. Fish Fish 16:230–239

Potter IC, Warwick RM, Hall NG, Tweedley JR (2015b) The physico-chemical characteristics, biota and fisheries of estuaries. In: Craig J (ed) Freshwater Fisheries Ecology. Wiley-Blackwell, USA

Potter IC, Veale LJ, Tweedley JR, Clarke KR (2016) Decadal changes in the ichthyofauna of a eutrophic estuary following a remedial engineering modification and subsequent environmental shifts. Estuar Coast Shelf Sci 181:345–363

Prince J, Potter I, Lenanton R, Loneragan N (1982) Segregation and feeding of atherinid species (Teleostei) in south-western Australian estuaries. Mar Freshw Res 33:865–880

Pyke G (2005) A review of the biology of *Gambusia affinis* and *G. holbrooki*. Rev Fish Biol Fish 15:339–365

Richardson MJ, Whoriskey FG, Roy LH (1995) Turbidity generation and biological impacts of an exotic fish *Carassius auratus*, introduced into shallow seasonally anoxic ponds. J Fish Biol 47:576–585

Rowe DK, Smith JP, Baker C (2007) Agonistic interactions between *Gambusia affinis* and *Galaxias maculatus*: implications for whitebait fisheries in New Zealand rivers. J Appl Ichthyol 23:668–674

Schoenherr AA (1981) The role of competition in the displacement of native fishes by introduced species. In: Naiman RJ, Stolz DL (eds) Fishes in North American Deserts. Wiley, New York, pp 173–203

Schofield PJ, Brown BE, Fuller PF (2006) Salinity tolerance of goldfish, *Carassius auratus*, a non-native fish in the United States. Fla Sci 69:258–268

Sheaves M, Baker R, Nagelkerken I, Connolly RM (2015) True value of estuarine and coastal nurseries for fish: incorporating complexity and dynamics. Estuaries Coasts 38:401–414

Strydom NA (2015) Patterns in larval fish diversity, abundance, and distribution in temperate South African estuaries. Estuaries Coasts 38:268–284

Tweedley JR (2011) The Relationships Between Habitat Types and Faunal Community Structure in Broke Inlet, Ph.D., Murdoch University, Western Australia

Tweedley JR, Warwick RM, Valesini FJ, Platell ME, Potter IC (2012) The use of benthic macroinvertebrates to establish a benchmark for evaluating the environmental quality of microtidal, temperate southern hemisphere estuaries. Mar Pollut Bull 64:1210–1221

Tweedley JR, Chambers JM, Paice R (2013) Sediment accumulation and resuspension in the Vasse-Wonnerup Wetlands and its relationship to internal nutrient cycling. Murdoch University, Perth

Tweedley JR, Warwick RM, Clarke KR, Potter IC (2014) Family-level AMBI is valid for use in the north-eastern Atlantic but not for assessing the health of microtidal Australian estuaries. Estuar Coast Shelf Sci 141:85–96

Tweedley JR, Warwick RM, Potter IC (2016) The contrasting ecology of temperate macrotidal and microtidal estuaries. Oceanogr Mar Biol Annu Rev 54:73–171

Tweedley JR, Warwick RM, Hallett CS, Potter IC (2017) Fish-based indicators of estuarine condition that do not require reference data. Estuar Coast Shelf Sci 191:209–220

Valesini FJ, Coen NJ, Wildsmith MD, Hourston M, Tweedley JR, Hallett CS, Linke TE, Potter IC (2009) Relationships between fish faunas and habitat type in south-western Australian estuaries. Project 2004/045. Draft Final Report for Fisheries Research and Development Corporation. Murdoch University, Perth

Valesini F, Tweedley J, Clarke K, Potter I (2014) The importance of regional, system-wide and local spatial scales in structuring temperate estuarine fish communities. Estuaries Coasts 37:525–547

Veale L, Tweedley JR, Clarke KR, Hallett CS, Potter IC (2014) Characteristics of the ichthyofauna of a temperate microtidal estuary with a reverse salinity gradient, including inter-decadal comparisons. J Fish Biol 85:1320–1354

Wetland Research and Management (2007) Ecological character description: Vasse-Wonnerup wetlands Ramsar site south-west Western Australia. Wetland Research and Management, Perth

Wildsmith MD, Rose TH, Potter IC, Warwick RM, Clarke KR, Valesini FJ (2009) Changes in the benthic macroinvertebrate fauna of a large microtidal estuary following extreme modifications aimed at reducing eutrophication. Mar Pollut Bull 58:1250–1262

12

Effects of temperature and salinity on larval survival and development of the western school prawn *Metapenaeus dalli*

Jason A. Crisp · Gavin J. Partridge · Frances M. L. D'Souza · James R. Tweedley · Navid R. Moheimani

Abstract The effects of temperature and salinity on the survival and development of larval *Metapenaeus dalli* were investigated in two experiments. First, the effects of four temperatures of 22.6, 25.8, 29.4 and 32.6 °C on survival and development time were examined from the Nauplius VI to the Mysis I sub-stage. Survival was significantly lower at 32.6 °C than at 22.6 and 25.8 °C. Development times differed significantly across all temperatures, decreasing linearly with temperature from 161.5 h at 22.6 °C to 74.8 h at 29.4 °C then decreased slightly to 71.4 h at 32.6 °C. Second, the combined effects of three temperatures (21.4, 26.5 and 29.7 °C) and three salinities (30, 35 and 40‰) on survival and development were quantified over a 48 h period from the Nauplius VI sub-stage. Only salinity was found to influence survival, with peak survival (77.7%) found to occur at the control salinity ($\sim$35‰). Any increase or decrease in salinity from this value resulted in a decrease in survival, with the lowest salinity tested (30‰) having a significantly negative effect on survival (58.4%) when compared to the control. Only temperature was found to influence the rate of development, with significant increases in development index values being recorded as temperature increased. The recommended conditions for optimal survival and development of *M. dalli* larvae as determined by this study are, therefore, 25.8 °C and 35‰.

Keywords Restocking · Penaeid · Crustacean · Abiotic · Shrimp

Introduction

The western school prawn *Metapenaeus dalli* was once the focus of a small commercial and iconic recreational fishery in the Swan-Canning Estuary in Western Australia (Potter et al. 1986). This species is known to complete its life-cycle within this estuary, with maturation and reproduction occurring in water temperatures of $\sim$22–26 °C and approximately marine salinities (Potter et al. 1986, 1989; Broadley et al. 2017). However, under continued fishing and environmental pressures a significant decline in *M. dalli* abundance occurred, leading to a cessation of commercial fishing activities in the 1970s and greatly reduced recreational fishing

J. A. Crisp (✉) · G. J. Partridge · J. R. Tweedley · N. R. Moheimani
School of Veterinary and Life Sciences, Murdoch University, Perth, Australia
e-mail: j.crisp@murdoch.edu.au

G. J. Partridge
Australian Centre for Applied Aquaculture Research, South Metropolitan TAFE, Fremantle, Australia

F. M. L. D'Souza
Department of Water, Perth, Australia

effort in the late 1990s (Maher 2002). Despite the reduced fishing pressure, stocks of *M. dalli* in the Swan-Canning Estuary have not recovered naturally (Smith et al. 2007). Thus, a restocking program was implemented to bypass the recruitment bottleneck during the early larval stages. As this species had never been cultured before, the development of hatchery techniques were initiated in 2012, with the aim of releasing post-larvae (PL) at sub-stage PL15-20; ~25–30 days post-hatch (Jenkins, G. I., Australian Centre for Applied Aquaculture Research, Fremantle, Western Australia, unpublished data).

Successful rearing of penaeid prawn larvae for stocking requires high survival rates to be attained, however, this must be achieved in conjunction with development rates that limit the time larvae spend in the hatchery to reduce operating costs (ASEAN 1978). Two significant environmental factors that influence the survival and development rates of larval crustaceans are water temperature and salinity (Kinne 1963, 1964). The complex early life stages of penaeid prawns are considered to be particularly vulnerable to temperature and/or salinity changes, with high mortality occurring during the larval stages under adverse conditions (Anger 2003). Of the three larval stages Nauplius, Protozoea and Mysis in penaeid prawns, Protozoea stage larvae appear to be the most sensitive to changes in temperature and salinity, and thereby provide an effective proxy for assessing suitable culture conditions for all larval and post-larval life stages (Preston 1985; Zacharia and Kakati 2004).

Studies of the effects of water temperature and salinity on larval penaeid prawns, have demonstrated that salinity has a greater effect on survival than temperature (e.g. Kumlu et al. 2000, 2001; Zacharia and Kakati 2004; Ch and Shailender 2013), although increases in mortality are known to occur at relatively high temperatures (Aktas and Cavdar 2012; Jackson and Burford 2003). In contrast, water temperature has a more pronounced effect on larval development than salinity (Parado-Estepa 1998; Jackson and Burford 2003), with growth rates only shown to be affected by salinity at brackish ranges (Ponce-Palafox et al. 1997; Kumlu et al. 2000). Temperature and salinity have also been show to interact, further exacerbating mortality rates when both factors are at the extremes of their ranges (Ponce-Palafox et al. 1997; Kumlu et al. 2000). For the establishment of rearing protocols, however, it is important to determine the temperature and salinity regime that provides the highest possible development rates without compromising larval survival.

Once preferred water temperatures and salinities for culture are established, they can be used to predict larval survival and growth rates for future planning. This is particularly important in larval stock management from the Protozoea stage, as this is the point at which penaeid larvae switch to exogenous feeding (Dall et al. 1990). It is at this point that hatchery feeding relies on the expensive cultivation of live feeds such as microalgae, which require planning days in advance to produce (D'Souza et al. 2000). Live zooplankton is often used from the Mysis and early post-larval stages with similar planning requirements (ASEAN 1978; Zacharia and Kakati 2004). Without predicting survival and development rates, feeding protocols cannot be calibrated, which may result in increased cultivation costs for live foods and potential negative effects on water quality as a result of over or under feeding.

The purpose of this study was to determine the effects of culture temperature and salinity on *M. dalli* Protozoea stage larvae, acting as proxy for all larval developmental stages, in an effort to maximize survival and optimize development rates for the purpose of increasing the number of hatchery-reared *M. dalli* post-larvae that can be restocked into the Swan-Canning Estuary.

Materials and methods

Broodstock collection

Female *M. dalli* were collected at night from the Swan-Canning Estuary in Perth (31°56′50″S 115°54′58″E), south-western Australia from December 2014 to March 2015, using a hand trawl net that was 1.5 m high, 4 m wide and constructed from 9 mm mesh. Water temperature and salinity during brood stock collection were recorded with a YSI 556 Handheld Multiparameter Instrument (http://www.ysi.com), with temperature and salinity ranges of 22.8–26.6 °C and 34.2–35.7‰ recorded over the course of this study. Maturity was determined as per descriptions by Tuma (1967) and Crisp et al. (2017), with mature prawns immediately transported to the aquaculture facility and placed in aerated holding tanks overnight (ASEAN 1978). Females were then stocked into 300 L conical base spawning tanks at a density of up to 40 per tank for two to four days. The tanks were filled with seawater at a salinity of ~35‰ drawn from a bore sunk through limestone rock, accessing near-shore marine water, and aerated constantly and maintained at a temperature of ~26 °C, to approximate conditions in the Swan-Canning Estuary at the time and place of capture. The base of each broodstock tank was fitted with a fine grate that allowed eggs to pass through, separating them from broodstock to prevent any potential cannibalism. Broodstock spawned naturally within 48 h of capture without eyestalk ablation. After spawning, eggs were collected on 48 and 63 μm screens and rinsed, sub-sampled for counting and re-suspended in 300 L of filtered (1 μm) seawater, under constant aeration. Egg quality was assessed by hatch rate, with a value of >75% survival post-hatch considered suitable for use in the study (ASEAN 1978).

Larval rearing system

After spawning, hatched larvae were held until they had entered into the Nauplius (henceforth denoted as 'N') VI sub-stage at ~48 h post-hatch. Morphological characters of each larval sub-stage were identified as per Fig. 1 and Crisp et al. (2016). Briefly, N VI larvae were identified as having two pairs of antennae, mandibles, an eyespot and a posterior end with two groups of spines with three plumose spines and three naked spines each. Protozoea sub-stages (PZ I, II and III) were defined as having a cephalothorax, abdomen and telson, with eye stalks appearing from PZ II and two pairs of rudimentary uropods from PZIII. Mysis I sub stage (M I) exhibits a dorsal spine as a precursor to the rostrum and a tail fan with a telson and two pairs of uropods almost fully formed.

Once larvae reached N VI, they were harvested and stocked at a density of 250 larvae L^{-1} (D'Souza and Kelly 2000) into 6 L flat bottom cylindrical glass culture vessels containing 4 L of 1 μm filtered seawater. Once larvae were stocked, experimental salinities and water temperatures were acclimated by lowering or increasing the salinity/temperature at a rate of 1‰ or 1 °C per 15 min, respectively. Salinity was adjusted by adding dechlorinated municipal supply water or hypersaline (60‰) sea water (Instant Ocean, USA), as per methods modified from Kumlu et al. (2000, 2001), while water temperature was adjusted by applying either heated or chilled water to water baths until target temperatures were reached. Water temperature for each experiment was maintained by housing culture vessels in 115 L temperature-controlled water baths, heated with pre-calibrated Eheim Jager 150 W aquarium water heaters and monitored with Thermocron TCS temperature loggers every 10 min. A constant salinity was maintained in the culture vessels and monitored daily using an ATAGO PAL-03S digital refractometer. Vessels were exposed to 3.5 μmol photons m^{-2} s^{-1} white light with 12:12 h light:dark photoperiod. The larval cultures were aerated from the base to provide vertical mixing as well as oxygenation.

Feed cultivation

Larval *M. dalli* were fed a diet comprising two microalgae species obtained from the Australian National Algal Culture Collection held by the Commonwealth Scientific and Industrial Research Organization (CSIRO) in Hobart, Tasmania. The chlorophyte *Tetraselmis suecica* (CSIRO strain number CS-187) and the diatom *Chaetoceros muelleri* (CSIRO strain number CS-176) were maintained in batch cultures under 14:10 h light:dark photoperiod with 180 μmol photons m^{-2} s^{-1} white fluorescent light, in 15 L culture vessels with ambient salinity (~35‰) and temperature (~25 °C). Each species was cultured in Guillard's F2 medium, with sodium metasilicate added at 30 g L^{-1} to the *C. muelleri* (Ryther and Guillard 1962). To avoid carbon limitation, food grade CO_2 was injected to maintain a pH of 7.4–7.7.

Feeding larvae

Larval feed in culture vessels was maintained daily at 3×10^4 cells mL^{-1} of *T. suecica* and 9×10^4 cells mL^{-1} of *C. muelleri*, as per feeding protocols used by the Australian Centre for Applied Aquaculture Research (Jenkins, G. I., Australian Centre for Applied Aquaculture Research, Fremantle, Western Australia, unpublished data). These species were used in a previous study of larval *M. dalli* by Crisp et al. (2016) and *Farfantepenaeus brasiliensis* Gaxiola et al. (2010) at similar feeding concentration for similar life stages. Feed intake was measured by counting residual cells of each microalgae species in larval culture vessels using a Neubauer haemocytometer, then replacing feed that had been consumed. Water in the culture vessels was

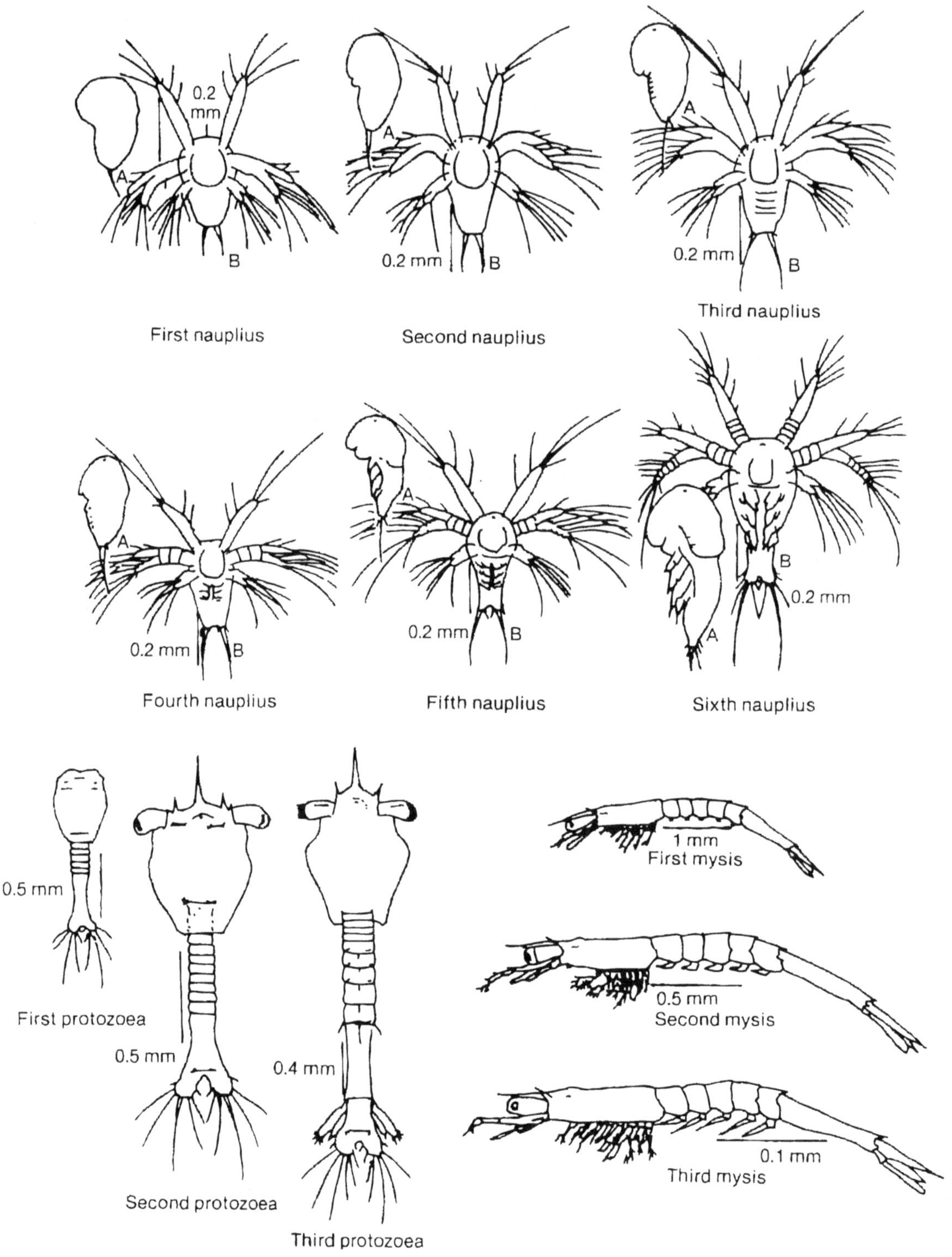

Fig. 1 Representation of the larval life-cycle of penaeid and metapenaeid prawns post hatching from nauplii stages I to VI, through protozoea stages I to III and on to mysis stages I to II (Motoh 1985)

topped up to 6 L daily with a combination of microalgal feed and fresh seawater, with no greater than 10% of culture water exchanged per day. Gut contents of the larvae were also briefly examined under a dissecting microscope to confirm that the algal cells were being ingested (D'Souza and Loneragan 1999).

Experimental design

Effect of temperature

The influence of four nominal water temperatures, i.e. 23, 26, 29 and 32 °C in the laboratory at a constant salinity of 35‰, were assessed on the development time (h) and percentage survival of larval *M. dalli* from N VI, through the protozoal stage (PZ I, II and III) to the Mysis I sub stage (M I). Six replicate culture vessels were used for each of the four temperatures. Development time was calculated as the time taken for 50% of the larvae in each vessel to reach the M I stage. Single 200 mL sub-samples were taken every 12 h from each vessel and each larva staged according to the descriptions in Crisp et al. (2016) and Fig. 1. When M I staged larvae were first observed, each vessel was then sub-sampled every 0.5–1 h. All sampled larvae were carefully returned to the vessels. Once each vessel was found to contain 50% of animals at the M I stage, time was recorded and triplicate 200 mL samples were taken and fixed in 5% tetraborate-buffered formaldehyde solution. All whole larvae that were fixed were assumed to be live at the time of fixation, as larvae are known to be rapidly decomposed by bacteria. Samples were then analyzed for rates of survival by recording the number of whole fixed individuals in each sample and comparing to the initial stocking rate of larvae. This method of determining survival was repeated for the second experiment.

Combined effects of temperature and salinity

The combined effects of three nominal water temperatures, i.e. 23, 26 and 29 °C, and three nominal salinities, i.e. 30, 35 and 40‰, on the development and survival of *M. dalli* larvae were quantified in a 3 × 3 fully factorial design over a 48 h period, after the temperature experiment was run. Each combination of salinity and temperature was conducted in two vessels per run, with an average of the vessels used to create a replicate value of each treatment regime. The experiment was run three times, creating three replicates for each treatment. Each run commenced from the N VI larval sub-stage, running for a fixed period of 48 h. This time period was chosen to examine the acute combined effects of temperature and salinity on the Protozoea stage of development and was informed by the results of the temperature experiment.

After 48 h, all culture vessels were sampled via triplicate 200 mL sub-samples and fixed with 5% tetraborate-buffered formaldehyde, from which survival was determined and the metamorphic sub-stage for each larva was recorded. A development index was calculated for each sub-sample based on that described by Villegas and Kanazawa (1979) and Gaxiola et al. (2010). Each larvae counted was assigned a value for its developmental sub-stage, where PZ I = 1, PZ II = 2 and PZ III = 3. A development index was then calculated, for each sub-sample, by multiplying the sub-stage value (i.e. 1, 2, or 3) by the number of larvae counted at that sub-stage. The product of each sub-stage is then added together then divided by the total number of larvae staged.

Development Index = sum (number of larvae at that stage × stage value)/total number larvae staged. The more larvae there are at higher stage numbers, the larger the index.

Statistical analysis

Effect of temperature

Analysis of the percentage values for survival indicated that data ranged both >80 and <20%. As a result this data was subjected to an arcsine square-root transformation as per Field (2009). To determine whether a transformation of the development data was required, the extent of the linear relationship between the $\log_e$ mean and $\log_e$ standard deviation of development time was calculated. Examination of the slope of the linear relationship indicated that a square-root transformation was necessary to meet test assumptions of homogeneity of variance, as per Clarke et al. (2014). Both survival and development time were analyzed separately using a one-way analysis of variance (ANOVA). When ANOVA detected a significant difference among temperatures, post hoc tests were conducted using Tukey's HSD to elucidate the pairs of temperature that were responsible for each of those differences. In this and all tests, a null hypothesis of no significant difference between a priori groups was rejected when $p < 0.05$.

Combined effects of temperature and salinity

Analysis of the percentage survival data indicated that there were values >80%, but not <20%. As a result, a square-root transformation of the data was performed as per Field (2009). Analysis of the relationship between the $\log_e$ mean and $\log_e$ standard deviation of the development index data indicated that no data transformation was required (Clarke et al. 2014). Separate two-way ANOVAs were used to determine whether development index and survival were affected by temperature and/or salinity, or the interaction term. Both factors were considered fixed. If a significant difference in any factor and/or interaction term was detected, post hoc analysis was conducted using Tukey's HSD.

All statistical analyses were conducted using SPSS version 22 software (IBM Corp.). Values for all results are given as mean ± 95% CL.

Results

Effect of temperature

Mean water temperatures achieved for the nominal temperatures of 23, 26, 29 and 32 °C were 22.6 ± 0.2, 25.8 ± 0.3, 29.4 ± 0.4 and 32.6 ± 0.4 °C, respectively. One-way ANOVA demonstrated that the percentage survival of *M. dalli* larvae differed significantly among temperatures ($p < 0.001$). Survival rates were greatest at 25.8 and 22.6 °C, i.e. 73.0 ± 17.8 and 66.7 ± 12.4%, respectively, and lowest at the highest temperature of 32.6 °C, i.e. 26.3 ± 12.4% (Fig. 2). Time taken for *M. dalli* larvae to develop from N VI to M I differed significantly among temperatures ($p < 0.001$). Development time decreased markedly from 161.5 ± 0.9 h at 22.6 °C to 74.8 ± 0.3 h at 29.4 °C, but only decreased marginally to 71.4 ± 0.1 h at 32.6 °C (Fig. 3).

Combined effects of temperature and salinity

Mean water temperatures achieved for each of the three nominal temperatures tested (23, 26 and 29 °C) were 21.4 ± 0.3, 26.5 ± 0.2 and 29.7 ± 0.4 °C, respectively, with the salinities obtained consistent with nominal salinities (30, 35 and 40‰). Two-way ANOVA showed no interacting effect of temperature and salinity on either survival ($p = 0.480$) or development rate ($p = 0.906$) of *M. dalli* larvae over a 48 h culture period. There were, however, two significant main effects, presented below. First, salinity had a significant effect on larval survival ($p = 0.010$), but not temperature ($p = 0.570$). Salinity of 35‰ was found to be the best for survival, resulting in a survival of 77.7 ± 5.0%. This survival was significantly higher than that measured at 30‰ (58.4 ± 8.4%). Survival at 40‰ (68.4 ± 7.8%) was not significantly different from either of the other two salinities (Fig. 4). Second, the reverse was found for larval development such that temperature had a significant effect ($p < 0.001$), but not salinity ($p = 0.774$). Development of larvae was significantly different

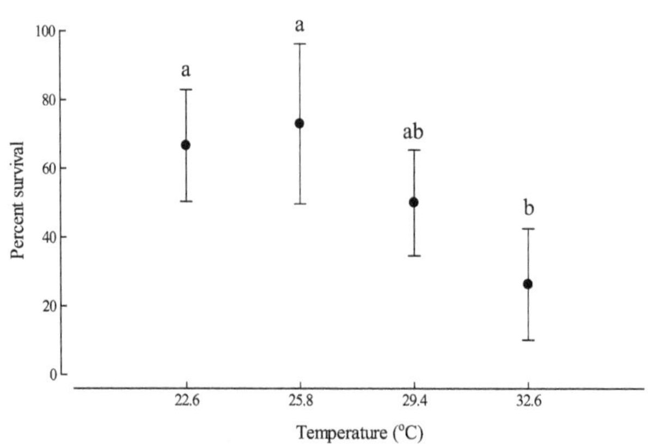

Fig. 2 Mean percentage survival (±95% CL) of *Metapenaeus dalli* larvae during development from Nauplius VI to Mysis I at four different water temperatures. *Letters* above *error bars* denote groups of samples identified by Tukey's HSD

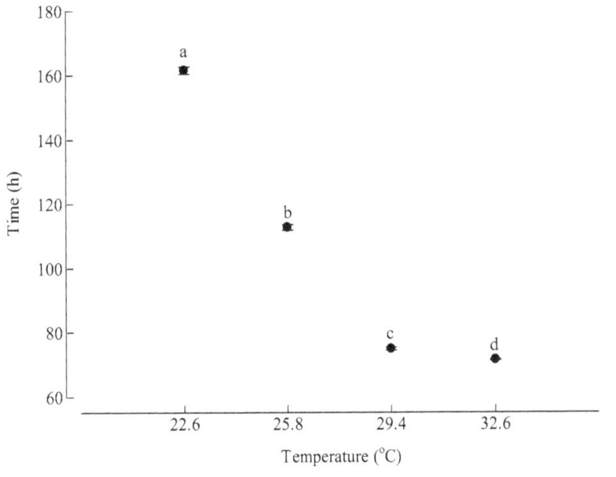

Fig. 3 Average time (±95% CL) taken for *Metapenaeus dalli* larvae to develop from Nauplius VI to Mysis I at four different water temperatures. *Letters* above *error bars* denote groups of samples identified by Tukey's HSD

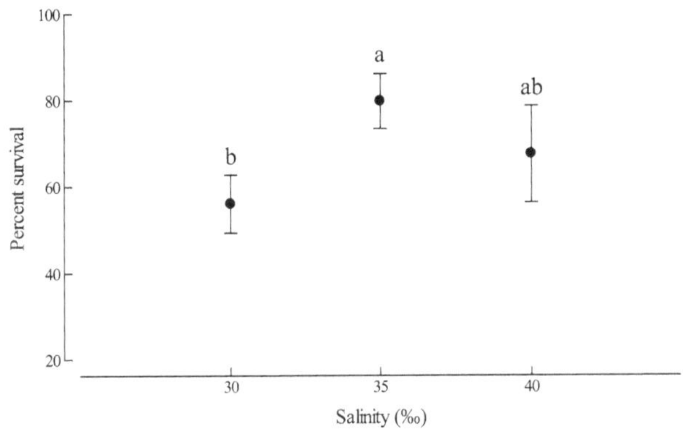

Fig. 4 Mean percentage survival (±95% CL) of *Metapenaeus dalli* larvae over a 48 h period from N VI sub-stage at three different salinities. *Letters* above *error bars* denote groups of samples identified by Tukey's HSD

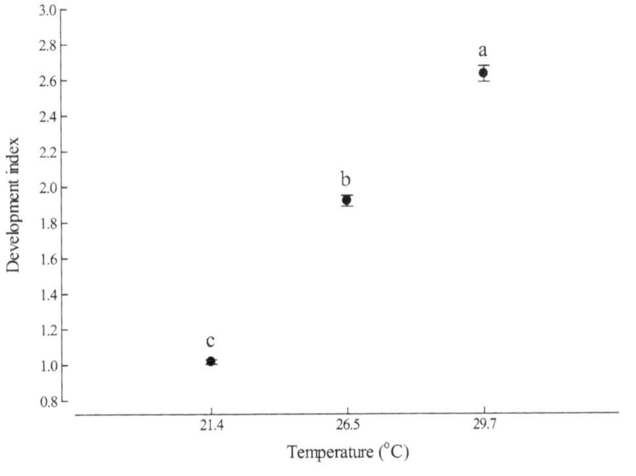

Fig. 5 Mean development index (±95% CL) of *Metapenaeus dalli* larvae over a 48 h period from N VI sub-stage at three different water temperatures. *Letters* above *error bars* denote groups of samples identified by Tukey's HSD

at all three temperatures. A stepwise increase in development rate occurred from the lowest temperature (1.01 ± 0) to the highest (2.61 ± 0.05), with some larvae observed reaching a peak development of PZ III in the highest two temperatures tested (Fig. 5).

Discussion

Effect of temperature

In this study, a significant negative impact on survival was detected at the highest temperature tested of 32.6 °C, but not at any other temperature tested, though there appeared to be reduced survival at 29.4 °C. This result is supported by a study by Kumlu et al. (2000), where temperatures above 30 °C were found to increase mortality in larval *Penaeus semisulcatus*. Furthermore, preliminary findings by Ponce-Palafox et al. (1997) on larval *Litopenaeus vannamei* demonstrated that temperatures above 30 °C, with similar salinities to those used in this study, appeared to have a negative effect on survival. The consistency of these results across larval penaeid prawns may indicate there is an upper thermal limit to their biochemical processes.

Results of this study demonstrate that the time taken by *M. dalli* to develop from N VI to M I differ significantly among temperatures. Rapid sequential decreases in development time to M I occurred between temperatures of 22.6 and 29.4 °C, with only a marginal decrease in development time at the highest temperature of 32.6 °C. These results are consistent with that of larval *P. semisulcatus* (Jackson and Burford 2003) and *Penaeus monodon* (Parado-Estepa 1998), where no significant difference was found in development rates between water temperatures of 29 and 32 °C during the Protozoea stages.

For commercially cultured crustacean larvae, the effect of water temperatures on larval growth rates are often incorporated into planning tools, such as degree-hours or degree-days calculations (Kittaka et al. 2001, 2002; Gendron and Ouellet 2009), which are used to determine larval duration under various temperatures from spawning to post-larval metamorphosis (Roberts et al. 2012). This allows for the determination of potential feed requirements, which is a particularly important factor in larval culture where forward planning is required for the cultivation of live foods (Stevens 1990). When applied to the temperature ranges tested, it can be assumed that culture temperature should be maintained between 25.8 and 29.4 °C provided high survival of larvae can be achieved. Reducing temperature by ~3 °C would increase development time by >2 days and thus feed and resource allocation requirements and increasing temperature by ~6 °C would increase mortality.

Combined effects of temperature and salinity

Salinity significantly affected survival, while temperature significantly affected the rate of development of larval *M. dalli*, without any interacting effects in the ranges tested. Unlike in the current study, interacting effects of temperature and salinity on survival have been found in other species such as *P. semisulcatus* (Jackson and Burford 2003) and *L. vannamei* (Ponce-Palafox et al. 1997). However, the results of the current study are consistent with these previous studies when similar temperatures and salinities were compared. In addition, any variation in salinity from 35‰ had an apparent negative effect on survival, with a 5‰ decrease having a greater effect than a 5‰ increase; noting that the increase in salinity to 40‰ did not significantly affect survival from that obtained at 35‰. This compares well with the effects of salinity on larval *Metapenaeus bennettae* (Preston 1985) and larval *Penaeus merguiensis* (Zacharia and Kakati 2004). However, these results contrast with those obtained for juvenile *L. vannamei* and larval *P. semisulcatus*, where although temperature was the primary factor in determining growth rates, salinities far lower than those used in the current study were also found to negatively impact on growth as well as survival (Ponce-Palafox et al. 1997; Kumlu et al. 2000). It may be the case that the energetic demands of osmoregulation under extreme conditions are greater than can be replenished by the larvae. It is only when the extreme ranges are tested that salinity and temperature appeared to interact, placing a combined burden on the energetic reserves of the larvae (Anger 2003).

Finally, the preferred temperatures (25.8 and 29.4 °C) and salinity (35‰) determined by this study appear to relate well to the natural spawning environments in the Swan-Canning Estuary, from which broodstock were obtained. Wild *M. dalli* are known to spawn in the middle and lower Swan-Canning Estuary during spring and summer, where the salinity in the river system is 35–38‰ and water temperature is 26–28 °C (Potter et al. 1986), indicating a close relationship between preferred larval hatchery conditions and those expected to influence larvae in the natural population. This relationship appears to be in line with findings on

M. bennettae, where the water temperatures found where broodstock were obtained, appeared to be the most suitable for larval culture in terms of survival (Preston 1985).

Conclusion

From this study, the most appropriate temperature and salinity to be used in the cultivation of *M. dalli* larvae were 25.8 °C and 35‰. It is also clear that any increase in temperature and/or variation in salinity may have a negative impact on survival. However, for short-term gains in growth rate, culture temperatures may be increased, providing they remain below 30 °C as this increase has a negligible effect on survival. Using this information, further research into the nutritional requirements of this species may now be conducted under ideal abiotic rearing conditions.

Acknowledgements Gratitude is expressed to Robert Michael, Bruno Pais and Lindsey Woolley at the Australian Centre for Applied Aquaculture Research (ACAAR), South Metropolitan TAFE, Fremantle, Western Australia for their assistance in the cultivation of microalgae, collection of broodstock and spawning of larvae in the laboratory.

Funding Funding for this project was supplied by the Recreational Fishing Initiatives Fund, a fund supported by Recfishwest and the Department of Fisheries, Western Australia; Fisheries Research and Development Corporation, Australia; Department of Parks and Wildlife, Western Australia; Australian Centre for Applied Aquaculture Research, Western Australia and Murdoch University, Western Australia.

References

Aktas M, Cavdar N (2012) The combined effects of salinity and temperature on the egg hatching rate, incubation time, and survival until protozoal stages of *Metapenaeus monoceros* (Fabricius) (Decapoda: Penaeidae). Turk J Zool 37:249–253

Anger K (2003) Salinity as a key parameter in the larval biology of decapod crustaceans. Inverteb Reprod Dev 43(1):29–45

ASEAN (1978) Manual on pond culture of penaeid shrimp. In: Philippines, A. N. C. A. O. T. (eds) Ministry of Foreign Affairs, Manila

Broadley AD, Tweedley JR, Loneragan NR (2017) Estimating biological parameters for penaeid restocking in an Australian temperate estuary. Fish Res 186(2):488–501

Ch SB, Shailender M (2013) Effect of salinity and temperature on larval growth and survival of black tiger shrimp *Penaeus Monodon* (Fabricius) in laboratory conditions. Int J Bio-Pharm Res 2:72–77

Clarke KR, Gorley RN, Somerfield PJ, Warwick RM (2014) Change in marine communities: an approach to statistical analysis and interpretation, 3rd edn. PRIMER-E, Plymouth, p 260

Crisp JA, Tweedley JR, D'Souza FML, Partridge GJ, Moheimani NR (2016) Larval development of the western school prawn *Metapenaeus dalli* Racek, 1957 (Crustacea: Decapoda: Penaeidae) reared in the laboratory. J Nat Hist 50(27–28):1699–1724

Crisp JA, D'Souza FML, Tweedley JR, Partridge GJ, Moheimani NR (2017) Quantitative determination of ovarian development in penaeid prawns. J Crustac Biol. doi:10.1093/jcbiol/ruw001

Dall W, Hill BJ, Rothlisberg PC, Staples DJ (1990) The biology of Penaeidae. Adv Mar Biol 27:1–489

D'Souza FML, Kelly GJ (2000) Effects of a diet of a nitrogen-limited alga (*Tetraselmis suecica*) on growth, survival and biochemical composition of tiger prawn (*Penaeus semisulcatus*) larvae. Aquaculture 181:311–329

D'Souza FML, Loneragan NR (1999) Effects of monospecific and mixed-algae diets on survival, development and fatty acid composition of penaeid prawn (*Penaeus* spp.) larvae. Mar Biol 133:621–633

D'Souza FML, Lecossois D, Heasman MP, Diemar JA, Jackson CJ, Pendrey RC (2000) Evaluation of centrifuged microalgae concentrates as diets for *Penaeus monodon* Fabricius larvae. Aquacult Res 31:661–670

Field A (2009) Discovering statistics using SPSS, 3rd edn. SAGE Publications Inc, London, pp 153–160

Gaxiola G, Gallardo P, Simões N, Cuzon G (2010) A red shrimp, *Farfantepenaeus brasiliensis* (Latreille, 1817), larvae feeding regime based on live food. J World Aquacult Soc 41(3):402–410

Gendron L, Ouellet P (2009) Egg development trajectories of early and late-spawner lobsters (*Homarus americanus*) in the Magdalen Islands, Québec. J Crustac Biol 29:356–363

Jackson CJ, Burford MA (2003) The effects of temperature and salinity on growth and survival of larval shrimp *Penaeus semisulcatus* (Decapoda: Penaeoidea). J Crustac Biol 23:819–826

Kinne O (1963) The effects of temperature and salinity on marine and brackish water animals: I. Temperature. Oceanogr Mar Biol Ann Rev 1:301–340

Kinne O (1964) The effects of temperature and salinity on marine and brackish water animals: II. Salinity and temperature–salinity combinations. Oceanogr Mar Biol Ann Rev 2:281–339

Kittaka J, Kudo R, Onoda S, Kanemaru K, Mercer JP (2001) Larval culture of the European spiny lobster *Palinurus elephas*. Mar Freshw Res 52:1439–1444

Kittaka J, Stevens B, Teshima S, Ishikawa M (2002) Larval culture of the king crabs *Paralithodes camtschaticus* and *P. brevipes*. Crabs in Cold Water Regions: Biology, Management, and Economics. Anchorage, pp 189–209

Kumlu M, Eroldogan O, Aktas M (2000) Effects of temperature and salinity on larval growth, survival and development of *Penaeus semisulcatus*. Aquaculture 188:167–173

Kumlu M, Eroldogan O, Aktas M, Sağlamtimur B (2001) Larval growth, survival and development of *Metapenaeus monoceros* (Fabricius) cultured in different salinities. Aquacult Res 32:81–86

Maher J (2002) Where have all the swan-canning prawns gone. West Fish 2002:20–21

Motoh H (1985) Biology and ecology of *Penaeus monodon*. In: Proceedings of the first international conference on the culture of penaeid prawns/shrimps, 4–7 December 1984, Iloilo City, Philippines. Aquaculture Department, Southeast Asian Fisheries Development Center, pp 27–36

Parado-Estepa FD (1998) Survival of *penaeus monodon* post larvae and juveniles at different salinity and temperature levels. Isr J Aquacult Bamidgeh 50:174–183

Ponce-Palafox J, Martinez-Palacois CA, Ross LG (1997) The effects of salinity and temperature on the growth and survival rates of juvenile white shrimp, *Penaeus vannamei*, Boone, 1931. Aquaculture 157:107–115

Potter IC, Penn JW, Brooker KS (1986) Life cycle of the western school prawn, *Metapenaeus dalli* Racek, in a Western Australian estuary. Aust J Mar Freshw Res 37:95–103

Potter IC, Baronie FM, Manning RJG, Loneragan NR (1989) Reproductive biology and growth of the western school prawn, *Metapenaeus dalli*, in a large Western Australian estuary. Aust J Mar Freshw Res 40:327–340

Preston N (1985) The combined effects of temperature and salinity on hatching success and the survival, growth, and development of the larval stages of *Metapenaeus bennettae* (Racek and Dall). J Exp Mar Biol Ecol 85:57–74

Roberts SD, Dixon CD, Anderacchio L (2012) Temperature dependent larval duration and survival of the western king prawn, *Penaeus* (Melicertus) *latisulcatus* Kishinouye, from Spencer Gulf, South Australia. J Exp Mar Biol Ecol 411:14–22

Ryther JH, Guillard RR (1962) Studies of marine planktonic diatoms: III. Some effects of temperature on respiration of five species. Can J Microbiol 8(4):447–453

Smith K, Lenanton R, Valesini F (2007) Preliminary survey to determine the abundance of river prawns (*Metapenaeus dalli*) in the Swan-Canning Estuary. Report to the Swan River Trust, Perth, p 11

Stevens BG (1990) Temperature-dependent growth of juvenile red king crab (*Paralithodes camtschatica*) and its effects on size-at-age and subsequent recruitment in the eastern Bering Sea. Can J Fish Aquat Sci 47:1307–1317

Tuma D (1967) A description of the development of primary and secondary sexual characters in the banana prawn, *Penaeus merguiensis* de Man (Crustacea: Decapoda: Penaeidae). Mar Freshw Res 18:73–88

Villegas CT, Kanazawa A (1979) Relationship between diet composition and growth rate of the zoeal and Mysis stages of *Penaeus japonicus* Bate. Fish Res J Philippines 4:32–40

Zacharia S, Kakati V (2004) Optimal salinity and temperature for early developmental stages of *Penaeus merguiensis* de Man. Aquaculture 232:373–382

13

Saccate thallus of the red alga *Halosaccion glandiforme* harbor diverse invertebrate fauna

Ritindra N. Bhaduri · Maya Wolf

Abstract Rocky intertidal zones are biologically diverse environments with numerous physical stressors; as such, close associations between organisms often develop to overcome such stressors and enhance survival of associates. As major components of rocky shores, macroalgae support numerous invertebrate species. In this study, we evaluated the habitat-providing role of the red alga, *Halosaccion glandiforme* (S.G. Gmelin) Ruprecht. This alga, also called sea sacs, is commonly found on rocky shores along the West Coast of North America. During emersion, this seaweed holds water within its saccate thallus, which can potentially serve as microhabitat for various organisms. Little is known about the composition of microfauna associated with this seaweed; as such, we documented richness and abundance of species found inside its thallus. Algal specimens were collected from Charleston, Oregon and transported to the laboratory for further analyses. Of the 119 thalli examined, 12 taxa were documented. Nematodes and copepods were the dominant taxa. Other groups represented included ostracods, turbellarians, halacarid mites, bivalves, rotifers, and three larval types (barnacle cyprids, copepod nauplii, mollusc veliger). Diatoms, crustacean molt remnants, detritus, and sand particles were also observed inside thalli. Species richness and abundance were positively correlated with thalli volume, as well as intra-thalli fluid volume. Several feeding and other behaviors of colonizers were noted; they included herbivory, predator–prey interactions, detritivory, and molting. Our findings demonstrate that *H. glandiforme* provide refuge for organisms from harsh environmental conditions during emersion periods, and also serve as feeding and nursery grounds for its diverse invertebrate fauna.

Keywords Sea sacs · Microhabitat · Invertebrate composition · Rocky intertidal community

Introduction

Rocky shores are dynamic environments inhabited by organisms that are able to withstand harsh abiotic and biotic conditions. Some of these organisms, especially macroalgae, provide habitats for many benthic species. These species depend on the physical structure and other resources provided by the seaweeds, thereby enhancing their survival and increasing diversity and abundance (Lilley and Schiel 2006; Wikström and Kautsky 2007).

R. N. Bhaduri (✉)
Department of Biological Sciences, California State University Stanislaus, One University Circle, Turlock, CA 95382, USA
e-mail: rbhaduri@csustan.edu

M. Wolf
Oregon Institute of Marine Biology, University of Oregon, Charleston, OR 97420, USA
e-mail: mwolf1@uoregon.edu

Size and morphological features of seaweeds can have a major bearing on the benefits they provide to other species. Numerous invertebrates are dependent on seaweeds with varying morphological features, including the complex three-dimensional *Macrocystis* (Graham 2004), the compact spongy *Codium* (Trowbridge 1998; Bulleri et al. 2006), the fan-like broad thallus of *Chondrus* (Janiak and Whitlatch 2012), the crustose coralline algae *Clathromorphum* (Chenelot et al. 2011) and the calcareous rhodoliths (Foster 2001). Steller et al. (2003) reported that the total number of organisms supported by the coralline red algal rhodoliths significantly increased with both complexity (branching density) and space available (thallus volume) when compared with adjacent sand community. Thus, thallus size/volume can be a factor determining species richness and abundance in associated organisms.

Another morphological trait found in some algae is the presence of saccate or sac-like thallus that retains sea water when exposed to air during low tide. The saccate red alga *Halosaccion glandiforme* grows in clumps in the rocky intertidal areas along the Pacific coast of North America, ranging from the northern Aleutian Isles to southern California (Mondragon 2003). The plant has a short stipe, a discoid holdfast, and thalli measuring 3–4 cm in diameter; it grows to a length of 10–20 cm over the course of a year (Johnson 1994). This seaweed is commonly referred to as the sea sac because its thallus is a cylindrical hollow sac, partially filled with sea water and about a cubic centimeter of gas. Each thallus possesses 5–15 irregularly shaped pores, 10–200 μm in diameter, which allow water into and out of the thallus during exposure and immersion associated with tidal cycles (DePamphilis 1978; Vogel and Loudon 1985). When tides recede, water inside the thallus is retained. Such water reserves reduce desiccation and heating of tissue during periods of emersion (DePamphilis 1978). It has been reported in some field guides that the water-filled sacs of *H. glandiforme* are inhabited by a variety of invertebrates (Mondragon 2003; Lamb and Hanby 2005).

The role of this "sea sac" as a microhabitat for assemblages of small invertebrates has been virtually unexplored. In this study, for the first time, we note the presence of fauna found inside the saccate thallus of *H. glandiforme*. The objectives of this study were to: (1) quantify species richness and abundance (2) determine relationship between species richness/abundance and thallus/seawater volume using multiple regression, and (3) observe miscellaneous ecological interactions between taxa.

Materials and methods

Fronds of *H. glandiforme* were collected from Cape Arago State Park in Charleston, Oregon (43.3062°N; 124.3935°W), USA, between July and September 2015, during low tide. During sampling, the short stipe was gently snipped using a scalpel while making sure the water-filled thalli were intact (Fig. 1). The seaweeds

Fig. 1 Saccate thallus of *Halosaccion glandiforme*, tapering to a short stipe and attached to rocks with small discoidal holdfast. Note the oblong sac partially filled with seawater

were transferred to plastic containers and transported to the laboratory at the Oregon Institute of Marine Biology for further analysis. Any damaged or dehydrated thalli were discarded and not analyzed in our study. All analyses were performed within 8 h of collection, which ensured quantifying and observing live fauna.

Using dissecting scissors, each thallus was carefully removed from the stipe. Assuming that each water-filled thallus was roughly columnar in shape, we measured thallus volume using the formula $V = \pi r^2 h$, where h was the length and r was the width at the widest point on the thallus. The thin seawater-filled sac was held vertically and the tip of the thallus was cut with a scissor. To measure seawater volume, the internal water was withdrawn using a calibrated 5 cc wide mouth syringe. The content of the syringe was transferred into a watch glass and examined for fauna under a binocular stereo microscope (Olympus 10X). Most specimens were identified as close to their taxonomic level as possible using available taxonomic keys (Light and Carlton 2007) and their abundance noted. 10 min were spent observing interactions (if any) between the colonizers for each thallus. Other recognizable intra-thallus components (e.g., phytoplankton, detritus, molt remnants, etc.) were also noted for documenting additional ecological associations.

Multiple regression analyses were performed to determine the relationship between the richness and abundance of the invertebrate fauna and the size of the sea sacs, which included thallus size and fluid volume within each sac. A Kolmogorov–Smirnov test showed the data were not normally distributed, and further transformation did not normalize the data. However, linear regression/correlation might still be robust to non-normality (McDonald 2014), so analyses were performed despite this violation. P values < 0.05 were considered statistically significant.

Results

We examined a total of 119 thalli of *H. glandiforme*. Table 1 lists the different types of invertebrates and their numbers associated with this alga. Of the 119 thalli inspected, 99 (83%) were colonized by 845 individuals

Table 1 Richness, abundance (A), % occurrence (O), mean occurrence (M.O.), range of invertebrate taxa, as well as copepod molts, within the saccate thallus of the red alga *Halosaccion glandiforme*

Invertebrate taxa	A	O (%)	M. O.	Range
Nematoda	598	52.9	9.5	1–107
Order Chromadorida, Fam. Chromadoridae				
O. Monhysterida, Fam. Xyalidae, *Theristus* sp.				
Platyhelminthes				
Class Turbellaria, *Neoplana* sp.	3	2.5	1	1
Arthropoda				
Crustacea				
Acartia sp., *Calanus* sp. (Copepoda: Calanoida)	183	29.4	5.2	1–65
Copepod molt	373	47.9	6.5	1–62
Paradoxostoma sp. (Ostracoda)	4	3.4	1	1
Mystacocaridada (Class Maxillopoda)	3	2.5	1	1
Nauplius larvae	45	17.8	2.8	1–14
Cyprid larvae	4	3.4	1	1
Arachnida	3	2.5	1	1
Marine mites (Fam. Halacaridae)				
Mollusca				
Veliger larva	1			
Unidentified bivalve	1			
Rotifera	nq	18.4		
Foraminifera	1			

Percentages do not add to 100, as occurrence refers to presence of each taxa inside thallus divided by total number of thalli examined ($N = 119$)

nq not quantified

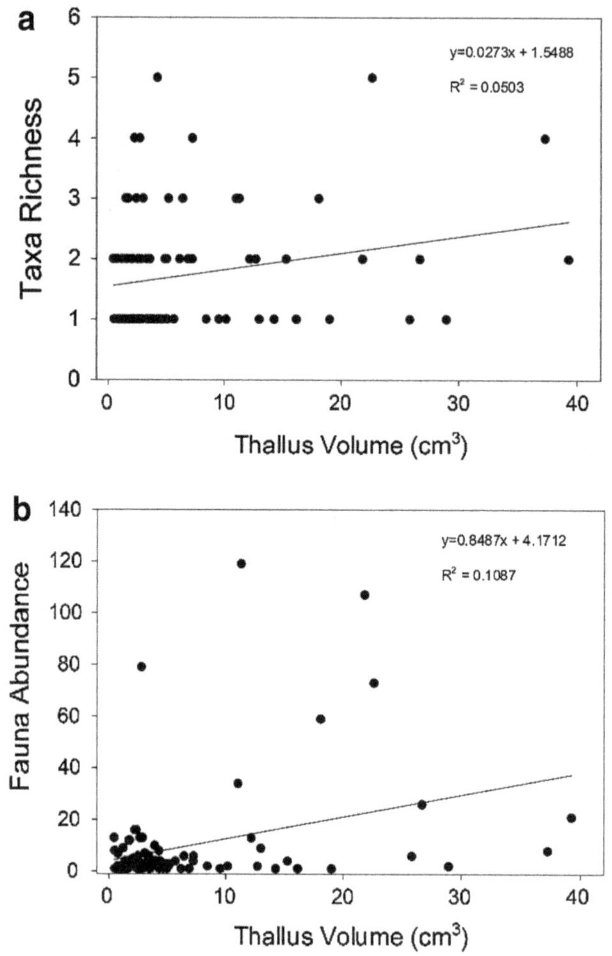

Fig. 2 Linear regression analysis of effect of thallus volume on **a** taxa richness, and **b** fauna abundance

belonging to 12 taxa in total. Up to 5 taxa and 119 individual invertebrates were found inside one sac. Table 1 also lists the percentage occurrence, mean occurrence, and range of all taxa associated with *H. glandiforme*.

Nematoda and Arthropoda were the dominant phyla; they made up over 92% of the total fauna. Nematodes were the most common inhabitants, occurring in 53% of thalli and constituting 70.8% of the total samples. Although the average number of nematodes per thallus was ~10, as many as 107 individuals were found in a single thallus. Copepods were the second most abundant group, occurring in 30% of thalli and comprising of 21.7% of all species. Nematodes and copepods were found to coexist in 26.1% of the samples. Copepods were occasionally found in various states of molting. The other lesser represented groups included ostracods, turbellarians, mites, and bivalves. Numerous copepod nauplii, a few barnacle cyprids, veliger larvae, rotifers, protozoans, diatoms, crustacean molt remnants, detritus, and sand particles were also found inside thalli.

Results of multiple regression analyses were separated in two groups: (a) thallus volume vs. taxa richness and abundance, and (b) fluid volume vs. taxa richness and abundance. The thallus volume and the intra-thallus fluid ranged from 0.42 to 39.27 cm^3 (mean 6.02) and 0.1–8.6 mL (mean 1.10), respectively. Overall, larger thallus volume and greater intra-thallus fluid volume supported higher taxa richness and species abundance. Taxa richness was positively correlated with thallus volume ($F = 4.66$, $P = 0.03$, $R^2 = 0.05$; Fig. 2a) as was species abundance ($F = 10.48$, $P = 0.002$, $R^2 = 0.11$; Fig. 2b). Furthermore, we found positive correlation between fluid volume and taxa richness ($F = 4.25$, $P = 0.04$, $R^2 = 0.05$; Fig. 3a) as well as fauna abundance ($F = 8.75$, $P = 0.004$, $R^2 = 0.11$; Fig. 3b).

A number of feeding interactions between colonizers were noted; they included herbivory, predator–prey interactions, and detritivory. Most of the invertebrates (e.g., copepods, ostracods, nematodes, mystacocaridads, rotifers) fed on diatoms and periphyton, some ingested other invertebrates (e.g., flatworm, nematodes,

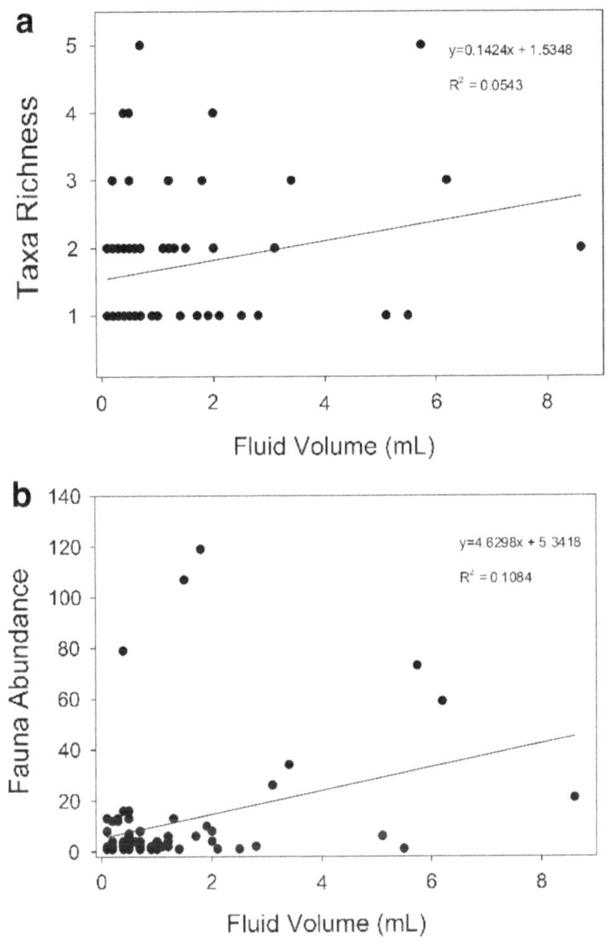

Fig. 3 Regression analysis of effect of fluid volume on **a** taxa richness **b** fauna abundance

copepods, mites), some consumed detritus (ostracods, bivalve, nematodes), and entrapped sediments (mostly nematodes). Nematodes were also found feeding on bacterial films on the copepod molts, which were common in our samples (~48% occurrence, with up to 201 molts in a single thallus). As crustacean molts were common and nematodes are known detritivores (Heip et al. 1985, and references within), it was unsurprising to find a significant positive relationship between the presence of nematodes and molt remains inside the thallus ($F = 56.77$, $P < 0.0001$, $R^2 = 0.40$; Fig. 4).

Discussion

The importance of seaweeds as habitats has been well studied for many rocky intertidal species; however, that has not been the case for the red alga *H. glandiforme*. Its fronds are annuals; new thalli appear in late winter and early spring, but degenerate after spore production during the following fall and winter (Johnson 1994). With pores as large as 200 μm, microfauna measuring <200 μm most likely enter the fronds via seawater medium and may leave the hollow sacs either during immersion or when the thalli degenerate at the end of the growing season. In this novel algal- invertebrate association study, we characterize this seaweed as a habitat-forming species as it shelters multiple invertebrate groups inside its saccate thallus, and provides multiple benefits for its diverse invertebrate fauna, which in turn, may increase their survivorship and fitness.

Macroalgal traits such as size and morphological complexity can have a large effect on local populations and communities. The presence of multiple pores on the thallus of *H. glandiforme* allows seawater into and out of thalli during tidal cycles and facilitate faunal colonization. Our results show that thallus volume and the

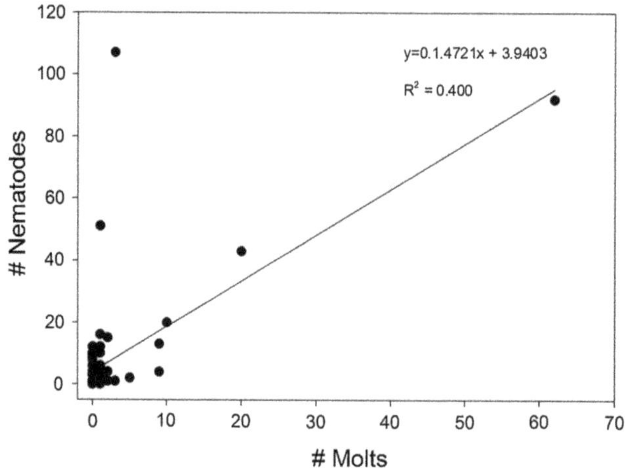

Fig. 4 Relationship between nematode abundance and presence of copepod molts

volume of seawater were important predictors of abundance and richness of invertebrates associated with this alga. Larger thalli possibly contained more resources, including sea water, which allowed them to shelter higher numbers of taxa and greater abundance of organisms. It is likely that bigger thalli have larger pores, which facilitate greater colonization opportunities. These thalli made the coexistence of species with different ecological necessities possible. Other studies have associated algal morphologies with species diversity and abundance. When given a choice between *Gracilaria vermiculophylla* and *Ulva rigida*, the architecturally more complex *G. vermiculophylla* supported higher species richness and diversity of associated macrofauna (Munari et al. 2015). Steller et al. (2003) also reported that the total organisms supported by rhodoliths, another group of benthic red alga, significantly increased with branching density and thallus volume.

We found nematodes to be the most abundant group in our samples. This is not surprising since they are considered to be among the most abundant organisms on earth. Marine roundworms inhabiting high-energy rocky shores are exposed regularly to changing tidal conditions. Being poor swimmers, they often remain suspended in the water column before re-entering the benthos as passive particles when tides recede; yet, nematodes are efficient in choosing algal habitats (Ullberg and Ólafsson 2003). Nematodes in our algal samples may have entered the saccate thallus via passive or active mechanisms.

The majority of the nematodes we found belonged to the family Chromadoridae (pers. comm. Ashleigh Smythe). In a related study at a Brazilian rocky coast, nematode biodiversity was related to the structural features of macrophyte habitats and the most abundant phytal nematodes were chromadorids (Da Rocha et al. 2006). Chromadorids are often associated with marine macrophytes (Trotter and Webster 1983; Pérez-García et al. 2015). Feeding experiments of the marine nematode *Chromadora* showed that diatoms and chlorophytes were preferred food items (Tietjen and Lee 1973). However, Moens and Vincx (1997) reported that many aquatic nematodes are in fact opportunistic feeders, which may change feeding strategies in response to available food. We also encountered nematodes belonging to the genus *Theristus* in our samples. This free-living marine nematode group has been found in diverse habitats, including degraded algal beds (Olafsson et al. 2013) and exposed sandy beaches (Lee and Riveros 2012). *Theristus* is known to ingest suitably sized food particles like microalgae cells (Moens and Vincx 1997). Diatoms have been shown to be a prominent food resource for *Theristus* (Moens et al. 2014). Diatoms were abundant in our samples and nematodes were frequently observed ingesting them as well as other items inside the thalli.

We found a positive association between the presences of nematodes and remnants of copepod molts. Nematodes were observed feeding on the distinct bacterial films covering these molts. As substrate consumers, nematodes are known to ingest organic substrates along with their associated microflora and microfauna (Moens et al. 2006). Tietjen and Lee (1973) reported that *Chromadora macrolaimoides*, isolated from the collections present on the green alga *Ulva intestinalis*, consumed different types of bacteria.

Copepods were the second most abundant group in our samples. This too is not unexpected as they are considered to be one of the most abundant groups in the marine realm. *Halosaccion glandiforme* belongs to class Rhodophyceae and, according to Fahrenback (1962), rhodophyceans show higher assemblage of copepods than other algal groups. Although the exact reasons why red algae are preferred over brown algae are unknown, Fahrenback (1962) postulates that some red algal fronds are thick enough to accommodate these crustaceans and prevent their dislodgement. Whereas Fahrenback (1962) observed one harpacticoid copepod species (*Diathrodes cystoecus*) in sea sacs collected from Moss Beach, California, we found two genera of planktonic calanoids, *Acartia* sp. and *Calanus* sp. within the algal thalli. They were frequently observed to feed on diatoms and rotifers, which were plentiful in our samples. As omnivores, they were found to graze both on phytoplankton (Paffenhöfer and Lewis 1989) and protozooplankton (ciliates, heterotrophic dinoflagellates, flagellates), and rotifers (Stoecker and Egloff 1987).

In addition to serving as feeding grounds, *H. glandiforme* may also serve as a nursery habitats as gravid copepods and various larval groups were frequently observed. Copepods were found in various states of molting, implying their growth and development inside thalli. The presence of other associated taxa suggested that sea sacs may act as a refuge for multiple species, especially during periods of emersion. According to Lilley and Schiel (2006), habitat-forming species increases spatial complexity and help ameliorate stressful environmental conditions, thereby promoting a diversified assemblage of animals. It is obvious that the use of *H. glandiforme* as a habitat for invertebrates is multifaceted, as it provides refuge from physical stresses and predation, facilitates trophic interactions, and serves as nursery ground. As such, this macroalgae should be addressed in studies dealing with coastal biodiversity conservation.

Acknowledgements We are grateful to A. Smythe for identification of a few taxa. We thank A. Bhaduri for assistance in the field and two anonymous reviewers for their helpful comments on the manuscript. Laboratory space was provided by the Oregon Institute of Marine Biology in Charleston, Oregon. This work was funded by California State University Stanislaus Biology Research Committee and Faculty Development Mini-Grant Awards. The primary author acknowledges the support of A. Kohlhaas, K. Schoenly, M. Gerson, M. Fleming, P. Kelly, and C. Jantz on this project.

References

Bulleri F, Airoldi L, Branca GM, Abbiati M (2006) Positive effects of the introduced green alga, *Codium fragile* ssp. *tomentosoides*, on recruitment and survival of mussels. Mar Biol 148:1213–1220

Chenelot H, Jewett SC, Hoberg MK (2011) Macrobenthos of the nearshore Aleutian Archipelago, with emphasis on invertebrates associated with *Clathromorphum nereostratum* (Rhodophyta, Corallinaceae). Mar Biodiv 41:413–424

Da Rocha CMC, Venekey V, Bezerra TNC, Souza JRB (2006) Phytal marine nematode assemblages and their relation with the macrophytes structural complexity in a Brazilian tropical rocky beach. Hydrobiologia 553:219–230

DePamphilis C (1978) The resistance to damage due to desiccation in the intertidal alga *Halosaccion glandiforme* (Gmelin) Ruprecht. MS., Library of Friday Harbor Laboratories, Friday Harbor

Fahrenback WH (1962) The biology of a harpacticoid copepod. La Cellule 62:301–376

Foster M (2001) Rhodoliths: between Rocks and Soft Places. J Phycol 37:659–667

Graham MH (2004) Effects of local deforestation on the diversity and structure of southern California giant kelp forest food webs. Ecosystems 7:341–357

Heip CHR, Vincx M, Vranken G (1985) The ecology of marine nematodes. Oceanogr Mar Biol Ann Rev 23:399–489

Janiak DS, Whitlatch RB (2012) Epifaunal and algal assemblages associated with the native *Chondrus crispus* (Stackhouse) and the non-native *Grateloupia turuturu* (Yamada) in eastern Long Island Sound. J Exp Mar Biol Ecol 413:38–44

Johnson LE (1994) Enhanced settlement on microtopographical high points by the intertidal red alga *Halosaccion glandiforme*. Limnol Oceanogr 39:1893–1902

Lamb A, Hanby BP (2005) Marine life of the Pacific Northwest. Harbour Pub, Madeira Park, BC

Lee MR, Riveros M (2012) Latitudinal trends in the species richness of free-living marine nematode assemblages from exposed sandy beaches along the coast of Chile (18–42°S). Mar Ecol 33:317–325

Light SF, Carlton JT (2007) The Light and Smith manual: intertidal invertebrates from central California to Oregon. Univ of California Press, Berkeley

Lilley SA, Schiel DR (2006) Community effects following the deletion of a habitat-forming alga from rocky marine shores. Oecologia 148:672–681

McDonald JH (2014) Handbook of biological statistics, 3rd edn. Sparly House Publishing, Baltimore

Moens T, Vincx M (1997) Observations on the feeding ecology of estuarine nematodes. J Mar Biol Assoc UK 77:211–227

Moens T, Traunspurger W, Bergtold M (2006) Feeding ecology of free-living benthic nematodes. Freshwater nematodes: ecology and taxonomy. CAB International publishing, Cambridge, pp 105–131

Moens T, Vafeiadou AM, De Geyter E, Vanormelingen P, Sabbe K, De Troch M (2014) Diatom feeding across trophic guilds in tidal flat nematodes, and the importance of diatom cell size. J Sea Res 92:125–133

Mondragon J (2003) Seaweeds of the Pacific Coast: common marine algae from Alaska to Baja California. Sea Challengers, Monterey, p 97

Munari C, Bocchi N, Mistri M (2015) Epifauna associated to the introduced *Gracilaria vermiculophylla* (Rhodophyta; Florideophyceae: Gracilariales) and comparison with the native *Ulva rigida* (Chlorophyta; Ulvophyceae: Ulvales) in an Adriatic lagoon. Ital J Zool 82:436–445

Olafsson E, Aarnio K, Bonsdorff E, Arroyo NL (2013) Fauna of the green alga *Cladophora glomerata* in the Baltic Sea: density, diversity, and algal decomposition stage. Mar Biol 160:2353–2362

Paffenhöfer G, Lewis KD (1989) Feeding behavior of nauplii of the genus *Eucalanus* (Copepoda, Calanoida). Mar Ecol Prog Ser 57:129–136

Pérez-García JA, Ruiz-Abierno A, Armenteros M (2015) Does morphology of host marine macroalgae drive the ecological structure of epiphytic meiofauna? J Mar Biol Oceanogr 4:1–7

Steller DL, Riosmena-Rodriguez R, Foster MS, Roberts CA (2003) Rhodolith bed diversity in the Gulf of California: the importance of rhodolith structure and consequences of disturbance. Aquat Conserv 13:5–20

Stoecker DK, Egloff DA (1987) Predation by *Acartia tonsa* Dana on planktonic ciliates and rotifers. J Exp Mar Biol Ecol 110:53–68

Tietjen JH, Lee JJ (1973) Life history and feeding habits of the marine nematode, *Chromadora macrolaimoides* Steiner. Oecologia 12:303–314

Trotter D, Webster JM (1983) Distribution and abundance of marine nematodes on the kelp *Macrocystis integrifolia*. Mar Biol 78:39–43

Trowbridge CD (1998) Stenophagous, herbivorous sea slugs attack desiccation-prone, green algal hosts (*Codium* spp.): indirect evidence of prey-stress models (PSMs)? J Exp Mar Biol Ecol 230:31–53

Ullberg J, Ólafsson E (2003) Free-living marine nematodes actively choose habitat when descending from the water column. Mar Ecol Prog Ser 260:141–149

Vogel S, Loudon C (1985) Fluid mechanics of the thallus of an intertidal red alga, *Halosaccion glandiforme*. Biol Bull 168:161–174

Wikström SA, Kautsky L (2007) Structure and diversity of invertebrate communities in the presence and absence of canopy-forming *Fucus vesiculosus* in the Baltic Sea. Estuar Coast Shelf 72:168–176

14

An indoor pelagic mesocosm facility to simulate multiple water-column characteristics

Ulf Båmstedt · Henrik Larsson

Abstract Mesocosms are important research tools in aquatic ecology because they close the gap between laboratory studies at the individual or lower organization level and field studies at the population and ecosystem level. However, most mesocosm studies regarding the pelagic environment do not consider the effects of physical factors like water-column stratification, turbulence and mixing. Neglecting such factors might bias the results compared to the natural system. Using a unique indoor mesocosm facility, we present results on how different water-column stratifications can be made and how they act as barriers for exchange between water layers. Turbulent mixing, simulated by vertically rotating incubation vessels, is shown to be of high importance for primary production, generating up to nine times higher production in humus-rich water than incubation vessels at fixed depths. Convective stirring is shown to be an attractive method for generating different turbulence conditions, and different temperature settings can be used to get turnover times from 84 h or more down to 17 min for a 5-m water parcel. We also demonstrate how an anoxic bottom layer can be achieved by stimulating heterotrophic bacteria through addition of bioavailable organic carbon.

Keywords Aquatic mesocosms · Turbulent mixing · Primary production · Pycnocline strength · Light spectrum · Bacterial oxygen depletion · Thermal convection

Introduction

Mesocosm-based research has become an increasingly important part of aquatic ecology by closing the gap between laboratory studies at the species, individual or lower organisation level and field studies at the population or ecosystem level (Stewart et al. 2013). Field studies in the marine environment often require large research ships with daily costs of 10,000€ or more and studies on natural systems, especially pelagic ones, also include much short-term variability. Thus, high cost of marine field studies combined with the inherent natural spatial and temporal variability heavily constrains the possibility of achieving statistically well-founded results solely based on field studies. Alternately, laboratory studies at the individual level can, for example, precisely define physiological responses to both abiotic (light, temperature, salinity, etc.) and biotic variables (food concentration and quality, crowding, gender, age, etc.). However, the complexity in the natural environment is not fully reflected in the relatively small volume of water used in laboratory experiments, and the time scales used are usually short. For example, Miko et al. (2015) compared results from ecotoxity tests on tadpoles in the lab and in outdoor mesocos which showed strikingly different results.

U. Båmstedt (✉) · H. Larsson
Umeå Marine Sciences Centre, University of Umeå, Norrbyn, 905 71 Hörnefors, Sweden
e-mail: ulf.bamstedt@umu.se

However, due to the reduction in complexity also present in a mesocosm, extrapolation of results from mesocosms to the natural environment should be done with caution. Spivak et al. (2011) concluded that results from mesocosm-scale experiments on micro-algae probably are relevant to large-scale processes, such as eutrophication. On the other hand, Williams et al. (2002) found that freshwater pond mesocosms showed lower biodiversity than corresponding natural ponds. Mesocosms in a mesotrophic and a eutrophic lake showed the same water chemistry as the surrounding water, but most biological variables diverged from the surrounding water (Dzialowski et al. 2014). The tendency of diverging succession in mesocos and natural habitats has been repeatedly shown by Schindler and colleagues (see, e.g., Schindler 1998 for a review). Another problem is variability between replicate mesocosms. Caquet et al. (2001) showed that water temperature, pH, dissolved oxygen and suspended solids exhibited a very low variability between replicate mesocosms whereas nutrients displayed elevated variability. They also found that biological parameters at the individual level were less variable than those at the community level.

Table 1 represents an attempt to categorize and compare different types of mesocosms and also compare with open field studies, thus giving a rough overview of which type of basic tool should be the best choice for a given research question. Costs, working conditions, range of environmental manipulation and measurement possibilities and replication possibilities have been considered. To cover as much as possible of the volume and depth ranges, the natural environment is of course number one, and going from sea/lake-based to land-based and indoor mesocosms gradually reduces the possible range in volume and depth. Field studies are also best if natural light conditions and irradiance profiles are of primary importance, although sea/lake-based mesocosms can provide the same light conditions as in nature; they have a practical limit in depth and thereby also a more restricted range in the vertical irradiance profile.

Sea-based mesocosms might yield more potential to use many replicates than land and indoor facilities, and therefore a better statistical evaluation of the results. Sea/lake-based mesocosms are often inexpensive, at least compared to the indoor variant, and provide better opportunities for a natural light environment. The negative characteristics of sea/lake-based mesocosms compared to land and indoor facilities are their sensitivity to weather conditions and very restricted potential to control light, temperature and turbulent mixing, whereas the ability to adjust nutrients, pH and food-web structure is approximately the same as for the land-based and indoor counterparts.

Table 1 Relative suitability of the basic research tool for doing ecological studies in pelagic environments

Function	A	B	C	D
Choice of volume	*****	***	**	*
Choice of depth range	*****	****	**	*
Sensitivity to weather conditions	***	*	****	*****
Access and working conditions	***	**	****	*****
Statistically based investigations	**	*****	****	****
Sensor-based online monitoring	*	**	*****	*****
Cost of a comparable research study	**	**	****	*****
Use of natural light conditions	*****	*****	****	***
Adjustment of light conditions	–	*	***	*****
Range in vertical irradiance profile	*****	***	****	****
Adjustment of nutrient conditions	*	****	*****	*****
Adjustment of temperature conditions	–	*	***	*****
Adjustment of salinity conditions	–	**	*****	*****
Adjustment of pH conditions	–	****	*****	*****
Control of turbulent mixing	–	*	***	*****
Density stratification of water parcel	*	**	***	*****
Manipulation of food-web structure	*	*****	*****	*****

A = Ship-based studies in the field; B = Sea-based mesocosms; C = Land-based mesocosms; D = Indoor mesocosms. The relative suitability scale goes from –, meaning not possible to *****, meaning highly suitable

Land-based and indoor mesocosms have a number of advantages, such as easy access, good working conditions, very low sensitivity to weather conditions, best possibilities for sensor-based online monitoring and especially that indoor mesocosms provide the best opportunity for manipulation of the physical environment. The number of land or indoor mesocosm units is usually restricted due to restricted space, and the number of replicate treatments and treatment levels will, therefore, also be restricted. Cost is an important factor in the choice of research tool, and land-based and indoor mesocosms often have a high initial construction cost but require limited operation budgets thereafter. A ship-based field study has full costs every time, and sea/lake-based mesocosm facilities often require frequent repair and renewal of parts.

A range of physical, chemical and biological shortcomings are obvious when trying to extrapolate mesocosm results to natural ecosystems (Schindler 1998). An enclosure of a water mass will inevitably differ in some conditions from the natural ones, and one factor that has received attention is the importance of dimensional relationships of the actual enclosure. Mesocosms with the same volume but different dimensional properties will often generate different results (e.g., Petersen et al. 1999; Petersen and Englund 2005). This might be explained by, for example, a changed surface/volume ratio or changes in the light environment and in air–water exchange processes. An obvious effect of an increased surface to volume ratio is a potential for build-up of periphyton biomass on the inside walls, which will act as an increasingly important factor for the flow of energy and material in the mesocosm during the time span of a study.

Most ecological studies using mesocosms lack information on physical characteristics of the experimental environment. Factors like mixing, turbulence, light spectrum and exchange rate through discontinuity layers are very important for biological processes and, if different from the natural environment, may cause biased results as compared to the natural environment. The natural environment in the sea and lakes is usually also stratified during most of the year, although thermoclines or pycnoclines are seldom part of a mesocosm-based ecological study. In this paper, we describe a unique mesocosm facility where the physical characteristics can be controlled and manipulated and give some examples of biological consequences of different physical conditions.

Materials and methods

For our functional tests and experimental evaluations, we used the indoor mesocosm facility at Umeå Marine Sciences Centre, University of Umeå, Sweden, situated at the northern Bothnian Sea (N63°34′; E19°50′) in the Baltic Sea.

The experimental facility

The facility is located in a 6-m-high temperature-regulated hall with a second floor about 120 cm below the top of the mesocosms, with a laboratory in direct connection to this upper floor (Fig. 1a). It consists of 12 insulated black cylindrical polyethylene tanks, 5 m high with an inner diameter of 0.73 m. There are three 1.4-m-high vertical sections in each mesocosm where the temperature is monitored and regulated (Fig. 1a), separated by two 0.4-m-high sections. The wall of each 1.4-m unit is divided into six interconnected concentric sections with holes in the dividing lamellae on opposite sides of two neighbouring sections (Fig. 1b), and with an inlet for the temperature control medium at the bottom and an outlet at the top section. The temperature control medium (40% propylene glycol, Dowcal 20) is, thus, forced to circulate around the wall (Fig. 1b). The outer part of the wall comprises an air-filled section that insulates the entire mesocosm tank (Fig. 1b). The actual temperature-control system consists of two units for circulation of the glycol solution, one with cold (− 6 °C) and one with warm (30 °C) solution. Both are connected to each of the three vertical sections on each tank (Fig. 1b) through computer-controlled valves that direct either cold or warm glycol solution according to the desired temperature (Honeywell AB, Johanneshov, Sweden and Gate AB, Umeå, Sweden). In general, the maximum deviation from the desired temperature is within ± 0.5 °C. The two intermediate sections of 0.4 m in height can either be switched off or connected in parallel with an adjacent 1.4-m temperature control section.

The technical design described above makes it possible to set up stratified water columns with a thermocline at 1.5–2 m in depth, at 3.2–3.7 m in depth, or at both depths. The temperature control system can also

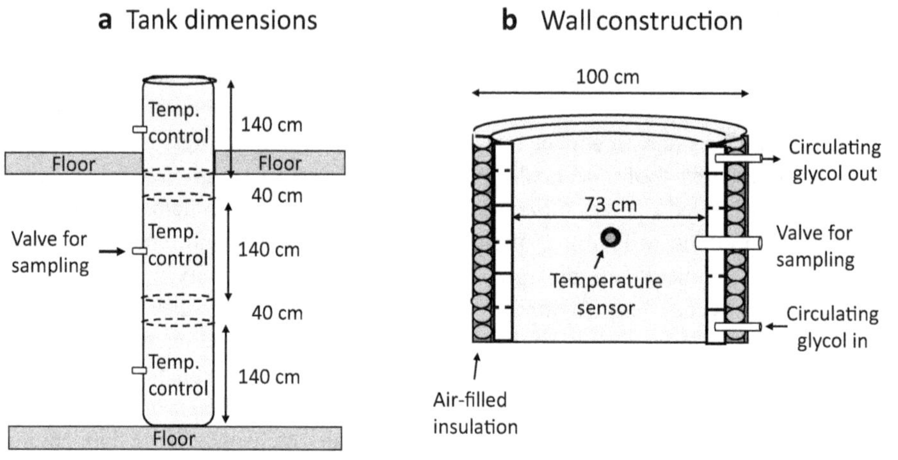

Fig. 1 Simplified view of a mesocosm tank. **a** Dimensions of the entire tank showing the positions of the three sections with temperature control and the two smaller sections between them. **b** Cross-view of the tank wall in one temperature-controlled section

be used to create convective stirring at different rates. A slight increase in the temperature settings in one or both of the two lower sections promotes convective stirring in the water column. It is also possible to maintain convective stirring above a cooler bottom layer by a similar slight offset in the temperature settings between the middle and upper sections. Faster convective stirring can also be maintained by maximum heating at the low section and maximum cooling at the top section, thus allowing the middle section to control the overall temperature in the water column. Further increase in pycnocline strength can be achieved by dissolving sea salt in the water. Each mesocosm tank is irradiated by an HID metal halogen lamp (Philips MH/CDM-T 150 W/930) with an emission spectrum resembling that of natural sunlight but with lower irradiance than sunlight below 400 nm. The position of the lamp in our experiments was adjusted to give 400 μmol quanta m^{-2} s^{-1} immediately below the water surface in all mesocosm tanks.

The mesocosm facility also includes an automatic system for water exchange, where a given percentage of water renewal per time unit can be set. The tanks can be filled either from the ordinary water-supply system, with intake approximately 800 m offshore, from 2 to 8 m in depth, or through a second pipe system, where water directly from the dock outside the lab or from a transport truck can be used. Water from the ordinary water-supply system is normally passed through a 300-μm filter; it can also be pre-flushed with air, or it can be used without any pre-treatments. A valve system ensures that all tanks are filled in parallel. There is a movable filtering system with 10 μm pre-filters (FSI Filter Spec. Int., Michigan City, Illinois, USA.) and patron filters down to 1 μm (Roki Techno Co. Ltd, Tokyo, Japan) which can be connected to the system. It takes around 3 h to fill all tanks without filtering and around 16 h to fill them with 1-μm-filtered sea water, including one necessary change of filters. A system is also available for collecting undisturbed bottom sediment from the field in shallow cylinders, fitted in size to the bottom of the mesocosm tanks. An automatic sampling device ensures reproducible sediment sampling over time (see Methods in Jonsson et al. 2014).

Sampling

For profiling of salinity, temperature, light (PAR, photosynthetic active radiation, 400–700 nm), oxygen saturation, chlorophyll a, turbidity and CDOM (colored dissolved organic matter), we used logging instruments, either an Aanderaa SeaGuard (see http://www.aanderaa.com) or Hanna HI 9828 multiparameter instrument with probe HI 769828 (see http://www.hannainst.com). The instrument was slowly lowered from the surface to the bottom with a measuring frequency of one per second (SeaGuard) or every 0.1–0.5 m in depth (Hanna). The chlorophyll-*a* concentration was also measured after ethanol extraction of pigments from cells filtered on a Whatman GF/C filter, using a Perkin Elmer LS 30 Luminescence spectrometer (Aminot and Rey 2002). The light profiles in the mesocosms were used to calculate the light attenuation coefficients (k in the equation $I_d = I_0 * e^{(-k*d)}$, where I_d is irradiance at d meter depth, I_0 is irradiance at the surface, and k is the slope in the exponential equation.

To study how materials on the inner walls with different optical properties influenced the light environment in the mesocosm tanks, we used thin glossy white PVC sheets (named White in the following) and thick aluminum foil (named Alu in the following) to cover the inside wall, in addition to the uncovered original black polyethene wall (named Black in the following). The mesocosm tanks were filled with tap water, the absolute irradiance was measured at 0.3, 1, 2, 3, 4 and 4.4 m in depth, the water-column irradiance was calculated by trapezoid depth integration, and the results shown are relative to those for Alu. The light spectra were measured with Ocean Optics™ radiometers USB 2000+ and USB 4000 (Ocean Optics, Winter Park, Florida, USA), with the fiber-optical cable mounted on a frame that held the tip of the fibre upward in a vertical position. To measure the highest irradiance levels, we used Teflon (polytetrafluoreten) tape as a neutral filter on the tip of the fibre. The instruments were calibrated using the Ocean Optics™ calibration unit, HAL 2000, before each series of measurements.

Water-column mixing and exchange through pycnoclines

The HID lamps emit heat that is transferred to the surface water. We recorded the temperature profiles in the water column with a temperature setting of 15 °C in all three sections of the mesocosms and with different mixing methods. Temperature profiles were measured in undisturbed conditions, in water bubbled with compressed air from 60 cm in depth, in water with an undulating fan blowing air on the surface, and in water with a 200-W electrical heater placed at the bottom. A further study of the exchange rate of water between the heated surface and deeper layers was also performed with the undisturbed conditions and with the 200-W electrical heater placed at the bottom of the mesocosm. This was done by injecting a solution containing 4 g of humic acid (Aldrich pnr: 536080) at 2.4 m in depth and profiling the content of humic acid in the upper part of the water columns over time using a Turner Designs Cyclops-7 CDOM sensor (Turner Designs, San Jose, California, USA).

Air bubbling and fan blowing were also compared in a study of the mixing efficiency through a thermocline with a 5 °C temperature gradient. Air bubbling was done by lowering a PVC tubing with an outlet diameter of 18 mm to pre-defined depths and regulating the air pressure to give a bubbling frequency of one bubble per second. This bubbling strategy creates stirring but it does not aerate like an air diffuser since the area of the bubbles at any given time is negligible compared to the area of the water surface.

When testing air blowing on the surface, we arranged the fan to blow at 45° towards the surface and undulate 45° left and right. To promote convective stirring, we used either a submersible 250-W heater or the three internal temperature control mantles that form parts of each mesocosm tank (Fig. 1). In these experiments, we used 160 mg of rhodamine in 50 ml of water, injected at 2.4 m in depth as described above.

Two different types of mixing rate experiments were performed. In one set of experiments, the rate of transport through thermoclines was studied. In another set of experiments, water-column mixing with different settings to promote convective stirring was studied. In both cases, we used a rhodamine solution that was slowly added through a small funnel with PVC tubing attached to the neck and with an air diffuser attached to the other end of the tubing. To ensure that all of the rhodamine was injected, the tubing was flushed with 100 ml of water after the injection. The rhodamine solution and flush water was kept at the same temperature as the surrounding water. In the thermocline experiment, the temperature in the bottom layer was set to, respectively, 5, 10 and 15 °C, and to 20 °C above this bottom layer in all three tanks. A rhodamine solution (160 mg in 50 ml of water) was injected at 4.5 m in depth, and a 250-W submersible heater was placed at 3.2 m in depth to ensure mixing in the upper water column. After adding the rhodamine, we lowered a small aquarium pump with horizontal in- and outlets down to the bottom and ran it for 1 min to get the bottom water well mixed with rhodamine. Vertical profiles of the rhodamine distribution were thereafter made over time, using a Turner Designs Databank handheld data logger with a pre-calibrated Turner Cyclops 7 model 2100-000 fluorimeter sensor (Turner Designs, San Jose, California, USA). After the profiling ended, the water column was mixed thoroughly, and the rhodamine concentration was recorded to get the equilibrium concentration. In the study of the transport rate through thermoclines, we divided the water column into four vertical sections, 0.2–1.8, 1.8–3.3, 3.3–3.6 m (main part of the thermocline), and 3.6–4.7 m (stagnant bottom layer where the rhodamine was added). The transport rate into the water above the bottom layer (above 3.6 m) was calculated by normalizing the final equilibrium concentration of rhodamine to 100% and then plotting the % values of equilibrium rhodamine concentration over time and fitting a linear regression equation to the data. This is a simplification of the actual process since, in theory, a given proportion of water is mixed per time unit, which would generate a logarithmic succession. The exchange rate of rhodamine from the bottom layer through the thermocline was calculated by setting the start value minus the equilibrium value as 100%, thereby normalizing the equilibrium rhodamine concentration to 0%, and then plotting the % values over time. To define the effect of thermocline strength, i.e., the density difference between the bottom and overlaying water, we plotted the transport rate versus thermocline strength (density difference), both for transport into the upper two layers and out from the bottom layer, and adapted the best fit regression equation to the data.

In the studies of water-column mixing, we defined homogenous mixing as the time it takes for the concentration of rhodamine to stabilize to within 95–105% of the final equilibrium concentration. This was tested with three temperature conditions: (1) temperature settings of 14, 15 and 16 °C in the low, middle and upper temperature-control sections, respectively, ensuring a continuous temperature increase from the bottom to the surface; (2) no temperature control, yielding a uniform temperature in the water column. (3) maximum heating in the low temperature-control section, a 15 °C setting in the middle section, and maximum cooling in the upper section, allowing for maximum convective stirring but with a constant temperature from the bottom to the surface. For these three experiments, we used 20 mg of rhodamine in 10 ml of water and injected it at 2.4 m in depth, followed by rinsing with 100 ml of water and rhodamine profiling over time, as described above.

Primary production under different mixing conditions

Primary production was measured in two experiments. In one experiment, comparing rotating incubations with fixed-depths incubations, we used 120-min incubations and glass incubation vials of 20 ml in volume, filled with water containing 40 mCi of $Na^{14}CO_3$ l^{-1}. A series of five replicates was fixed to each of two submersed rotating rubber bands driven by an electrical motor, one band rotating from the surface to 1.5 m in depth, the other one from the surface to 4.5 m in depth, both with a speed of 0.97 m min^{-1}. Duplicate vials were used for the rope incubations with positions at 0.2, 0.4, 0.6, 1.0 and 1.4 m in depth. After incubation, 5.0 ml of the content was transferred to a pre-labelled plastic scintillation vial, 300 μl of 3 M HCl was added, and the samples bubbled with air for 30 min. A scintillation cocktail (Optiphase HiSafe 3, 15 ml) was thereafter added, and the content was mixed and measured in a Beckman 6500 scintillation counter, which gave the value for DPM_{sample} (DPM = disintegrations per minute). For each experimental series, we prepared a standard series from the labelled water by pipetting out triplicates of 100, 200, 300 and 400 μl of the labelled water into 5 ml of distilled water. Immediately thereafter, 15 ml of scintillation cocktail was added, the content was mixed and measured as above. This gave the radioactivity in the incubation water, expressed as DPM_{total}. The coefficient of variation of the average value was always less than 2% for all standard series. Carbon assimilation was then calculated from the formula below:

$$\text{Prim. prod.}\ (\mu\text{mol C l}^{-1}\text{h}^{-1}) = \text{DPM}_{\text{sample}}/\text{DPM}_{\text{total}} \times 244.86/t$$

where t is the incubation time in minutes and the constant 244.86 is a product of a factor of 1.10 (sum CO_2), a factor of 1.05 for slower uptake of ^{14}C compared to ^{12}C, a factor of 1.06 for correction of respiration loss during incubation, and a factor of 200 for conversion from production in 5 ml to 1 l. See Colijn and Edler (1998) for further description.

Primary production per m^2 and hour for rotating incubations was obtained as the calculated primary production per m^3 and hour times the rotating depth (i.e., 1.5 or 4.5 m). Primary production per m^2 and hour for the rope incubations was given by first calculating the exponential regression equation of primary production m^{-3} h^{-1} versus depth and then integrating over 0–1.5 m, respectively, for 0–4.5 m in depth. The experiments were done with unfiltered brackish water containing added nutrients in surplus and with brackish water containing either 4 or 8 mg l^{-1} of dissolved humic acid (Aldrich pnr: 536080) in addition to the nutrients. In the second experiment, where we studied oxygen succession in a carbon-enriched environment, primary production was determined using 3-h incubations at mid-day with 100 ml of water in Nunc culture bottles to which 3.2 μCi of $Na^{14}CO_3$ where added. We used a light:dark cycle of 16:8. Single bottles were incubated at 0.1, 0.2, 0.4, 0.6, 0.8, 1.2, 1.6 and 2.0 m in depth. A pycnocline at 3–3.5 m in depth separated the upper mixed water parcel (18 °C, 4.41 g kg^{-1} salinity) from the bottom layer (10 °C, 5.07 g kg^{-1} salinity). An airlift system consisting of a 5-cm-wide pipe with its lower end at 3.2 m in depth, its upper end at the surface and tubing that delivered air bubbles into the lower end of the pipe ensured continuous mixing of the water above the pycnocline.

Further laboratory procedures and calculation of primary production per m^2 was done as defined for rope incubation above. Conversion to production per day from production per hour was done by multiplying by 16 (the daily light period).

Oxygen depletion and bacterial production

Bacterial production was measured in the oxygen succession experiment. We used 1-h incubations at in situ temperature with ^{3}H-labeled thymidine (Amersham®, 1 mCi ml^{-1}, 80 Ci $mmol^{-1}$) at a final concentration of 25 nM using a conversion factor of 1.4×10^{18} cells mol^{-1} (Wikner and Hagström 1999). Triplicate incubations and two controls (cold TCA added to 5% of the final concentration) were incubated for each sample. Calculations of bacterial production per m^2 and day were made by integrating over depth, assuming homogenous production within the upper and lower parts of the water column and uniform production during the 24 h of the day. To estimate bacterial respiration, we used information from Van den Meersche et al. (2004), who modelled bacterial uptake of carbon and nitrogen and the corresponding bacterial growth and estimated that approximately 19% of the carbon uptake goes into production. Assuming that all of the assimilated carbon that is not bound in growth is respired, approximately 4.3 times more carbon is used in respiration than in growth. Using the molar relationship of 12/32 between carbon and oxygen, the conversion factor from bacterial growth in mg carbon to respiration in mg oxygen will be $4.3 \times 32/12 = 11.5$.

Results

Light environment and wall surface

Aluminum-covered walls usually gave the highest irradiance levels at all depths, followed by white walls and then black walls, which absorbed most light (Fig. 2). The wavelength spectrum for the three wall surfaces had a similar shape at all depths (Fig. 2). The white walls reduced the irradiance compared to the aluminum-covered walls, especially between 580 and 700 nm (Fig. 3a), and in the upper meter of the water parcel, this corresponded to a reduction by 30–70%. In absolute terms, irradiance differences decreased with depth (Fig. 2a–c), whereas relative differences between both Alu and White and Alu and Black could reach between 200 and 300% in the region 500–650 nm in the deeper part of the water column. The difference between White and Black was highly significant in the upper meter of the water parcel, corresponding to a relative difference of 50–100% higher values for White in the range 450–700 nm. The effect of the different wall surfaces for the irradiance in the entire water parcel was a reduction to 68% over the entire spectrum with white compared to aluminium walls and a reduction to 46% with black compared to aluminium walls (Fig. 3d).

Thermal-lid effects

The 150-W HID metal-halogen lamps produced heat that was transferred to the water, causing an increased temperature that was significant down to 40 cm in depth (Fig. 4a). A gradual aggregation of organic material on the surface, including bacterial, algae and dead material, takes place as long as the surface is not mixed. Bubbling with ca. 1 bubble s^{-1} at 2 cm in depth eliminated visible surface aggregation, but the surface water was still heated by the lamp as without bubbling. Both bubbling from 60 cm in depth (Fig. 4b) and fan blowing (Fig. 4c) eliminated the surface temperature gradient, and the variation in the temperature in the upper meter of the water column was less than 0.3 °C. Convective stirring from the bottom, using a 200-W heater, eliminated most of the temperature increase in the surface layer (Fig. 4d). Without any mixing of the

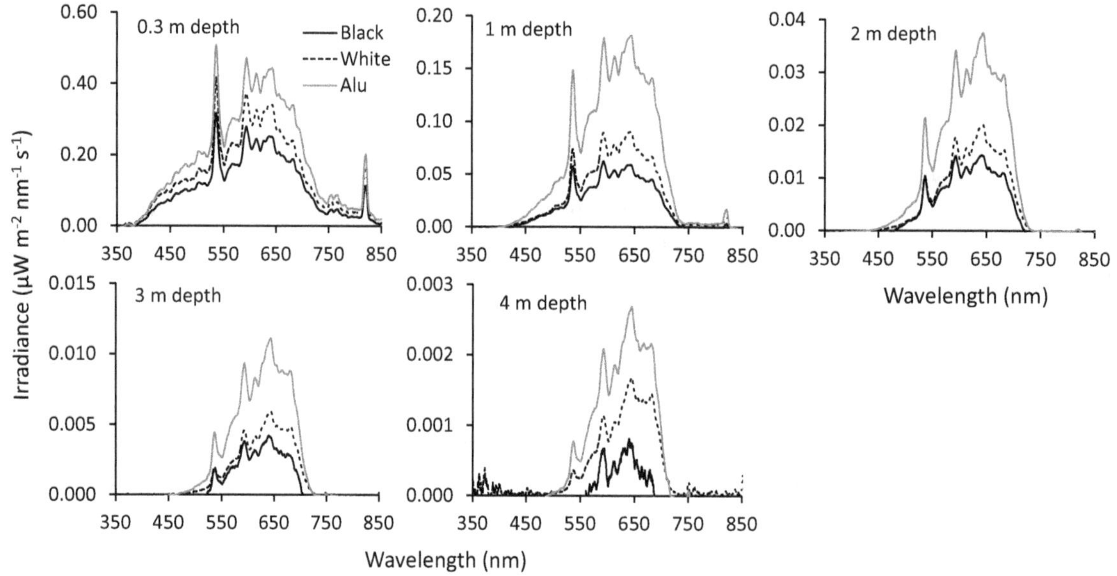

Fig. 2 Absolute irradiance spectrums in mesocosm tanks with three different surfaces on the inside wall and at five different depths. Black is the original polyethene surface. White is a glossy white PVC sheet, and Alu is a tube of aluminium foil kept in place towards the wall by stiff ring-formed plastic tubings

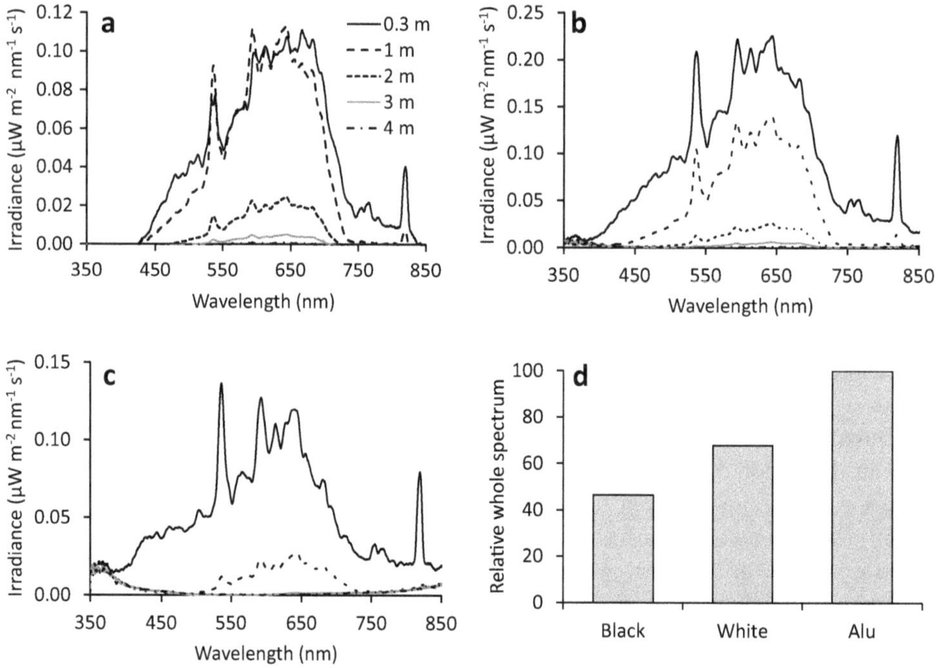

Fig. 3 Differences in absolute irradiance spectrum between mesocosm tanks with three different surfaces on the inside walls. **a** Alu minus White walls. **b** Alu minus Black walls. **c** White minus Black walls. **d** Relative irradiance values compared to Alu for the entire spectrum (350–850 nm) and 0–4.4 m in depth

surface water, as in Fig. 4a, the transport of material up to the surface was slow, and only 60% of the equilibrium concentration was reached in the upper 10 cm within 42 h (Fig. 4e). With convective stirring, as in Fig. 4d, the transport up to the surface increased, and the equilibrium concentration in the entire water column was reached within 31 h (Fig. 4f).

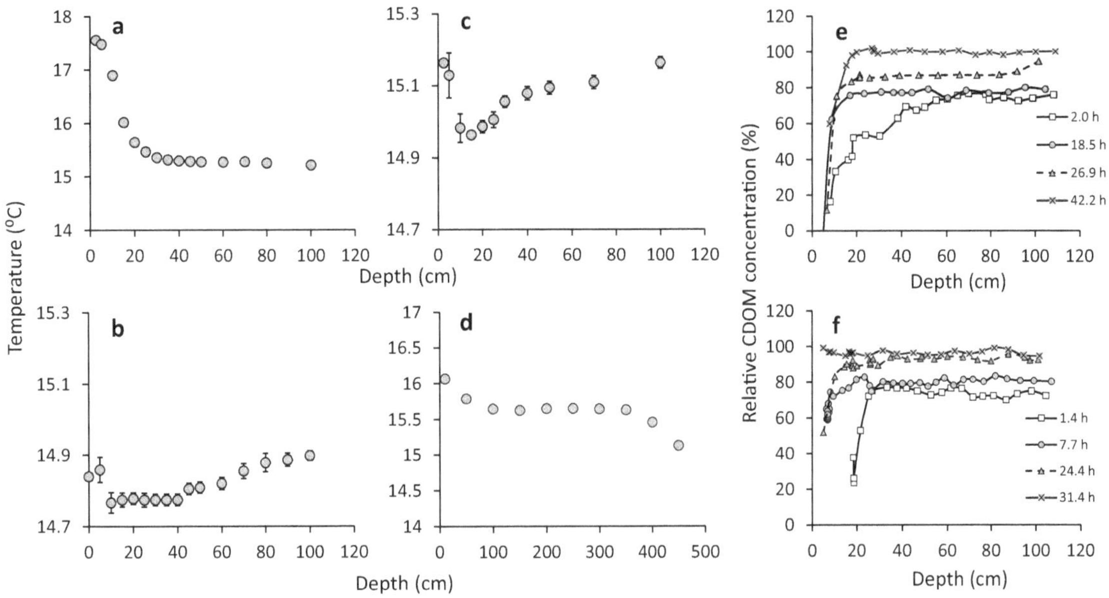

Fig. 4 Temperature profiles (**a–d**) and transport affected by a thermal lid (**e**, **f**) in mesocosms with a 150-W HID metal halogen lamp approximately 50 cm above the surface, giving an irradiance of ca. 400 μmol quanta m^{-2} s^{-1} below the surface. **a** No mixing in the water column. **b** Bubbling from 60 cm in depth. **c** An undulating fan (± 45° side movements) blowing at 45° on the surface. **d** A 200-W heater placed close to the bottom of the mesocosm tank. **e** Succession in relative CDOM concentrations following an injection of humus solution in the middle of the tank and with conditions as in **a**. **f** Succession in relative CDOM concentrations with the conditions in **d**. Note the different scales on the *Y*-axis in **a–d**. Vertical lines in **a–d** denote 95% confidence intervals, based on 13 measurements along two perpendicular lines of the tank diameter

Transport through thermoclines

Transport from a stagnant bottom-water layer through thermoclines of 15, 10 and 5 °C temperature gradients was studied while maintaining convective stirring above the thermocline with a 250-W aquarium heater. The temperature profiles differed markedly below 3 m in depth (Fig. 5a) and generated the observed differences in transport. In the upper 3.3 m of the water parcel, there was a linear increase in rhodamine concentration, with a rate constant of 0.05, 0.08 and 0.17% h^{-1}, respectively, of equilibrium concentration for the temperature gradients 15, 10 or 5 °C at 0.2–1.8 m and a rate constant of 0.10, 0.11 and 0.24% h^{-1} at 1.8 – 3.3 m (Fig. 5b, d, f). Within the thermocline, the rhodamine concentration showed a logarithmic increase over time, whereas the bottom layer showed a logarithmic decrease (Fig. 5c, e, g). A plot of transport rate versus the magnitude of the density gradient for the two upper vertical levels showed a relationship best described by a power function (Fig. 6). Transport to the upper layer (0.2–1.8 m) was slower than to the middle layer (1.8–3.3 m) and also decreased more with an increase in the density gradient, as shown by the exponent − 0.71 compared to − 0.52 (Fig. 6). Using the data from 22 to 138 h for the bottom layer (3.6–4.7 m) in calculation of the linear regression coefficient gave the data for transport rate versus density gradient for this layer, which showed a similar relationship as the other two layers (Fig. 6).

In an experiment with an upper warmer mixed layer of ca. 1.5 m in depth, overlaying a stagnant layer with a 5 °C lower temperature, we used either mixing by bubbling at 60 cm in depth or fan blowing as a way of getting rid of heat from the lamp. The resulting temperature profiles are displayed in Fig. 7a. The bubbling gave a constant temperature of 15.5 °C in the upper 1.7 m, a very sharp thermocline, and a constant temperature of 10–11 °C below the thermocline (Fig. 7a). The fan blowing gave a surface temperature of ca. 14 °C in the upper 60 cm, thereafter continuously decreasing to a constant temperature of approximately 10 °C below 230 cm in depth (Fig. 7a). The rhodamine injected at 240 cm in depth showed a linear change in concentration over time in the three depth intervals recorded. For the mesocosm with bubbling, the mixing rates at 0–100 and 100–150 cm in depth were identical, with an hourly increase in rhodamine concentration of 0.27% and a decrease below 200 cm of 0.17% per hour (Fig. 8b). Fan blowing yielded a faster mixing, with an hourly increase rate of 0.38 and 0.30% in the two upper layers and a decrease rate below 2 m in depth of

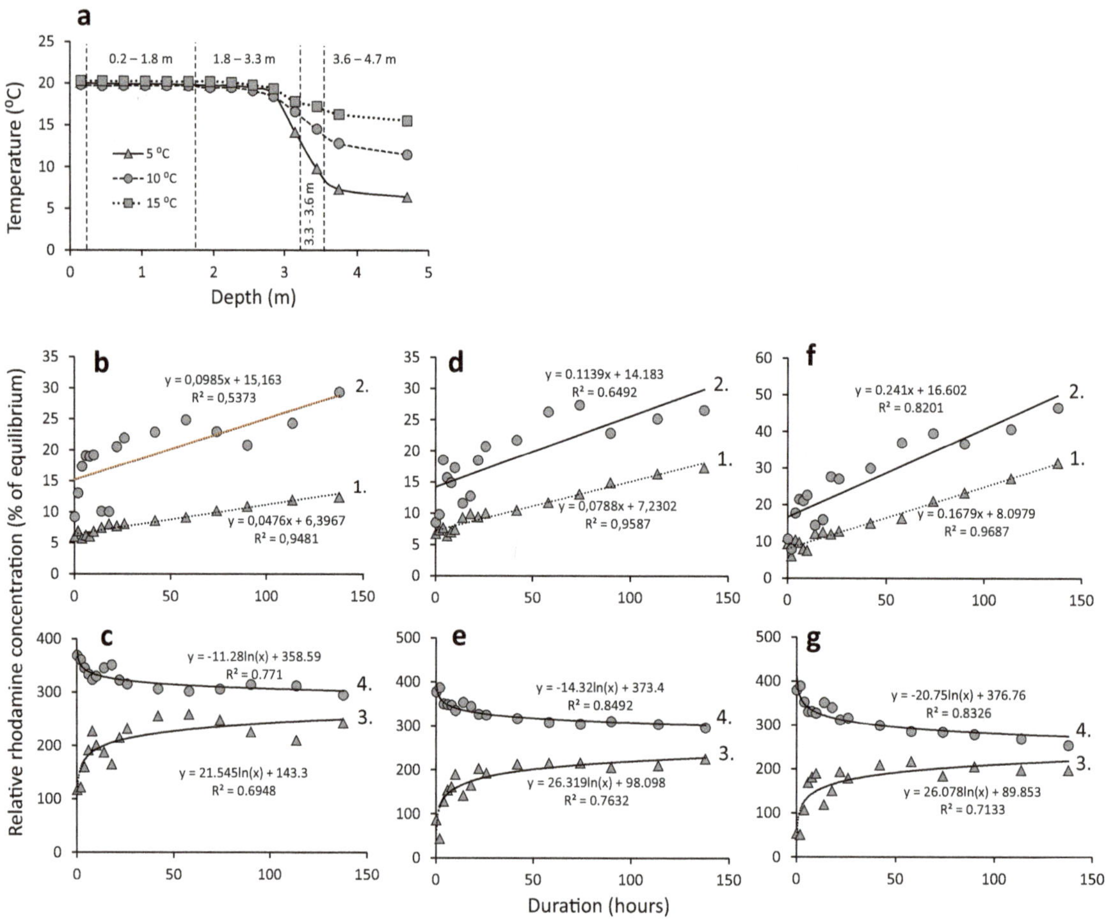

Fig. 5 **a** Temperature profiles in the studies of transport rates through three different temperature gradients, made by temperature settings in the bottom layer of 5, 10 and 15 °C, and 20 °C above the bottom layer. **c–g** Succession in relative rhodamine concentration in different vertical levels of the water column. The bottom temperature in **b** and **c** was 5 °C, in **d** and **e** was 10 °C, and in **f** and **g** was 15 °C. The plots represent average values within each of four vertical levels, 0.2–1.8 m (1.), 1.8 – 3.3 m (2.), 3.3–3.6 m (3.), and 3.6–4.7 m (4.), and rhodamine was added below the thermocline at the start. Values are given as % of the equilibrium concentration in layers 1, 2 and 3, i.e., the rhodamine concentration if the entire water column was completely mixed. In layer 4, 100% is defined as the starting concentration after the equilibrium concentration has been subtracted. Regression equations best fitting the plots are given in the graphs

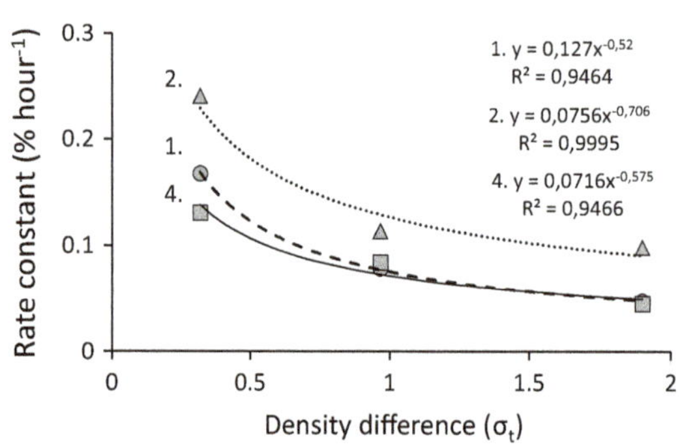

Fig. 6 Transport-rate constant ($Y = \%$ of rhodamine concentration per hour) versus density gradient, i.e., the difference in σt (kg m^{-3}) between the bottom layer and the water parcel above. Data from layers 1, 2 and 4 (see Fig. 5) have been used (for 1 and 2, with the regression coefficients from the linear regression equations). Regression coefficients based on data for 22–138 h and linear regression calculations where used for layer 4. Plots and equations are given that best fit the data

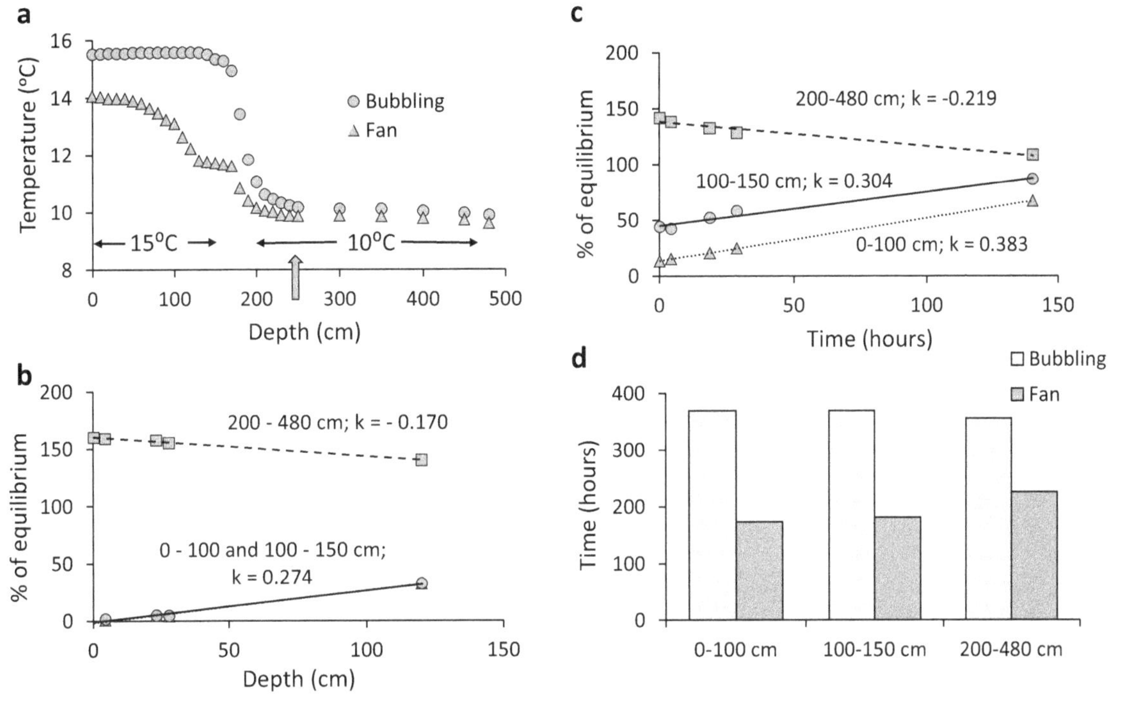

Fig. 7 Temperature profiles with bubble mixing at 60 cm in depth and fan mixing from the surface. **b** Mixing rate constant (% of equilibrium rhodamine distribution) in three different depth intervals and with bubble mixing. **c** Mixing rate constant with fan mixing. **d** Time to reach the equilibrium distribution based on linear regression and the estimated mixing-rate constant. Temperature setting as given in the figures. Rhodamine was injected at 2.40 m in depth (see the arrow in **a**)

0.22% per hour (Fig. 7c). To reach equilibrium concentrations in all three sections, bubble mixing required between 356 and 369 h, whereas the fan treatment required between 173 and 226 h to reach equilibrium (Fig. 7d).

Non-stratified water mixing

Temperature settings ranging from 16 °C at the top down to 14 °C at the bottom gave a gradual decrease of the temperature in the water column from 16.4 at the surface down to 13.9 °C at 4.5 m in depth, and with a strong drop further down due to cooling from the bottom (Fig. 8a). There was considerable variation in rhodamine concentrations following injection between the six replicates during the first 36 h but very uniform levels thereafter, and the equilibrium level was reached after 84 h (Fig. 8b). With both the temperature control and light switched off, the temperature in the water column was uniform at 16.4 °C except close to the bottom, where it was 16.2 °C (Fig. 8c). Equilibrium in rhodamine distribution was reached here within 24 h (Fig. 8d). Maximum convective stirring with the light switched on gave a temperature increase from 40 cm in depth to the surface of 15.0–16.1 °C, whereas the temperature below 40 cm in depth varied less than 0.1 °C from the temperature of 15.0 °C for the middle section (Fig. 8e). The rhodamine distribution reached equilibrium within 17 min at 100 cm in depth and below (Fig. 8f), whereas it took 34 min to reach equilibrium at 30 cm in depth and 78 min at 10 cm in depth (not shown) as a result of the temperature gradient caused by the lamp above (see also the thermal-lid effects in Fig. 4).

Primary production, turbulence and humus

In our test of the significance of mixing for primary production, we found a dramatic effect when the irradiance in the water parcel was reduced by the addition of humus. The light attenuation coefficient increased from 1.07 to 1.96 and 2.61 for the 4 and 8 mg l^{-1} additions, respectively. The primary production based on traditional fixed-depth incubation was successively reduced with added humus. A decrease of more

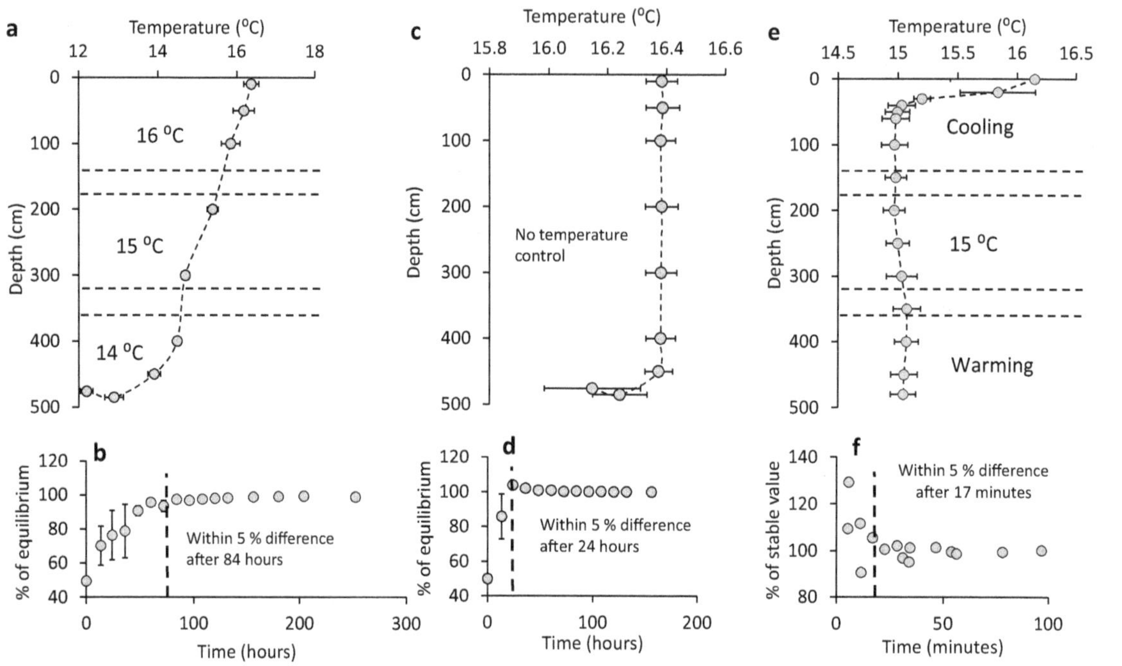

Fig. 8 **a** Temperature profile with temperature settings of 14, 15 and 16 °C from the bottom up; **b** succession of the rhodamine concentration for the condition in **a**, given as % of equilibrium; **c** temperature profile with temperature control switched off. **d** Succession of the rhodamine concentration for the condition in **c**, given as % of equilibrium. **e** Temperature profile with maximum convective stirring by constant heating the lowest section, maintaining 15 °C in the middle section and constant cooling in the upper section, and with the lamp above switched on. **f** Succession of the rhodamine concentration for the condition in **e**, given as % of equilibrium. Vertical lines denote 95% confidence intervals. Time of complete mixing given in the inserted text in **b**, **d**, and **f**

than 80 and 60% for the 8 mg l^{-1} additions could be observed for the 0- to 1.5- and 0- to 4.5-m sections, respectively (Fig. 9a). Without any addition, rotation incubation gave a 20% lower primary production than fixed-depth incubation in the upper 1.5 m and a 48% lower production between 0 and 4.5 m in depth (Fig. 9b). With the addition of 4 mg humus l^{-1}, the primary production in rotating incubation corresponded to roughly 300% of the primary production based on fixed-depth incubation, both for the shallow (0–1.5 m) and deep (0–4.5 m) water parcel (Fig. 9b). A further addition of 4 mg l^{-1} of humus caused the rotating incubation to reach levels 6.7 times higher in the 0- to 1.5-m interval and 8.7 times higher in the 0- to 4.5-m interval compared to fixed-depth incubations (Fig. 9b).

Anoxia through microbial respiration

The strong pycnocline in this experiment (Fig. 10a) reduced the transport of oxygen to the bottom layer, and the continuous addition of glycol stimulated bacterial production (Fig. 10b). Primary production reached its highest value in day 4, with 55 mg C m^{-2} day^{-1}, thereafter ranging between 16 and 42 mg C m^{-2} day^{-1} (Fig. 10b), whereas bacterial production increased heavily after day 4, with the highest value at day 22, when it was 32 times higher than the primary production (Fig. 10b). The theoretical oxygen consumption by bacterial respiration (see "Materials and methods") increased linearly over time with all oxygen respired for the upper water layer (18 °C) within 20 days and for the bottom layer (10 °C) within 23 days, assuming no oxygenation of the water (Fig. 10c). The actual measured oxygen level in the bottom layer had reached down to a 2% saturation level at day 19 (Fig. 10d). Although the water above the pycnocline was mixed by an air-lift system (see "Materials and methods"), the oxygen saturation level was also reduced here, from over 130% in the beginning to between 58 and 81% after day 15 (Fig. 10d).

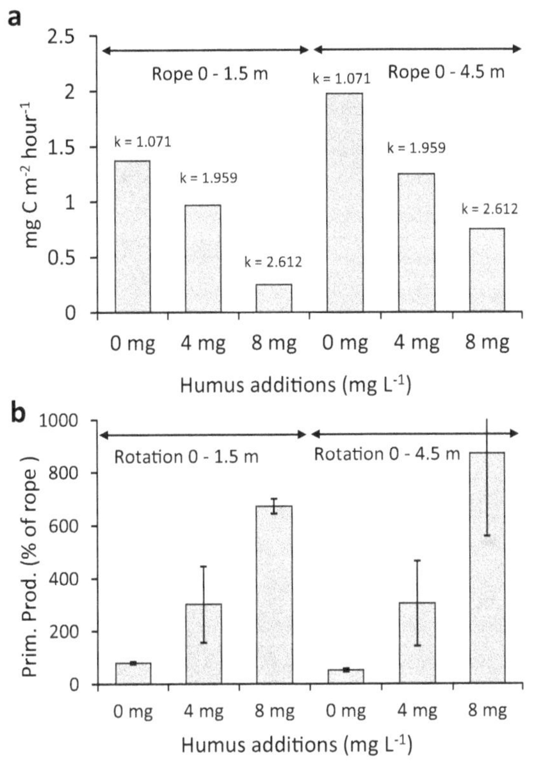

Fig. 9 Primary production based on rope incubations at five depths (**a**) and primary production using rotating incubations, either between the surface and 1.5 m in depth or down to 4.5 m in depth, expressed as relative production compared to rope incubation (**b**). Rope and rotating incubations were made simultaneously and at three different occasions with different humus content: no humus addition, 4 and 8 mg l^{-1}. Rotating speed = 97 cm min^{-1}. Vertical lines in **b** denote 95% confidence intervals

Discussion

Ecological studies where mesocosms are used to simulate natural conditions usually do not include measurements that describe all physical conditions of importance for the succession in the ecosystem. Irradiance as PAR (400–700 nm in wavelength), salinity and temperature are commonly included variables, whereas spectral distribution and level of turbulent mixing are seldom measured. The ultra-violet radiation from sunlight is known to affect microorganisms and especially bacteria negatively, but Ogbebo and Ochs (2008) showed that this effect was significant only with added inorganic nutrients, and then with a stronger effect on phytoplankton production (45% reduction) compared to bacterioplankton production (21% reduction). In addition, sunlight also causes photolysis of DOC (dissolved organic carbon), which will transform carbon compounds and make them more bioavailable for bacterioplankton (Bertilsson et al. 1999; Tranvik and Bertilsson 2001). Such effects of sunlight will, therefore, have a strong impact on the overall production, as well as the trophic balance of the system. By mesocosm enclosures with different types of inside material and different light sources, both quantity (PAR irradiance) and quality (spectral distribution) of the light climate in the water column can be manipulated, as shown in our Figs. 2 and 3.

One feature of solid mesocosm tanks is that the tank walls do not let light in from the sides into the water column. This is in contrast to in situ mesocosms, which are usually made of a thin plastic material, although of variable opacity. The light penetrating the water column is absorbed by the water itself, as well as by dissolved and particulate material. Light hitting the wall of a solid tank is partly absorbed and partly reflected, with the relative proportion dependent on the surface material and also how clean the surface is. One important factor for indoor mesocosms is the light source, where the beam angle is a key factor. In nature, the light beam from the sun is essentially parallel, while an electric light source over a mesocosm tank delivers a diverging light beam. The wider the beam angle from the light source, the higher the attenuation through absorption on the wall. In our mesocosm experiments, we recorded light attenuation coefficients ranging from 0.521 (pure tap

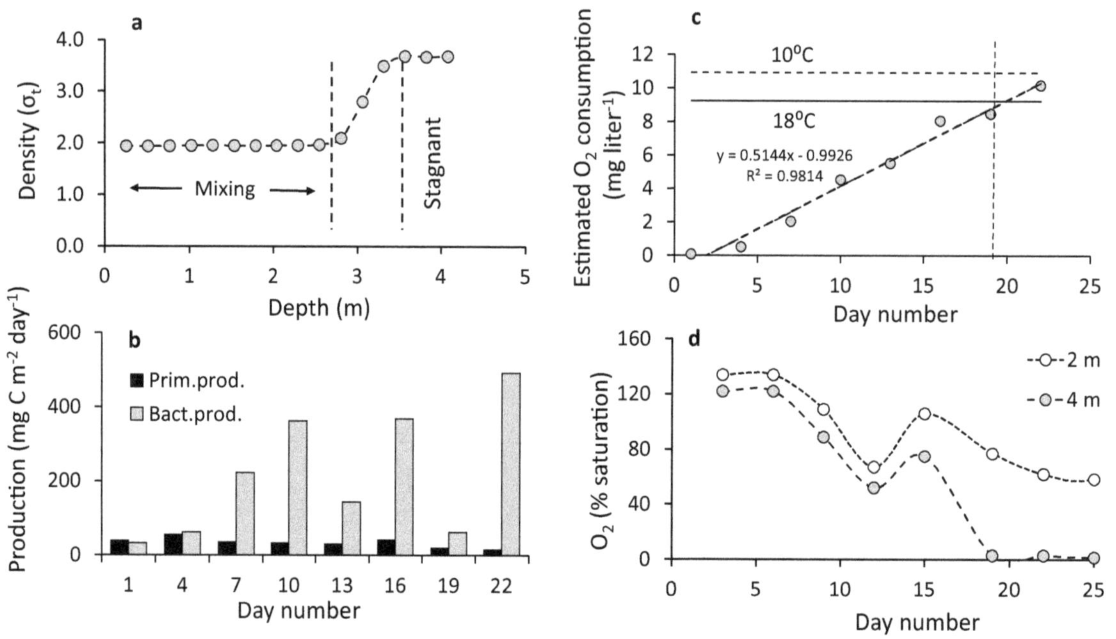

Fig. 10 Experiment forcing anoxic conditions in the bottom water through changes in microbial production balance. Glycol as an organic carbon source was used to stimulate bacterial production. **a** Vertical profile of density. **b** Succession in primary production and bacterial production. **c** Estimated succession in oxygen consumption through bacterial respiration without water exchange through the pycnocline. **d** Succession in oxygen saturation level in the water above the stagnant bottom water (2 m in depth) and within the stagnant bottom water (4 m in depth). Airlift mixing (see "Materials and methods") from 3.5 m in depth explains the oversaturation of oxygen during part of the experiment. The two horizontal lines in **c** denote the level of saturated oxygen content above (18 °C) and below (10 °C) the pycnocline

water) up to 2.612 [natural water with a high allochthonous dissolved organic carbon (ADOC) content], which can be compared to coefficients measured in the northern Bothnian Sea between June and December, where the average coefficient was 0.595, ranging from 0.300 up to 1.425 (own unpublished data). The negative consequence of a higher-than-natural light attenuation coefficient is that the same light-depth relationship as in nature cannot be reproduced. However, at the same time, it might be an advantage, since the irradiance profile in a natural water column can be reproduced on a compressed depth scale. For example, in the experiment by Båmstedt and Wikner (2016), the average compensation depth was 3.2 m without added ADOC and 2.6 m with added ADOC, whereas field measurements gave a range from 3.2 to 15.4 m, with an average of 7.7 m.

Pycnocline strength and turbulent mixing are physical properties that are of the utmost importance in, for example, supplying nutrients from deeper water to the euphotic zone or supplying oxygen from above to benthic organisms. The effect of variable pycnocline strength and turbulent mixing on ecosystem components can easily be studied in mesocosm experiments, but mesocosms of the type we describe here are also suitable tools for direct physical studies of how different combinations of pycnocline strength and turbulent mixing govern transport rates, as exemplified in our Figs. 5 and 6.

When using mesocosms to simulate a natural ecosystem, one major challenge is to create mixing, since the natural mechanisms of mixing will be altered in a closed water parcel. In our study, we found that mixing in the uppermost few cm of the water parcel is necessary to eliminate a gradually increasing surface aggregation of organic material. Mixing from 60 cm in depth was also necessary with our illumination system to eliminate the thermal-lid effect caused by heat from the lamps. The exchange of material between the surface and deeper layers is clearly hampered by the existence of a surface section that is warmer than the water below (Fig. 4e, f). The combination of a low mixing rate in the water column and temperature stratification at the surface may influence the exchange of gases between the water column and the atmosphere and the transport of nutrients from deeper water to the surface layer. In the worst case, this can lead to mesocosm environments that differ significantly from the prevailing natural conditions. Because different lamps produce different amounts of heat, it is necessary to investigate the actual condition for each mesocosm facility. Both bubbling with air from

60 cm in depth and blowing air on the surface with a fan eliminated thermal stratification at the surface (Fig. 4b, c). The use of a fan, however, caused a significantly higher loss of water due to evaporation and a lower surface temperature as compared to bubbling (Fig. 7a). The explanation for this effect is as follows: the enthalpy of water evaporation is 2260 kJ per kg (see textbooks on thermodynamics), and this energy is taken from the water, thereby causing a cooling of the surface layer and a different temperature profile as compared to air bubbling. The logic of the difference in temperature profile is a difference in transport rate, clearly illustrated in our Fig. 7d. The results are a strong reminder that the same methods should always be used if the results from different experiments are to be compared.

Striebel et al. (2013) compared the effect of three methods of mesocosm mixing, following a plankton community over 19 days and using a disc that moves up and down, an air-lift system and air bubbling. Mechanical mixing with the disc generated the highest zooplankton biomass, and an air-lift system generated a lower proportion of small flagellates, but otherwise, all methods gave the same community composition of phytoplankton and zooplankton. From mesocosm studies in Lake Thaihu, Zhou et al. (2016a) showed that the turbulence level had a strong impact on the structure of the zooplankton community, with a shift from the dominance of large-sized crustaceans to small-sized rotifers with increased turbulence. High turbulence also suppressed the zooplankton growth and biomass. Zou et al. (2016b) also reported a strong positive influence of turbulent mixing on the production of autochthonous chromophoric dissolved organic matter (CDOM). We have mainly used convective stirring as a standard method of mixing, since it corresponds better to field conditions than mixing by pumps, propellers or air bubbles (e.g., Paul et al. 2012; Sjöqvist et al. 2014). This is done either by slightly different temperature settings for the temperature-control mantles, or by submerging an electrical heater, or by both. The maximum convective stirring settings correspond to only 17 min for complete mixing of the mesocosm water (Fig. 8f) with still a rather uniform temperature profile (Fig. 8e). These results show that high stirring rates can be achieved and that the overall temperature is precisely controlled by the setting of the middle section under these circumstances. The range in mixing rate, exemplified in Fig. 8, with relatively small differences in temperature profiles, shows that mixing by convective stirring is a versatile and effective method of creating different turbulent environments. However, to achieve a given mixing rate for a certain type of mesocosm, calibration measurements are necessary. We have used rhodamine injections for this purpose, with the criterion that mixing is achieved when the concentration of rhodamine has stabilized to within 95–105% of the equilibrium value.

Our results of primary production from rotating incubations (Fig. 9) also indicate a dramatic importance of turbulent mixing in water with a high content of humus, which causes brownification of the water, with an increased production ratio between rotation/fixed-point incubations with increased humus. Because primary production with a radioactive tracer technique is measured either by in situ incubations at fixed depths or in an incubator with a series of fixed light levels, such an effect from turbulent mixing will not be accounted for. Almost 40 years ago, Marra (1978) noted that variation in incubation depth yielded higher primary production than incubations at fixed depths. In addition, Gieskes et al. (1979) noted from their studies in the tropical open ocean that the primary production in the upper part of the euphotic zone was underestimated, probably caused by more rapid bleaching of chlorophyll in algal cells incubated at fixed depths compared to cells freely circulating in the vertically mixed surface layer. This important topic obviously needs more attention. If further studies show this issue to be of general significance, future measurements of the primary production in situ and in incubators should be made in better accordance with the prevailing turbulent mixing conditions.

Turbulent mixing is also of profound importance for the diffusive boundary layer above the bottom sediment and for the transportation of material and properties through a pycnocline, such as supplying the photic zone with nutrients from the deeper layers and for oxygen-rich surface water to reach the deeper layers. Lucas et al. (2016) used an oscillating grid mesocosm in which the turbulence could be varied and found that the thickness of the oxygen-diffusive boundary layer was inversely related and that the oxygen penetration depth in different sediments was positively related to the turbulence intensity. Our study on oxygen depletion in the bottom layer (Fig. 10) showed the effect of enrichment of bioavailable DOC for the oxygen concentration in the bottom layer. However, the estimated theoretical consumption of oxygen through bacterial respiration indicated that 23 days would be required for the observed reduction (Fig. 10c). The density difference between the upper and lower water layer was 1.737 g kg^{-1}, and if we use Eq. 4 in Fig. 7, we see that the transport rate from the bottom layer would be 0.053% per hour. This means that 1.27% per day of the bottom water is replaced by more oxygenated water from above, which would then somewhat reduce the effect of bacterial respiration. Bacterial abundance in the mesocosm with glycol addition was 35–66% higher than mesocosms without glycol addition, and bacterial production was between 150 and 168% higher (data not shown), clearly showing the stimulating effect of glycol on the bacterial community. We can only speculate about the difference between the theoretical and observed oxygen succession, and one aspect may be a lower theoretical growth efficiency than the 19% we used (see "Materials and methods"). The fact that the bacterial production increased by 150–168% with glycol but the bacterial abundance only increased by 35–66% might indicate that glycol was mainly used for respiration. Higher-resolution measurements, both in time and depth, would probably also provide a better basis for explanation. Here, our primary goal was to show that an anoxic environment can be achieved in a mesocosm enclosure via natural biological processes.

Final remark

We have described a unique aquatic mesocosm facility that is highly versatile regarding physical conditions in the water column. Such an experimental tool is of utmost importance in simulating environmental changes like, e.g., the ongoing climate change, and for studying the effects on the ecosystem. Ignoring factors like turbulent mixing, density stratification and light spectrum might give results that diverge from the natural environment and which thereby support wrong environmental management decisions.

Acknowledgements The work was conducted with support from the Nordic Marine Academy, the European Union Seventh Framework Program (FP7/2007–2013) under Grant agreement no. 228224, (MESOAQUA), and the Strategic Marine Environmental Research program Ecosystem dynamics in the Baltic Sea in a changing climate perspective (ECOCHANGE). We thank Umeå marine sciences centre for good working facilities and necessary technical support during all parts of the study.

References

Aminot A, Rey F (2002) Chlorophyll *a*: determination by spectroscopic methods. ICES Tech Mar Environ Sci 30:1–18

Båmstedt U, Wikner J (2016) Mixing depth and allochthonous dissolved organic carbon: controlling factors of coastal trophic balance. Mar Ecol Prog Ser 561:17–29

Bertilsson S, Stepanauskas R, Cuadros-Hansson R, Graneli W, Wikner J, Tranvik L (1999) Photochemically induced changes in bioavailable carbon and nitrogen pools in a boreal watershed. Aquat Microb Ecol 19:47–56

Caquet T, Lagadic L, Monod G, Lacaze JC, Coute A (2001) Variability of physicochemical and biological parameters between replicated outdoor freshwater lentic mesocosms. Ecotoxicology 10:51–66

Colijn F, Edler L (eds) (1998) Working manual and supporting papers on the use of a standardised incubator—technique in primary production measurements. In: Report of the working group on phytoplankton ecology. CM 1998/C: 3. ICES, Copenhagen, pp 49–57

Dzialowski AR, Rzepecki M, Kostrzewska-Szlakowska I, Kalinowska K, Palash A, Lennon JT (2014) Are the abiotic and biotic characteristics of aquatic mesocosms representative of in situ conditions? J Limnol 73:603–612

Gieskes WWC, Kraay GW, Baars MA (1979) Current C-14 methods for measuring primary production—gross underestimates in oceanic waters. Neth J Sea Res 13:58–78

Jonsson S, Skyllberg U, Nilsson MB, Lundberg E, Andersson A, Björn E (2014) Differentiated availability of geochemical mercury pools controls methylmercury levels in estuarine sediment and biota. Nat Commun 5:10

Lucas S, Moulin F, Guizien K (2016) Oscillating grid mesocosm for studying oxygen dynamics under controlled unsteady turbulence. Limnol Oceanogr Methods 14:1–13

Marra J (1978) Phytoplankton photosyntthetic response to vertical movement in a mixed layer. Mar Biol 46:203–208

Miko Z, Ujszegi J, Gal Z, Imrei Z, Hettyey A (2015) Choice of experimental venue matters in ecotoxicology studies: comparison of a laboratory-based and an outdoor mesocosm experiment. Aquat Toxicol 167:20–30

Ogbebo FE, Ochs C (2008) Bacterioplankton and phytoplankton production rates compared at different levels of solar ultraviolet radiation and limiting nutrient ratios. J Plankton Res 30:1271–1284

Paul C, Reunamo A, Lindehoff E, Bergkvist J, Mausz MA, Larsson H, Richter H, Wängberg SÅ, Leskinen P, Båmstedt U, Pohnert G (2012) Diatom derived polyunsaturated aldehydes do not structure the planktonic microbial community in a mesocosm study. Mar Drugs 10:775–792

Petersen JE, Englund G (2005) Dimensional approaches to designing better experimental ecosystems: a practitioners guide with examples. Oecologia 145:216–224

Petersen JE, Cornwell JC, Kemp WM (1999) Implicit scaling in the design of experimental aquatic ecosystems. Oikos 85:3–18

Schindler DW (1998) Replication versus realism: the need for ecosystem-scale experiments. Ecosystems 1:323–334

Sjöqvist C, Kremp A, Lindehoff E, Båmstedt U, Egardt J, Gross S, Jönsson M, Larsson H, Pohnert G, Richter H, Selander E, Godhe A (2014) Effects of grazer presence on genetic structure of a phenotypically diverse diatom population. Microb Ecol 67:83–95

Spivak AC, Vanni MJ, Mette EM (2011) Moving on up: can results from simple aquatic mesocosm experiments be applied across broad spatial scales? Freshw Biol 56:279–291

Stewart RIA, Dossena M, Bohan DA, Jeppesen E, Kordas RL, Ledger ME, Meerhoff M, Moss B, Mulder C, Shurin JB, Suttle B, Thompson R, Trimmer M, Woodward G (2013) Mesocosm experiments as a tool for ecological climate-change research. Adv Ecol Res 48(Global Change in Multispecies Systems, Pt 3):71–181

Striebel M, Kirchmaier L, Hingsamer P (2013) Different mixing techniques in experimental mesocosms-does mixing affect plankton biomass and community composition? Limnol Oceanogr Methods 11:176–186

Tranvik LJ, Bertilsson S (2001) Contrasting effects of solar UV radiation on dissolved organic sources for bacterial growth. Ecol Lett 4:458–463

Van den Meersche K, Middelburg JJ, Soetaert K, van Rijswijk P, Boschker HTS, Heip CHR (2004) Carbon-nitrogen coupling and algal-bacterial interactions during an experimental bloom: modelling a C-13 tracer experiment. Limnol Oceanogr 49:862–878

Wikner J, Hagström Å (1999) Bacterioplankton intra-annual variability: importance of hydrography and competition. Aquat Microb Ecol 20:245–260

Williams P, Whitfield M, Biggs J, Fox G, Nicolet P, Shillabeer N, Sherratt T, Heneghan P, Jepson P, Maund S (2002) How realistic are outdoor microcosms? A comparison of the biota of microcosms and natural ponds. Environ Toxicol Chem 21:143–150

Zhou J, Han XX, Qin BQ, Casenave C, Yang GJ (2016a) Response of zooplankton community to turbulence in large, shallow Lake Taihu: a mesocosm experiment. Fundam Appl Limnol 187:315–324

Zhou YQ, Zhou J, Jeppesen E, Zhang Y, Qin B, Shi K, Tang X, Han X (2016b) Will enhanced turbulence in inland waters result in elevated production of autochthonous dissolved organic matter? Sci Total Environ 543:405–415

15

Inhibition of biofilm bacteria and adherent fungi from marine plankton cultures using an antimicrobial combination

Vanessa Ochi Agostini · Alexandre José Macedo · Erik Muxagata

Abstract The presence of organic matter in plankton cultures will lead to 10- to 1000-fold increases in bacterial density in less than 24 h. The presence of bacteria and fungi can damage cultivated phytoplankton and zooplankton. These microorganisms can also inhibit experiments investigating the role of these microorganisms in the community and in biological and ecological laboratory studies. The aim of this study was thus to evaluate the effect of penicillin + streptomycin + neomycin (antibiotics) in combination with nystatin (antifungal) to select an antimicrobial combination for the inhibition of biofilm bacteria and adherent fungi that is effective and also non-toxic to marine phytoplankton and zooplankton. *Acartia tonsa* was exposed to different antimicrobial treatments and application routes (culture medium, culture food, both) to evaluate the survival and egg and fecal pellet production endpoints. The same treatments were also applied to measure *Amphibalanus improvisus* survival and settlement and *Conticribra weissflogii* growth endpoints. We selected the most sensitive experimental organism and exposed it to some novel antimicrobial combinations. To evaluate the inhibition potential, biofilm bacteria and adherent fungi were exposed to the treatments that were safe for the bioindicator species. *A tonsa* was considered the most sensitive of all tested organisms. The treatment composed of 0.025 g L^{-1} penicillin G potassium + 0.08 g L^{-1} streptomycin sulfate + 0.04 g L^{-1} neomycin sulfate showed the best results for *A. tonsa* and *C. weissflogii* cultures. No differences were observed for *A. improvisus* between the treatments. *A. tonsa* survival rates showed no differences from the

V. O. Agostini (✉) · E. Muxagata
Laboratório de Zooplâncton - Instituto de Oceanografia da Universidade Federal do Rio Grande (FURG), Caixa Postal, 474, CEP: 96203-900 Rio Grande, RS, Brazil
e-mail: nessaochi@gmail.com

V. O. Agostini
Programa de Pós-graduação em Oceanografia Biológica (PPGOB), Bolsista do Conselho Nacional de Desenvolvimento Científico e Tecnológico (CNPq), Rio Grande, Brazil

Present Address:
V. O. Agostini
Laboratório de Microcontaminantes Orgânicos e Ecotoxicologia Aquática, CAPES Postdoctoral fellow at Programa de Pós-Graduação em Oceanografia Física, Química e Geológica, Instituto de Oceanografia, Universidade Federal do Rio Grande, Rio Grande, Brazil

A. J. Macedo
Laboratório de Biofilmes e Diversidade Microbiana - Faculdade de Farmácia e Centro de Biotecnologia da Universidade Federal do Rio Grande do Sul (UFRGS), Av. Ipiranga, 2752, Bairro Azenha, 90610-000 Porto Alegre, RS, Brazil

control at nystatin concentrations up to 0.005 g L^{-1} in combination with the antibiotics. The biofilm bacterial density decreased up to 94% under this treatment, and fungal growth was prevented. Species of other planktonic groups should also be tested to improve our understanding of the effectiveness of the treatments proposed.

Keywords Antibiotics · Antifungal · Bioassay · Holoplankton · Meroplankton · Phytoplankton

Introduction

Different groups of planktonic organisms, such as phytoplankton and zooplankton, have been cultivated in semi-intensive and intensive systems for ecological and biological studies as live food in aquaculture, as well as for toxicological tests (Gilles et al. 2013; Edwards et al. 2016; Lopes et al. 2018). Unicellular marine algae are commonly grown as food for several commercially valuable marine organisms (e.g., fishes, crustaceans, mollusks) (Riquelme and Avendaño-Herrera 2003; Gilles et al. 2013). Copepods and barnacle larvae are used as live food in aquaculture for fishes (Støttrup 2000; López et al. 2010). Microalgae, copepods, and barnacles are also increasingly used as bioindicator species in ecotoxicology (Raisuddin et al. 2007; Buttino et al. 2012; Cid et al. 2012; Piazza et al. 2012; Lopes et al. 2018) and as models to predict the impact of chemicals on communities. These organisms are also cultivated to investigate species and population characteristics (growth, development, interactions, production, settlement, and life cycles) (Thiyagarajan et al. 2003a, b; Lee et al. 2013; Freckelton et al. 2017).

The presence of small amounts of organic matter in cultures can lead to a 10- to 1000-fold increase in bacteria in less than 24 h (D'Agostino 1975). Such rapid microbial growth is often accompanied by marked decreases in pH, dissolved oxygen, and concentrations of dissolved nutrients, as well as the release of potentially toxic bacterial metabolites (Fitt et al. 1992; Carvalho and Fernandes 2010). Bacteria associated with diatoms are usually very tenaciously attached, and some may penetrate the gelatinous sheaths and harm phytoplankton (Subhash et al. 2004; Hamdan and Jonas 2007). Bacteria have the potential to negatively affect the growth and reproduction of crustaceans (Wang 2011; Edlund et al. 2012) and are associated with failures in the development of zooplankton cultures (Agostini et al. 2016).

Pathogenic microbes (i.e., virus, bacteria, fungi) generally invade cultures by three principal routes: culture medium, broodstock, and food. Numerous studies have investigated chemical (antibiotics, hypochlorite, iodine, hormones, phenol and detergents) and physical (membrane filtration, autoclave, ultraviolet light, and sonication) treatments aiming to inhibit these microorganisms without negatively affecting the cultivation of eukaryotic organisms (i.e., algae, mollusks, crustaceans) (Spencer 1952; McCracken 1989; Skjermo and Vadstein 1999; Hamdan and Jonas 2007; Liu et al. 2012; Scholz 2014; Agostini et al. 2016). Most studies used different chemicals (mainly the antibiotics penicillin, streptomycin and neomycin) to inhibit bacterial activity without preliminary tests to ensure that the chemical would not affect experimental organisms (Spencer 1952; Droop 1967; McCracken 1989; Skjermo and Vadstein 1999). Most of these studies also used one antibiotic alone and assumed inhibition of the target microorganisms without previously testing the efficiency of these chemicals.

The importance of antibiotics in controlling bacterial growth in marine cultures has been recognized since the 1940s, when they were used in protozoan cultures (Ithaca and Mahmoud 1944). The advantage of antimicrobials over all other methods is their efficiency in microbial inhibition combined with the availability of several different types of antimicrobials. It remains uncertain, however, whether the experimental animals (non-target organisms) would be affected by exposure to these antibiotics. Antibiotics combinations have broad-spectrum bacterial inhibition, but some authors have emphasized that the niche provided by these prokaryotes may be occupied by filamentous fungi and/or yeasts. Application of antifungals would thus inhibit these fungi (Provasoli et al. 1951; Tang et al. 2006; Agostini 2014; Agostini et al. 2016).

The application of antibiotics and antifungals (i.e., antimicrobials) in planktonic cultures can prevent disease and competition (Agostini et al. 2016). This allows ecological studies of the role of bacteria and fungi in aquatic communities (Trottet et al. 2011) and biological (genetic, taxonomy, development, and growth) studies without bacterial or fungal interference (Howes et al. 2014). The goal of this study was thus to evaluate the effect of a combination of penicillin, streptomycin and neomycin with an antifungal agent to select the most effective concentration for preventing bacterial and fungal growth without negatively affecting phytoplankton or zooplankton organisms. The novelty of the present study is the evaluation of the effect of antimicrobial (prokaryote and eukaryote inhibitors) combinations on target and experimental organisms.

Materials and methods

This study consisted of experiments to evaluate the effects of different antibiotics (penicillin, streptomycin and neomycin) at different concentrations (Agostini et al. 2016; DeLorenzo et al. 2001) administered in combination with nystatin (Finley 2012) (Table 1) to plankton cultures. The experimental organisms evaluated were *Acartia tonsa* Dana 1849 (Copepoda: Calanoida), *Amphibalanus improvisus* (Darwin 1854) (Crustacea: Cirripedia), and *Conticribra weissflogii* (Grunow) Stachura-Suchoples & Williams (Ochrophyta: Bacillariophyceae). These organisms represent zooplankton (holoplankton and meroplankton) and phytoplankton, respectively.

The animals were obtained from zooplankton samples collected by means of horizontal tows performed at the surf zone of Cassino Beach, Rio Grande, RS, Brazil. All samples were collected with a conventional 200-μm plankton net with a 30-cm mouth diameter. The microalgae *C. weissflogii* were obtained from cultures kept at the Marine Phytoplankton and Microorganisms Laboratory of the Federal University of Rio Grande in F/2 medium (Guillard 1975) at salinity 30, temperature 25 ± 1 °C and a 12hL:12hD (Light:Dark) photoperiod under 70 mol photons s^{-1} m^{-2} artificial illumination inside BOD incubators (Marconi 403).

Copepods and barnacles were identified in the laboratory (Sabatini 1990; Jones and Crisp 1954, respectively). Healthy specimens of *A. tonsa* and *A. improvisus* were manually picked using Pasteur pipettes under a stereoscopic microscope (Olympus SZ40). All specimens were classified according to the development stage (i.e., copepodites VI males and females of *A. tonsa* and nauplius VI of *A. improvisus* to obtain cyprids of the same age after metamorphosis). Before the experiments were performed, the organisms were submitted to a 30-h acclimation under similar field conditions (e.g., salinity 30, temperature 25 °C and a 14hL:10hD photoperiod) to remove any organism that had been injured or debilitated by stress inherent in collection and sorting. Adults of *Acartia tonsa* and nauplius VI of *A. improvisus* were acclimated separately.

After acclimation, eight replicates containing two copepods, one male and one female, were placed in 50-mL experimental units (EUs) (density of 1 org 25 mL^{-1}). Each EU was fitted with a 140-μm mesh to separate the copepod eggs and fecal pellets produced (Runge and Roff 2000). Copepods were exposed to different antimicrobial treatments (Table 1) following different application routes: antimicrobial applications only in the copepod culture medium (M), only in the culture medium of their food (F), and in the culture media of both the copepods and their food (M/F). The experiment was static; in this way, the treatments were applied once per culture (time 0). Recently molted cyprid larvae were deposited in each EU individually (1 org 20 mL^{-1}) under the same acclimation conditions, with six replicates per treatment for barnacles. Cyprids were then exposed to different antimicrobial treatments, as indicated in Table 1, and the experiment was static.

The copepod and barnacle cultures were observed every 24 h for 96 h. Eggs and fecal pellets produced by copepods were counted only at the end of the experiment (96 h) to evaluate the effects of the treatments on reproduction and feeding, respectively. Copepods were fed daily with the diatom *C. weissflogii* at a

Table 1 Summary of the treatments tested on experimental (copepod, barnacle, microalgae) organisms

Abbreviature	Composition
Control	Untreated
T_A	0.025 g L^{-1} penicillin G potassium + 0.08 g L^{-1} streptomycin sulfate + 0.04 g L^{-1} neomycin sulfate (Agostini et al. 2016)
$T_{A+nystatin}$	0.025 g L^{-1} penicillin G potassium + 0.08 g L^{-1} streptomycin sulfate + 0.04 g L^{-1} neomycin sulfate + 0.05 g L^{-1} of nystatin
T_D	0.025 g L^{-1} penicillin G potassium + 0.04 g L^{-1} streptomycin sulfate + 0.08 g L^{-1} neomycin sulfate (DeLorenzo et al. 2001)
$T_{D+nystatin}$	0.025 g L^{-1} penicillin G potassium + 0.04 g L^{-1} streptomycin sulfate + 0.08 g L^{-1} neomycin sulfate + 0.05 g L^{-1} nystatin

concentration of 20,000 cells mL^{-1} (Teixeira et al. 1896). Because cyprids do not feed at this stage, no food was added to their EU.

In the experiments with the microalgae *C. weissflogii*, the initial concentration was 31 cells mL^{-1} in 200 mL of medium, with six replicates per treatment. The same growing conditions were maintained, with the exception of the photoperiod, which was changed to 14hL:10hD to follow the same conditions of the experiments with the zooplankton. Due to the absence of aerators, the EUs were shaken every 6 h to prevent sedimentation of the cells. The experiment was static. To follow the growth of *C. weissflogii*, 2 mL of each culture was sampled at 24, 96, and 168 h at the same time and after homogenization of the EU. This subsample was deposited in Eppendorf tubes and preserved with lugol (1%). Cell counts were performed in Neubauer chambers under a microscope (40 × – Olympus BH-2) to quantify growth (density per exposure time) and yield (final cell density – initial cell density).

After the first set of experiments, the antifungal nystatin was added at different concentrations (based on Lopes 2014) to the antibiotic combinations that produced the highest survival and production rates of *A. tonsa* compared to the control (Table 1). This new set of treatments (Table 2) was performed in two steps. In the first step, the survival of the calanoid copepod *A. tonsa* under the treatments was assessed (Table 2). In the second step, the efficacy on the inhibition of biofilm bacteria and adhered fungi in marine cultures and the half-life of the best antimicrobial combinations obtained in the first step were investigated.

Individuals of *A. tonsa* were collected from zooplankton tows and acclimated under the same conditions as described above. After acclimation, adult organisms were separated without sex distinction (ISO 14669 1999) and placed in pairs in each EU with 40 mL of culture medium (1 org 20 mL^{-1}) (Lopes et al. 2018), representing seven treatments with ten replicates each (Table 2). The cultures were kept under the same condition as in the previous experiments, with a 14hL:10hD photoperiod. *C. weissflogii* at 20,000 cells mL^{-1} (Teixeira et al. 1896) was added daily to each EU as food. The experiment was semi-static, with the culture medium renewed daily. To evaluate the impact of treatments on copepod survival, each EU was observed every 24 h for 96 h.

To evaluate the antimicrobial efficacy and half-life, eight substrates of wood (20 × 10 mm) that had been previously cleaned with 70% ethyl alcohol (Caixeta et al. 2012; Agostini et al. 2017) were placed in each EU (120 mL), with three replicates per treatment. To prevent biofilm colonization on only one side of the wooden substrates, the EUs were shaken five times a day. All EUs were kept under the same conditions as in the previous experiments inside a BOD incubator (Marconi 403). To assess the initial effects of the antibiotics as well as the half-life in artificial seawater, the experiment was static, and antimicrobials were applied only once (time 0).

To estimate the biofilm bacterial density, three aliquots from three substrates were taken after 3, 6, 9, 12, 15, 18, 21, 24 and 168 h of exposure (based on Lopes 2014) and placed in sterile saline solution (20 mL). The microorganisms had to be detached from the wooden substrates using three pulses of 20 kHz for 15 s on each side of the substrate (Oliveira et al. 2006) using a Cole-Parmer® ultrasound (series 4710). The detached biofilm solution was fixed with 4% sterile formaldehyde, placed inside Eppendorf tubes, and stored in the dark at 8 °C until analysis. Biofilm bacterial density (org cm^{-2}) was estimated in a flow cytometer (BD FACS-Verse™) at the Faculty of Pharmacy of the Federal University of Rio Grande do Sul (UFRGS) (Agostini et al. 2016, 2017).

Table 2 Summary of the treatments tested on target (bacteria and fungi) and experimental (copepod, barnacle, microalgae) organisms

Abbreviature	Composition
Control	Untreated
T_A	0.025 g L^{-1} penicillin G potassium + 0.08 g L^{-1} streptomycin sulfate + 0.04 g L^{-1} neomycin sulfate (Agostini et al. 2016)
$T_A + n_1$	T_A + 0.0025 g L^{-1} nystatin
$T_A + n_2$	T_A + 0.005 g L^{-1} nystatin
$T_A + n_3$	T_A + 0.01 g L^{-1} nystatin
$T_A + n_4$	T_A + 0.015 g L^{-1} nystatin
$T_A + n_5$	T_A + 0.02 g L^{-1} nystatin

To evaluate the microbial community, biological material (of water and substrate) obtained from the EUs was filtered on polycarbonate filters (darkened with irgalan black), stained with acridine orange (1%) and viewed under an epifluorescence microscope (Zeiss Axioplan) at 1000× magnification. The presence or absence of fungi was also recorded, and bacterial morphotypes were classified (Zaritski 1975). The size (μm) and cell complexity were evaluated using the mode Forward Light Scatter (FSC-A) and Light Side Scatter (SSC-A) parameters using 6-μm latex beads (Molecular Probes®) as the standard (Herzenberg et al. 2006; Bouvier et al. 2011; Picot et al. 2012). The cell size (μm) estimated by flow cytometry was compared with measurements performed over epifluorescence photos (100 bacteria manually measured). The bacterial cell biovolume (μm^3) estimation (Sun and Liu 2003) and conversion to bacterial cell biomass (pg C $cell^{-1}$) were made using the allometric conversion factor ($0.09^*biovolume^{0.09}$) (Norland 1993).

General linear model (GLM) analysis was performed for *A. tonsa* survival and production, *A. improvisus* survival and settlement, and *C. weissflogii* and bacterial growth at each exposure time using a statistical computing package (R Development Core Team 2016). The model used was adapted to the data with a binomial distribution (survival and settlement) with a "logit" link function and a Poisson distribution (growth and production) with a "log" link function. Post hoc Tukey's test was performed after the analyses. After analysis, antimicrobials were stored in sealed containers and collected by the Engineering and Environmental Services, SANIPLAN©.

Results

Significant survival rates of *Acartia tonsa* among treatments were observed. Significant differences occurred among the antimicrobial treatments applied in M after 48 h of exposure. At 48 h, 100, 100, 38, 88 and 25% survival rates were observed in the control, T_A, $T_{A+nystatin}$, T_D and $T_{D+nystatin}$, respectively. $T_{D+nystatin}$ was significantly different from the control ($p < 0.028$) and T_A ($p < 0.028$). At 72 h of exposure, $T_{A+nystatin}$ and $T_{D+nystatin}$ showed lower survival rates (13 and 0%, respectively), being different from all treatments ($p < 0.029$). At 96 h of exposure, no significant difference was observed between the control and the T_A and T_D treatments ($p > 0.985$); copepod survival rates were 88, 88 and 75%, respectively (Fig. 1a). When antimicrobials were applied only in F, no significant differences were observed ($p > 0.441$) (Fig. 1b). At 96 h of exposure, the average survival was 88, 75, 50, 63 and 63% for the control, T_A, $T_{A+nystatin}$, T_D and $T_{D+nystatin}$, respectively. When antimicrobials were applied in M/F, differences were observed from 48 h of exposure, where 100, 88, 0, 75 and 38% survival were recorded in the control, T_A, $T_{A+nystatin}$, T_D and $T_{D+nystatin}$, respectively. $T_{A+nystatin}$ ($p < 0.002$) was significantly different from the control, as well as from T_A ($p < 0.007$) and T_D ($p = 0.027$). At 72 h, $T_{A+nystatin}$ and $T_{D+nystatin}$ showed no survivors, differing from all treatments ($p < 0.028$). At the end of the experiment (96 h), no significant difference was observed between the control (88%) and the T_A (88% survival) and T_D (75% survival) treatments (Fig. 1c).

Significant differences in the egg production of *A. tonsa* ($female^{-1}$ day^{-1} at 96 h) between treatments ($p < 0.001$) were observed (Fig. 1d). T_A showed higher average egg production than the control (20 ± 3), with significant differences when antimicrobials were applied in M ($p < 0.001$), F ($p < 0.048$), and M/F ($p < 0.001$) at 49 ± 6, 27 ± 7, and 63 ± 24, eggs $female^{-1}$ day^{-1}, respectively. In M/F and M, $T_{A+nystatin}$ and $T_{D+nystatin}$ also presented differences from the control ($p < 0.001$), as the copepods under those antimicrobial treatments did not produce any eggs. $T_{A+nystatin}$ showed egg production only in F (16 ± 8), which was statistically equivalent to the control ($p = 0.088$). No differences in T_D between M (13 ± 8) and F (8 ± 5) and the control ($p = 0.088$) were observed. In M/F, 36 ± 21 eggs were produced, which was significantly higher than the control ($p < 0.001$). $T_{D+nystatin}$ only presented egg production in F (7 ± 1), with significant differences from the control ($p < 0.044$). T_A showed higher average egg production (46 ± 11 eggs $female^{-1}$ day^{-1}). Among the different antimicrobial application routes, M/F presented the best average results (25 ± 11 eggs $female^{-1}$ day^{-1}).

Fecal pellet production by *A. tonsa* ($copepod^{-1}$ day^{-1} at 96 h) in those routes presented significant differences between treatments ($p < 0.001$) (Fig. 1e). T_A in M/F showed a higher average pellet count (60 ± 16), followed by M (60 ± 26) and F (22 ± 17); but these values did not statistically differ from the control (49 ± 18). $T_{A+nystatin}$ showed fecal pellet production only in F (28 ± 10), equivalent to the control. No differences in T_D between the different application methods and the control were observed. When

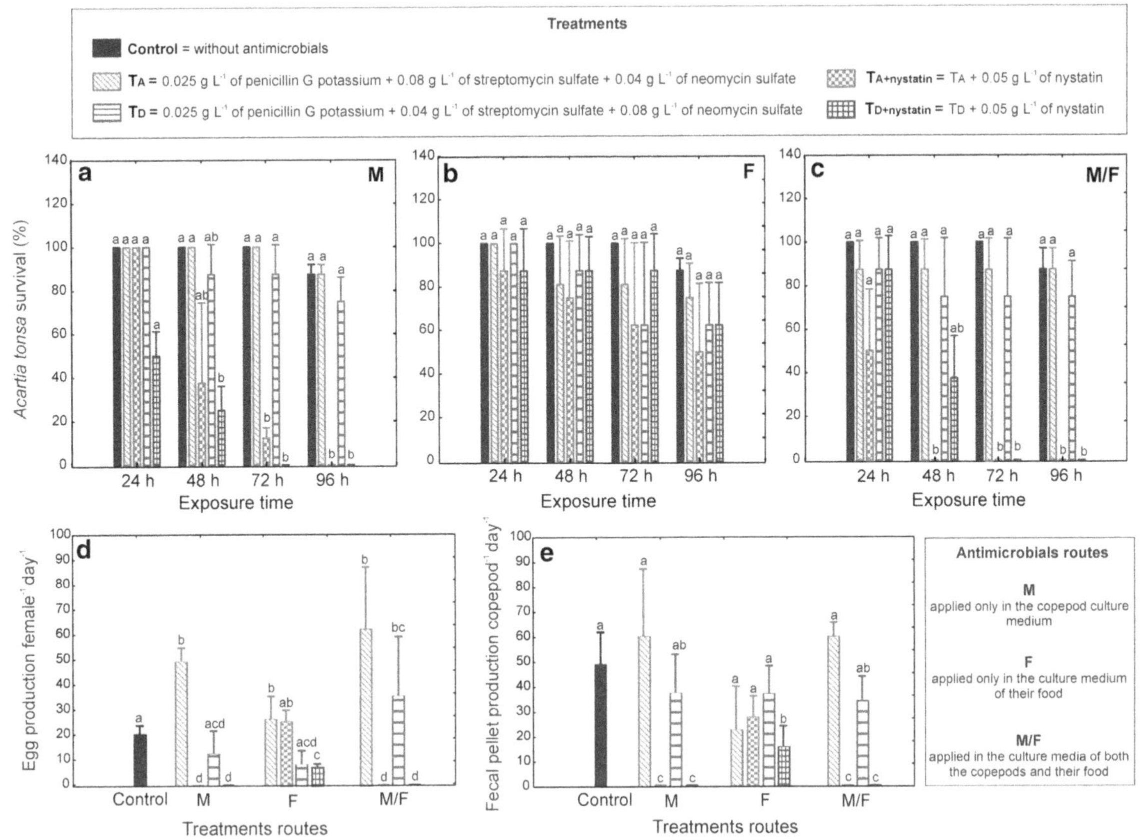

Fig. 1 *Acartia tonsa* copepod survival and egg and fecal pellet production under different treatments and application routes. **a** Survival under antimicrobial exposure in the culture medium; (M) **b** survival under antimicrobial exposure through food (F); **c** survival under antimicrobial exposure in the culture medium and through food (M/F); **d** egg production (female^{-1} day^{-1} at 96 h); **e** fecal pellet production (copepod^{-1} day^{-1} at 96 h). The vertical lines denote 95% confidence intervals (standard error*1.96), and lowercase letters indicate statistical similarities or differences between treatments for each time or situation evaluated

antimicrobials were administered in M, 38 ± 17 pellets were produced, followed by F (37 ± 12) and M/F (35 ± 10). $T_{D+nystatin}$ only presented fecal pellet production in F (16 ± 8), with significant differences from the control ($p < 0.001$). Among the antimicrobial application methods, T_A applied in M/F and M showed the best average results.

No differences were observed in *A. improvisus* survival ($p > 0.991$) (Fig. 2a) or settlement ($p > 0.692$) (Fig. 2b) between treatments. $T_{A+nystatin}$ and T_D showed a lower survival average (83%). $T_{A+nystatin}$ also presented a lower settlement percentage at 96 h of exposure (50%). A higher settlement percentage was observed in the control (83%), but $T_{A+nystatin}$ only settled at 48 h. Significant differences in *C. weissflogii* growth rates between treatments were also found at 96 h ($p < 0.001$) (Fig. 2c). At 24 h, no significant differences were observed between all treatments with antimicrobials and the control (109 ± 76 cells mL^{-1}), although T_A presented higher growth rates (198 ± 61 cells mL^{-1}). At 96 h of exposure, $T_{A+nystatin}$ (109 ± 21 cells mL^{-1}) and $T_{D+nystatin}$ (78 ± 42 cells mL^{-1}) had a lower average growth rate than the control (266 ± 100 cells mL^{-1}) ($p < 0.001$). At approximately 168 h of exposure, only T_A (536 ± 134 cells mL^{-1}) and T_D (365 ± 83 cells mL^{-1}) were different from the control (270 ± 52 cells mL^{-1}) ($p < 0.001$), with a higher cell concentration. T_A showed the highest yield (505 cells mL^{-1}), followed by T_D (333 cells mL^{-1}), $T_{A+nystatin}$ (193 cells mL^{-1}), the control (177 cells mL^{-1}), and $T_{D+nystatin}$ (162 cells mL^{-1}).

In the resistance tests with *A. tonsa*, *A. improvisus* and *C. weissflogii*, the treatments with the antifungal nystatin (0.05 g L^{-1}) were the most harmful to non-target organisms, as was the antibiotic concentration T_D. Subsequent tests (Table 2) thus applied antimicrobials only in the culture medium, the condition in which the copepod *A. tonsa* was the most sensitive in the current study. The results presented significant survival rates of

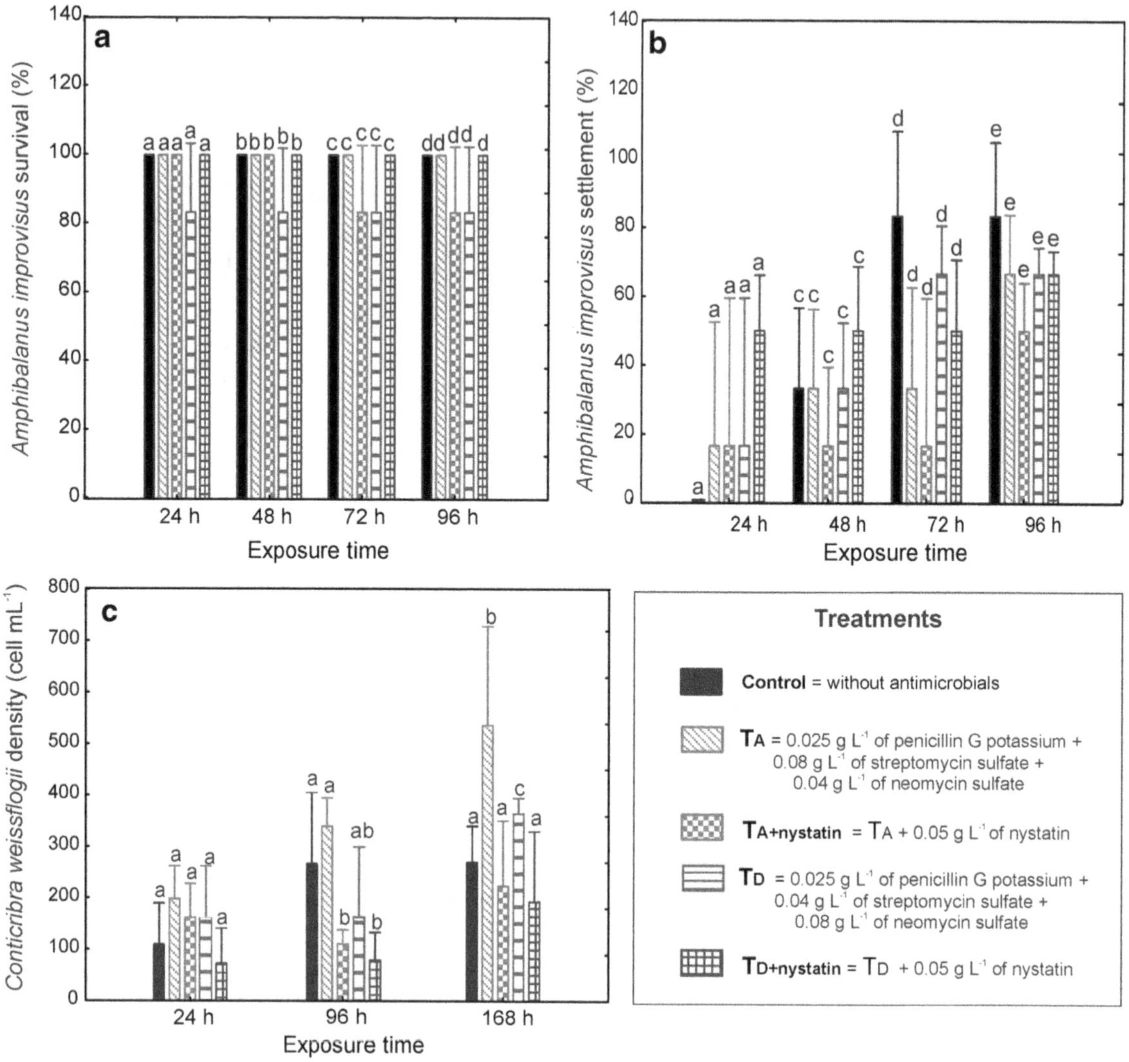

Fig. 2 *Amphibalanus improvisus* cyprid **a** survival and **b** settlement at 24, 48, 72 and 96 h; **c** *Conticribra weissflogii* diatom growth (cells mL^{-1}) at 24, 96, and 168 h of exposure. The vertical lines denote 95% confidence intervals (standard error*1.96), and lowercase letters indicate statistical similarities or differences between treatments for each time evaluated

A. tonsa between treatments after 48 h of exposure. $T_A + n_4$ and $T_A + n_5$ were lethal to copepods, different from the control (90% survival) and the other antimicrobial treatments ($p < 0.003$). The same situation occurred at 72 h of exposure; however, the survivorship in the control was 80% ($p < 0.009$). At 96 h, no significant differences ($p = 0.960$) between the control (80% survival) and T_A (100% survival), $T_A + n_1$ (70% survival), and $T_A + n_2$ (65% survival) were observed. $T_A + n_3$ (20% survival) ($p < 0.009$), $T_A + n_4$ (0% survival) ($p < 0.010$) and $T_A + n_5$ (0% survival) ($p < 0.034$), in contrast, showed the lowest survival rates when compared to the control.

The bacterial density (org cm^{-2}) decreased in treatments with antimicrobials (T_A, $T_A + n_1$ and $T_A + n_2$), as expected, compared to the control at 3 ($p < 0.007$), 6 ($p < 0.001$), 9 ($p < 0.003$), 12 ($p < 0.001$), 15 ($p < 0.001$), 18 ($p < 0.001$), and 21 ($p < 0.001$) h of exposure (Fig. 3a). In the first 3 h of the experiment, a reduction in the bacterial density (> 26%) was observed compared with the control. Between 9 and 15 h, an inhibition of up to 94% of the biofilm bacterial density was recorded compared to the control. After 15 h, bacterial densities increased in all antimicrobial treatments. The bacteria found in the control and antimicrobial treatments were in the coccus form with similar sizes.

All treatments with antimicrobials had the same cell size (FSC-A), biovolume and biomass values as the control at 12 h of exposure: 1.11 μm, 0.72 μm^3 and 0.09 pg C $cell^{-1}$, respectively. All treatments showed just

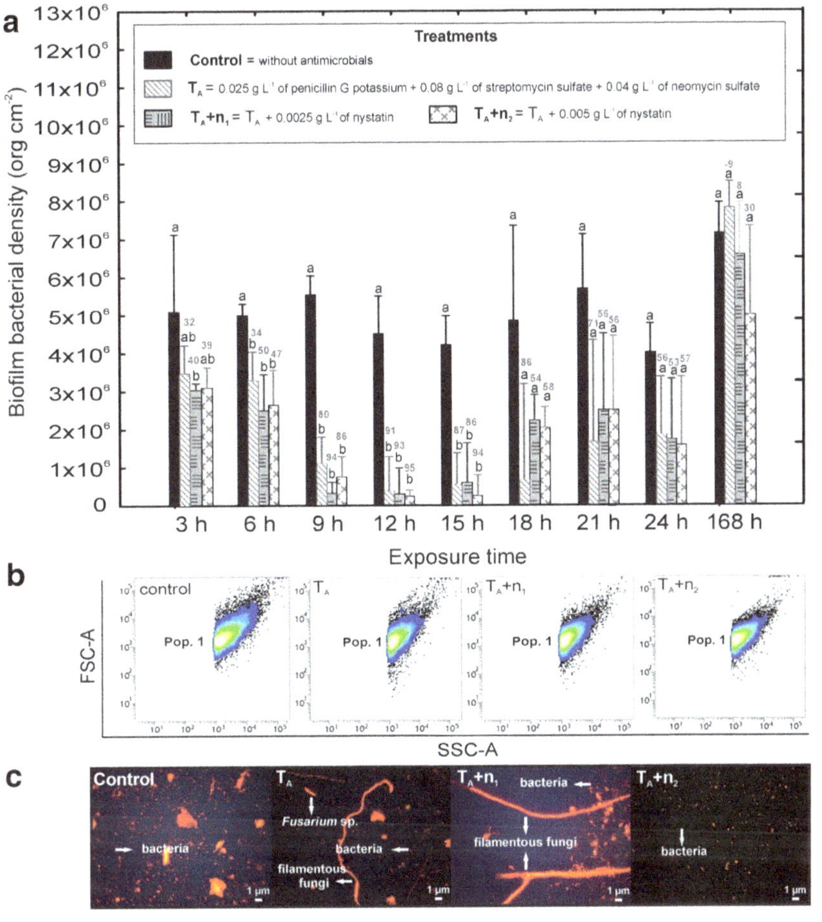

Fig. 3 **a** Biofilm bacteria density from 3 to 168 h of exposure. The vertical lines denote 95% confidence intervals (standard error*1.96). The lowercase letters indicate statistical similarities or differences between treatments at each time evaluated. The numbers above the columns denote the bacteria inhibition percentage (%) compared to the control. **b** biofilm bacteria populations by flow cytometer (BD FACSVerse^{TM}) after 10,000 acquired events at 12 h. Lighter colors are related to higher density cells; **c** microbial community under epifluorescence microscope (×1000) stained with acridine orange with emphasis on fungi presence/absence at 168 h

one population and the same cell complexity (SSC-A) (Fig. 3b). The following size, biovolume and biomass were verified independent of treatment using an epifluorescence microscope at 12 h: 0.61 μm, 0.12 μm^3, and 0.07 pg C $cell^{-1}$, respectively. Overall, all antimicrobial treatments were effective in reducing bacterial density, but only $T_A + n_2$ prevented fungal colonization. No fungi were observed in the control. The antimicrobial treatment without nystatin (T_A) showed filamentous fungi and *Fusarium* sp., while $T_A + n_1$ showed only a filamentous form (Fig. 3c).

Discussion

Our findings indicated that the application of T_D (0.025 g L^{-1} penicillin G potassium + 0.04 g L^{-1} streptomycin sulfate + 0.08 g L^{-1} neomycin sulfate) proposed by DeLorenzo et al. (2001) resulted in a lower survival rate, with lower egg and fecal pellet production for the copepod *Acartia tonsa* when compared with T_A (0.025 g L^{-1} penicillin G potassium + 0.08 g L^{-1} streptomycin sulfate + 0.04 g L^{-1} neomycin sulfate) proposed by Agostini et al. (2016). The differences between these two treatments are the streptomycin and neomycin concentrations. The higher concentration of neomycin sulfate (0.08 g L^{-1}) proposed by DeLorenzo et al. (2001) caused a greater mortality of *A. tonsa* than did the same concentration of streptomycin proposed by Agostini et al. (2016).

Antibiotics such as penicillin, streptomycin and neomycin have been applied in cultures to inhibit bacterial contamination without negative effects in microalgae (Molina-Cárdenas et al. 2016), protozoans (Divan and Schnoes 1982), fish (Forberg et al. 2011), mollusks (Howes et al. 2014), crustaceans (Agostini et al. 2016) and ecological assays with the aquatic community (Middelburg and Nieuwenhuize 2000a, b; DeLorenzo et al. 2001; Veuger et al. 2004; Cozzi and Cantoni 2006; Trottet et al. 2011). The concentration of these antibiotics, however, needs to be evaluated before application in cultures. In this study, T_A and T_D, composed only of antibiotics, showed different responses in non-target organisms due to differences in the concentrations of some compounds.

Acartia tonsa, *A. improvisus* and *C. weissflogii* are possibly more sensitive to high concentrations of neomycin than to streptomycin. Droop (1967), for example, used 0.04 g L^{-1} neomycin in combination with other antibiotics in diatom cultures. Jones et al. (1973) applied only 0.0000001 to 0.000016 g mL^{-1} mixed with other prokaryotic inhibitors in cyanobacteria cultures, while Green et al. (1967) purified seaweed cultures with 0.0002 g L^{-1} neomycin mixed with other antibiotics. In addition to the lower resistance to T_D compared to T_A, T_{A+N} and T_{D+N} had lower survival of non-target organisms, probably due to the nystatin concentration used (0.05 g L^{-1}).

Kaminski and Montú (2005) reported *A. tonsa* egg production of 14–34 eggs $female^{-1}$ day^{-1} when fed with *Nannochloropsis oculata* (Droop) Hibberd and *Chaetoceros calcitrans* (Paulsen) Takano in excess. Teixeira et al. (1896) observed an average of 28 eggs $female^{-1}$ day^{-1} when fed with *C. weissflogii* at the same concentrations as used this study. The data presented here indicate a beneficial effect of those antibiotics in copepod cultures; our results recorded an average of 63 eggs $female^{-1}$ day^{-1} in T_A in M/F and 49 eggs $female^{-1}$ day^{-1} in T_A in M. In the control, we found 20 eggs $female^{-1}$ day^{-1}. The number of fecal pellets produced by copepods can be used as an additional tool to evaluate the nutritional quality of the food (Ianora et al. 1995), although food quality, concentration and size must be considered. Shaw et al. (1994) observed that the culture medium of the copepod *Tigriopus californicus* (Baker, 1912) without antibiotics had a production rate of 1.6 pellets h^{-1}, while cultures with antibiotics had a production rate of 1.9 pellets h^{-1} with the same amount of food. The current study confirms the results of Shaw et al. (1994) because T_A presented higher production of pellets in M/F and M (2.5 pellets h^{-1}) than the control (2.0 pellets h^{-1}). The effect of the antimicrobials on fecal pellets and egg production should be analyzed carefully because production was observed during short periods and the antimicrobials were not replaced, decreasing their effect over the duration of the experiment. The benefit of using antibiotics in cultures was also reported by Tighe-Ford et al. (1970), who used 100 μL L^{-1} Crystamycin® (0.3 g of penicillin G sodium and 0.5 g of streptomycin sulfate in 2 mL of distilled water) in barnacle cultures of *Austrominius modestus* (Darwin 1854) to virtually eliminate bacterial growth and increase the rate of survival of larvae (Tighe-Ford et al. 1970; Harms 1987). The survival of *Amphibalanus improvisus* in all treatments indicates that these antimicrobials can be applied in cultures of this species without causing lethal damage. Thus, we assume that barnacles are less susceptible than copepods and microalgae. Laboratory cultures allow the study of the growth, development, metamorphosis, fertility and nutritional needs of species as a means of research into the role of organisms in nature (D'Agostino 1975). Thus, the application of $T_A + n_2$ in marine plankton cultures can have different objectives.

We expected that the antimicrobial treatments would lead to lower *A. improvisus* settlement rates, because the presence of bacterial biofilms has been reported to condition larval settlement of a number of fouling benthic species (Unabia and Hadfield 1999; Lau et al. 2002; Hung et al. 2005; Bao et al. 2010; Agostini et al. 2017), such as barnacles. This relationship was only observed from 72 h of exposure, when the control reached the highest settlement rate. Because the antibiotics were applied only at the beginning of the cultures, we believe that at 24 h, the effect of the treatments had decreased, resulting again in an increase in the bacterial load.

The use of $T_{A+nystatin}$, T_D and $T_{D+nystatin}$ in cultures of the diatom *C. weissflogii* did not produce differences when compared to the control. This indicates the possibility of applying these treatments for purification of microalgae cultures. T_A presented better results than the control, being the most recommended treatment for intensification of microalgal growth. The use of antibiotics in microalgae culture has been reported to be beneficial because bacteria actively compete for the same resources necessary for the survival of phytoplankton (Spencer 1952; Subhash et al. 2004; Hamdan and Jonas 2007). It should be considered, however, that this study was performed only with *C. weissflogii* and that the half-life of antimicrobials reduces the effect over time. In addition, bacteria can mediate a variety of harmful or beneficial interactions with eukaryotic organisms. Beneficial interactions would be the acquisition of vitamin B_{12} through symbiotic relationships with bacteria (Croft et al. 2005). In contrast, the disruption of the microflora would likely delay the development of some juvenile crustaceans (Edlund et al. 2012). Antimicrobial substances can thus affect symbiont-mediated interactions in hosts, negatively so if this inhibition continues over time. Lethal effects on microalgae are the lowest when using penicillin compared to other antibiotics (Youn and Hur 2007), although the efficiency of the antibiotics and their concentrations for axenic cultures vary with microalgal species (Lai et al. 2009). The establishment of Bacillariophyceae and Dinophyceae axenic cultures is more difficult than that of Chlorophyceae and Haptophyceae because of their complicated external morphology (Youn and Hur 2007).

Our results indicate that the copepod *A. tonsa* is more sensitive than *A. improvisus* and *C. weissflogii*, especially when treatments are applied only in the culture medium. This result was somewhat expected because this copepod species is very sensitive and is indicated for acute ecotoxicological tests by the International Standardization Organization (ISO 14669 1999). *A. tonsa* was therefore selected to test the other treatments with different concentrations of nystatin. The results obtained in the settlement of *A. improvisus* prompted us to re-apply the antimicrobial every 24 h to maintain the culture medium throughout the experiment. Testing different concentrations of nystatin in combination with T_A revealed that T_A, $T_A + n_1$ and $T_A + n_2$ could be applied to *A. tonsa* cultures. According to Lopes (2014), it is possible to apply up to 0.02 g L^{-1} nystatin in combination with antibiotics in cultures without lethal effects on *A. tonsa* for up to 48 h. There was a higher mortality of copepods in the treatments that showed the antifungal nystatin, probably because nystatin is a eukaryotic inhibitory substance (Groll et al. 1999). The use of an antifungal, however, is essential for avoiding fungal contamination due to the ecological niche provided by bacteria (Agostini et al. 2016). Diseases in cultures of invertebrates are caused by fungi of the genera *Aspergillus*, *Penicillium*, and *Fusarium* (Santos et al. 2016).

T_A, $T_A + n_1$ and $T_A + n_2$ showed inhibition in biofilm bacterial density of up to 94% compared to the control at 9–15 h of exposure. In this time interval, the microbial community started to restore, and a new dosage of antibiotics was necessary (Agostini et al. 2016). Lopes (2014) obtained adherent bacterial reduction of up to 95% using the same combination of antimicrobials. Trottet et al. (2011) tested six different antibiotics under laboratory conditions. These authors found that penicillin and streptomycin showed the greatest effectiveness in inhibiting bacterial density. Penicillin targets Gram-positive bacteria; streptomycin targets Gram-negative and Gram-positive bacteria, and neomycin targets Gram-negative bacteria and some Gram-positive bacteria. This combination thus inhibits various bacterial species (i.e., *Pseudomonas*, *Vibrio*, *Agrobacterium*, *Xanthomonas*, *Achromobacter*, *Flavobacterium*, *Micrococcus*, and *Staphylococcus*) (Berland and Maestrini 1969).

Fungal proliferation could only be avoided by using $T_A + n_2$. To eliminate bacterial and fungal contamination in marine cultures without causing lethal damage to non-target organisms, we suggest the application of $T_A + n_2$. To maintain the culture medium with a minimum bacterial load for more time, antibiotics should be replaced every 12 h. Further studies are therefore necessary to evaluate the effects of these antimicrobials on non-target organisms and to examine alternatives to these substances in culture medium. The antifungal nystatin may be reapplied over a longer time interval, considering that in 7 days (168 h) of observation, the presence of adherent fungi was not recorded in $T_A + n_2$.

Conclusion

We found an absence of negative effects on experimental organisms and efficiency in the inhibition of biofilm bacteria using a combination of 0.025 g L^{-1} penicillin G potassium + 0.08 g L^{-1} streptomycin sulfate + 0.04 g L^{-1} neomycin sulfate diluted in seawater. Nevertheless, the absence of adherent fungi was only observed using 0.005 g L^{-1} nystatin antifungal together with the antibiotics. This concentration of nystatin did not show significant differences from the control regarding *A. tonsa* survival, but this substance should be avoided if fungal contamination is not a concern because this substance is a eukaryote inhibitor. Species of other planktonic groups should also be tested at this concentration to improve our understanding of the effectiveness of the treatments proposed.

Acknowledgements The authors acknowledge the support of FAPERGS-PRONEM 16/2551-000244-4, Conselho Nacional de Desenvolvimento Científico e Tecnológico (CNPq-141217/2014-6, 408578/2013-0, 303353/2016-3), and Marine Phytoplankton and Microorganisms Laboratory of the Federal University of Rio Grande and Faculty of Pharmacy of Universidade Federal do Rio Grande do Sul. We thank the veterinarian M Sc. Alice Meirelles Leite for supplying the prescriptions for the antimicrobials, Dr. Lucélia Borges and Biol. Waldemar Appolinário Amaral for their support in the development of the experiments, Mr. Camilo D'Amore for help with cytometer analyses, and M. Sc. Graciéle Cunha Alves for fungi identification.

References

Agostini VO (2014) Avaliação dos efeitos do uso de antimicrobianos em cultivos de plâncton marinho. Dissertation, Universidade Federal do Rio Grande, Brasil. 124 pp

Agostini VO, Macedo AJ, Muxagata EM (2016) Evaluation of antibiotics as a methodological procedure to inhibit free-living and biofilm bacteria in marine zooplankton culture. An Acad Bras Cienc 88(1):733–746. https://doi.org/10.1590/0001-3765201620150454

Agostini VO, MdoN Ritter, Macedo AJ, Muxagata E, Erthal F (2017) What determines sclerobiont colonization on marine mollusk shells? PLoS One 12(9):e0184745. https://doi.org/10.1371/journal.pone.0184745

Bao W-Y, Lee O-O, Chung H-C, Li M, Qian P-Y (2010) Copper affects biofilm inductiveness to larval settlement of the serpulid polychaete *Hydroides elegans* (Haswell). Biofouling 26:119–128. https://doi.org/10.1080/08927010903329680

Berland BR, Maestrini SY (1969) Action de quelques antibiotiques sur le developpement de cinq diatomlees en culture. J Exp Mar Bio Ecol 3:62–75

Bouvier T, Troussellier M, Anzil A, Courties C, Servais P (2011) Using light scatter signal to estimate bacterial biovolume by flow cytometry. Cytometry 44:188–194

Buttino I, Ianora A, Buono S, Vitiello V, Malzone MG, Rico C, Langellotti A, Sansone G, Gennari L, Miralto A (2012) Experimental cultivation of the Mediterranean calanoid copepod *Temora stylifera* and *Centropages typicus* in a pilot re-circulation system. Aquaculture 43:247–259. https://doi.org/10.1111/j.1365-2109.2011.02822.x

Caixeta DS, Scarpa TH, Brugnera DF, Freire DO, Alves E, Abreu LR, Piccoli RH (2012) Chemical sanitizers to control biofilms formed by two Pseudomonas species on stainless steel surface. Ciênc Tecnol Aliment 32(1):142–150. https://doi.org/10.1590/S0101-20612012005000008

Carvalho CCCR, Fernandes P (2010) Production of metabolites as bacterial responses to the marine environment. Mar Drugs 8(3):705–727. https://doi.org/10.3390/md8030705

Cid A, Prado R, Rioboo C, Suarez-Bregua P, Herrero C (2012) Use of microalgae as biological indicators of pollution: looking for new relevant cytotoxicity endpoints. In: Johnsen MN (ed) Microalgae: biotechnology, microbiology and energy. Nova Science Publishers, New York, pp 311–323

Cozzi S, Cantoni C (2006) Partition nitrogen uptake in phytoplankton and bacteria using bactericidal agents and light dependent incubations. Period Biol 108:145–150

Croft MT, Lawrence AD, Raux-Deery E, Warren MJ, Smith AG (2005) Algae acquire vitamin B_{12} through a symbiotic relationship with bacteria. Nature 438:90–93. https://doi.org/10.1038/nature04056

D'Agostino A (1975) Antibiotics in cultures of invertebrates. In: Smith WL, Chanicy MH (eds) Culture of marine invertebrate animals. Plenum Press, New York, pp 109–133

DeLorenzo ME, Lewitus AJ, Scott GI, Ross PE (2001) Use of metabolic inhibitors to characterize ecological interactions in an estuarine microbial food web. Microb Ecol 42:317–327. https://doi.org/10.1007/s00248-001-0004-1

Divan CL, Schnoes HK (1982) Production of axenic *Gonyaulax* cultures by treatment with antibiotics. Appl Environ Microbio 44:250–254

Droop MR (1967) A procedure for routine purification of algal cultures with antibiotics. Br Phycol Bull 3:295–297

Edlund A, Ek K, Breitholtz M, Gorokhova E (2012) Antibiotic-induced change of bacterial communities associated with the copepod *Nitocra spinipes*. PLoS One 7(3):e33107. https://doi.org/10.1371/journal.pone.0033107

Edwards KF, Thomas MK, Klausmeier CA, Litchman E (2016) Phytoplankton growth and the interaction of light and temperature: a synthesis at the species and community level. Limnol Oceanogr 61:1232–1244. https://doi.org/10.1002/lno.10282

Finley D (2012) Biofiles: cell culture contamination. Sigma-Aldrich 7(1):24

Fitt WK, Hesling GA, Watson TC (1992) Use of antibiotics in the mariculture of giant clams (F. Tridacnidae). Aquaculture 104:1–10. https://doi.org/10.1016/0044-8486(92)90132-5

Forberg T, Arukwe A, Vadstein O (2011) A protocol and cultivation system for gnotobiotic Atlantic cod larvae (*Gadus morhua* L.) as a tool to study host microbe interactions. Aquaculture 315:222–227. https://doi.org/10.1016/j.aquaculture.2011.02.047

Freckelton M-L, Nedved BT, Hadfield MG (2017) Induction of invertebrate larval settlement; different bacteria, different mechanisms? Sci Rep 7:42557. https://doi.org/10.1038/srep42557

Gilles S, Fargier L, Lazzaro X, Baras E, De Wilde N, Drakidès C, Amiel C, Rispal B, Blancheton JP (2013) An integrated fish-plankton aquaculture system in brackish water. Animal 7(2):322–329. https://doi.org/10.1017/S1751731112001279

Green B, Heilporn V, Limbosch S, Bolonkhere M, Brachet I (1967) The cytoplasmic DNAs of *Acetabularia mediterranea*. Proc Natl Acad Sci USA 58:1351–1358

Groll AH, Michkiene D, Werner K, Piscitelli S, Walsh TJ (1999) High-performance liquid chromatographic determination of liposomal nystatin in plasma and tissues for pharmacokinetic and tissue distribution studies. J Chromatogr B 735:51–62

Hamdan L, Jonas R (2007) The use of antibiotics to reduce bacterioplankton uptake of phytoplankton extracellular organic carbon (EOC) in the Potomac River estuary. J Exp Mar Biol Ecol 342:242–252. https://doi.org/10.1016/j.jembe.2006.10.054

Harms J (1987) Energy budget for the larval development of *Elminius modestus* (Crustacea: cirripedia). Helgolinder Meeresun 41:45–67. https://doi.org/10.1007/BF02365099

Herzenberg LA, Tung J, Moore WA, Herzenberg LA, Parks DR (2006) Interpreting flow cytometry data: a guide for the perplexed. Nat Immunol 7(7):681–685. https://doi.org/10.1038/ni0706-681

Howes EL, Bednarsek N, Budenbender J, Comeau S, Doubleday A, Gallager SM, Hopcroft RR, Lischka S, Maas AE, Bijma J, Gattuso J-P (2014) Sink and swim: a status review of thecosome pteropod culture techniques. J Plankton Res 36(2):299–315. https://doi.org/10.1093/plankt/fbu002

Hung OS, Thiyagarajan V, Wu RSS, Qian PY (2005) Effect of ultraviolet radiation on biofilms and subsequent larval settlement of *Hydroides elegans*. Mar Ecol Prog Ser 304:155–166. https://doi.org/10.3354/meps304155

Ianora A, Poulet SA, Miralto A (1995) A comparative study of the inhibitory effect of diatoms on the reproductive biology of the copepod *Temora stylifera*. Mar Biol 121:533–539

ISO-International Organization for Standardization (1999) ISO 14669:1999, Water quality—determination of acute lethal toxicity to marine copepods (Copepoda, Crustacea). Geneva

Ithaca NY, Mahmoud AH (1944) Isolation of *Trichomonas foetus* (Riedmuller, 1928) in bacteria-free cultures by the use of penicillin. Ann Trop Med Parasit 38:219–222

Jones LWG, Crisp DJ (1954) The larval stages of the barnacle *Balanus improvisus* Darwin. Proc Zool Soc Lond 123:765–780

Jones AK, Rhodes ME, Evans SC (1973) The use of antibiotics to obtain axenic cultures of algae. Brit Phycol J 8:185–196

Kaminski SM, Montú MA (2005) Produção de ovos dos copépodes costeiros *Acartia tonsa*, *Temora stylifera* e *Temora turbinata*, da Praia do Cassino—Rio Grande—RS. Atlântica 27(2):103–111

Lai HT, Hou HH, Su CI, Chen CL (2009) Effects of chloramphenicol, florfenicol, and thiamphenicol on growth of algae *Chlorella pyrenoidosa*, *Isochrysis galbana*, and *Tetraselmis chui*. Ecotox Environ Safe 72:329–334. https://doi.org/10.1016/j.ecoenv.2008.03.005

Lau SCK, Mak KKW, Chen F, Qian PY (2002) Bioactivity of bacterial strains isolated from marine biofilms in Hong Kong waters for the induction of larval settlement in the marine polychaete *Hydroides elegans*. Mar Ecol Prog Ser 226:301–310

Lee K-W, Dahms H-U, Park HG, Kang J-H (2013) Population growth and productivity of the cyclopoid copepods *Paracyclopina nana*, *Apocyclops royi* and the harpacticoid copepod *Tigriopus japonicus* in mono and polyculture conditions: a laboratory study. Aquacult Res 44:836–840. https://doi.org/10.1111/j.1365-2109.2011.03071.x

Liu H, Zhou Y, Xiao W, Ji L, Cao X, Song C (2012) Shifting nutrient-mediated interactions between algae and bacteria in a microcosm, evidence from alkaline phosphatase assay. Microbiol Res 167:292–298. https://doi.org/10.1016/j.micres.2011.10.005

Lopes LFP (2014) Aplicação de antimicrobianos no cultivo de *Acartia tonsa* (Copepoda:Calanoida): uma ferramenta para a inibição de bactérias e fungos. Universidade Federel do Rio Grande, Rio Grande, Trabalho de Conclusão de Curso, p 38

Lopes LFP, Agostini VO, Muxagata E (2018) Could some procedures commonly used in bioassays with the copepod *Acartia tonsa* Dana 1849 distort results? Ecotoxicol Environ Saf 150:353–365. https://doi.org/10.1016/j.ecoenv.2017.12.004

López DA, López BA, Pham CK, Isidro EJ, Mde Girolamo (2010) Barnacle culture: background, potential and challenges. Aquacult Res 41:367–375. https://doi.org/10.1111/j.1365-2109.2010.02508.x

McCracken IR (1989) Purifying algal cultures a review of chemical methods. Proc Ns Inst Sci 38:145–168

Middelburg JJ, Nieuwenhuize J (2000a) Uptake of dissolved nitrogen in turbid, tidal estuaries. Mar Ecol Prog Ser 192:79–88

Middelburg JJ, Nieuwenhuize J (2000b) Nitrogen uptake by heterotrophic bacteria and phytoplankton in the nitrate-rich Thames estuary. Mar Ecol Prog Ser 203:13–21

Molina-Cárdenas CA, MdelP Sánchez-Saavedra, Licea-Navarro AF (2016) Decreasing of bacterial content in *Isochrysis galbana* cultures by using some antibiotics. Rev Biol Mar Oceanogr 51(1):101–112. https://doi.org/10.4067/S0718-19572016000100010

Norland S (1993) The relationship between biomass and volume of bacteria. In: Kemp PF, Sherr BF, Sherr EB, Cole JJ (eds) Handbook of methods in aquatic microbial ecology. Lewis Publishers, USA, p 777

Oliveira SS, Wasielesky JRW, Ballester ELC, Abreu PC (2006) Caracterização da assembléia de bactérias nitrificantes pelo método "Fluorescent in situ Hybridization" (FISH) no biofilme e água de larvicultura do Camarão-rosa *Farfantepenaeus paulensis*. Atlântica 28(1):33–45

Piazza V, Ferioli A, Giacco E, Melchiorre N, Valenti A, Delprete F, Biandolino F, Dentone L, Frisenda P, Faimali M (2012) A standardization of *Amphibalanus* (*Balanus*) *amphitrite* (Crustacea, Cirripedia) larval bioassay for ecotoxicological studies. Ecotoxicol Environ Saf 79:134–138. https://doi.org/10.1016/j.ecoenv.2011.12.014

Picot J, Guerin CL, Le Van Kim C, Boulanger CM (2012) Flow cytometry: retrospective, fundamentals and recent instrumentation. Cytotechnology 64:109–130. https://doi.org/10.1007/s10616-011-9415-0

Provasoli L, Pintner I, Packer L (1951) Use of antibiotics in obtaining pure cultures of algae and protozoa. Proc Am Soc Protozool 2:6

Raisuddin S, Kwok KWH, Leung KMY, Schlenk D, Lee JS (2007) The copepod *Tigriopus*: a promising marine model organism for ecotoxicology and environmental genomics. Aquat Toxicol 83:161–173. https://doi.org/10.1016/j.aquatox.2007.04.005

Riquelme CE, Avendaño-Herrera R (2003) Microalgae and bacteria interaction in the aquatic environment and their potential use in aquaculture. Rev Chil Hist Nat 76:725–736

Runge JA, Roff JC (2000) The measurement of growth and reproductive rates. In: Harris RP, Wiebe PH, Lenz J, Skjoldal HR, Huntley M (eds) Zooplankton methodology manual. Academy Press, USA, p 684

R Core Team (2016) R: a language and environment for statistical computing. R Foundation for Statistical Computing, Vienna, Austria. https://www.R-project.org/

Sabatini ME (1990) The developmental stages (Copepodids I to VI) of *Acartia tonsa* Dana, 1849 (Copepoda Calanoida). Crustaceana 59:53–61

Santos A, Hauser-Davis RA, Santos MJ, De Simone SG (2016) Potentially toxic filamentous fungi associated to the economically important *Nodipecten nodosus* (Linnaeus, 1758) scallop farmed in southeastern Rio de Janeiro, Brazil. Mar Pollut Bull 115:75–79. https://doi.org/10.1016/j.marpolbul.2016.11.058

Scholz B (2014) Purification and culture characteristics of 36 benthic marine diatoms isolated from the Solthörn tidal flat (Southern North Sea). J Phycol 50(4):685–697. https://doi.org/10.1111/jpy.12193

Shaw BA, Harrison RJ, Andersen RJ (1994) Evaluation of the copepod *Tigriopus californicus* as a bioassay organism for the detection of chemical feeding deterrents produced by marine phytoplankton. Mar Biol 121:89–95

Skjermo J, Vadstein O (1999) Techniques for microbial control in the intensive rearing of marine larvae. Aquaculture 177:333–343. https://doi.org/10.1016/S0044-8486(99)00096-4

Spencer CP (1952) On the use of antibiotics for isolating bacteria-free cultures of marine phytoplankton organisms. J Mar Biol Assoc UK 31:97–106

Støttrup J (2000) The elusive copepods: their production and suitability in marine aquaculture. Aquacult Res 31:703–711. https://doi.org/10.1046/j.1365-2109.2000.318488.x

Subhash SK, Lipton AP, Paul Raj R (2004) Antibiotic exposure to minimize microbial load in live feed *Isochrysis galbana* used for larval rearing of Indian pearl oyster *Pinctada fucata*. Curr Sci 87:25

Sun J, Liu D (2003) Geometric models for calculating cell biovolume and surface area for phytoplankton. J Plankton Res 25(11):1331–1346. https://doi.org/10.1093/plankt/fbg096

Tang KW, Freund CS, Schweitzer CL (2006) Occurrence of copepod carcasses in the lower Chesapeake Bay and their decomposition by ambient microbes. Estuar Coast Shelf Sci 68:499–508. https://doi.org/10.1016/j.ecss.2006.02.021

Teixeira PF, Sonia MK, Avila TR, Cardozo AP, Bersano JGF, Bianchini A (2010) Diet influence on egg production of the copepod *Acartia tonsa* (Dana 1896). An Acad Bras Cienc 82(2):333–339

Thiyagarajan V, Harder T, Qian PY (2003a) Combined effects of temperature and salinity on larval development and attachment of the subtidal barnacle *Balanus trigonus* Darwin. J Exp Mar Bio Ecol 287:223–236. https://doi.org/10.1016/S0022-0981(02)00570-1

Thiyagarajan V, Harder T, Qiu JW, Qian PY (2003b) Energy content at metamorphosis and growth rate of the early juvenile barnacle *Balanus amphitrite*. Mar Biol 143:543–554

Tighe-Ford DJ, Power MJD, Vaile DC (1970) Laboratory rearing of barnacle larvae for antifouling research. Helgoland Wiss Meer 20:393–405

Trottet A, Fouilland E, Leboulanger C, Lanouguère E, Bouvy M (2011) Use of inhibitors for coastal bacteria and phytoplankton: application to nitrogen uptake measurement. Estuar Coast Shelf S 93:151–159

Unabia CRC, Hadfield MG (1999) Role of bacteria in larval settlement and metamorphosis of the polychaete *Hydroides elegans*. Mar Biol 133:55–64. https://doi.org/10.1007/s002270050

Veuger B, Middelburg J, Boschker H, Nieuwenhuize J, Rijswijk P, Rochelle-Newall E, Navarro N (2004) Microbial uptake of dissolved organic and inorganic nitrogen in Randers Fjord. Estuar Coast Shelf 61:507–515. https://doi.org/10.1016/j.ecss.2004.06.014

Wang W (2011) Bacterial diseases of crabs: a review. J Invertebr Pathol 106(1):18–26. https://doi.org/10.1016/j.jip.2010.09.018

Youn J-Y, Hur SB (2007) Antibiotics and their optimum concentration for axenic culture of marine microalgae. Algae 22(3):229–234. https://doi.org/10.4490/algae.2007.22.3.229

Zaritski A (1975) On dimensional determination of rod-shaped bacteria. J Theor Biol 54:243–248

16

Yolk utilization and growth during the early larval life of the Silver Perch, *Bidyanus bidyanus* (Mitchell, 1838)

Sulaeman · Ravi Fotedar

Abstract The aim of this research was to investigate the yolk sac and oil globule utilization by silver perch (*Bidyanus bidyanus*) larvae produced from domesticated broodfish. The larvae were kept unfed in the holding tank, sampled, and investigated by image analysis software to determine various characteristics, such as the diameters of ova, water-hardened eggs, yolk-sac, oil globules, and the total length of larvae. The research illustrated that, with the exception of oil globule diameter, all other morphometric parameters were significantly lower ($P < 0.05$) when compared to the larvae from the wild broodfish. The yolk sac was completely absorbed at 96 h post-hatching (hph) and the oil globule was visible until 240 hph. The larvae exhibited predatory movements and tried to catch rotifer at 4 days post hatching (dph). However, the onset of feeding took place at 5 dph, while 100% of feeding occurred at 6 dph. During the first 96 h (h), larvae grew significantly faster than the next 144 h. Larvae encountered low mortalities (<10%) during the first 96 hph, before increasing significantly in the next 24 h and no unfed larvae survived post 240 h. The results also suggested that the exogenous feed should be available at 96 hph, which is well after the yolk sac is completely depleted. In addition, although most of eggs and larval performance from domesticated broodfish were inferior compared to the wild one, it has larger oil globule that could make longer of its mixed feeding period and therefore could have better in viability.

Keywords Yolk utilization · Larvae · *Bidyanus bidyanus*

Introduction

The yolk stages (yolk-sac and oil globule) of most of the teleost with early-life planktonic stages and lack of parental care, are characterized by reduced capacities in swimming, vision, and feeding ability (Jug-Dujaković et al. 1995). The knowledge about the role of early fish ontogeny is particularly essential for better understanding of larval dynamics and mechanisms of adapting to variable environmental conditions. The essential events in larviculture are not only the period of these susceptible ontogenetic stages, but also a diversity of vital physical appearances such as the hatching size, onset of feeding, and the efficacy of yolk utilization

Sulaeman · R. Fotedar
Curtin Aquaculture Research Laboratory, Department of Environment and Agriculture, Curtin University, 1 Turner Avenue, Bentley, WA 6102, Australia

Sulaeman (✉)
Research Institute for Coastal Aquaculture (RICA), Jl. Makmur Dg. Sitakka 129, Maros 90511, Indonesia
e-mail: Sulaiman1@student.curtin.edu.au; bagren_brpbap@yahoo.com

(Koumoundouros et al. 2001; Polo et al. 1991). Some fish species are more highly prone to mass mortalities during the early feeding time than during the advanced development phase (Dou et al. 2005; Gisbert and Williot 1997; Kailasam et al. 2007; Mookerji and Rao 1999; Paperna 1978; Qin et al. 1997). Fish larvae are also highly vulnerable during the period of transition from consuming endogenous reserves of food to exogenous nutrient consumption. Failure to perform first feeding response immediately when the mouth has opened often causes morphological deformities, abnormal moving behaviour, and an inability to swim and feed (Gwak and Tanaka 2001; Kjørsvik et al. 1991), leading to high mortality (Dou et al. 2002; Houde 1974).

Generally, the rate of survival of fish larvae is poor during the early life stages (Bisbal and Bengtson 1995; Yang 2007). The availability of food at the right time is crucial because starvation can lead to high larval mortality (Bisbal and Bengtson 1995; Gisbert et al. 2004; Kohno et al. 1997; Peña and Dumas 2005). Some other factors such as larval size, yolk and oil quantity and absorption rate, feeding behaviour, and the time of initial feeding may also affect mortality (Blaxter 1974; Dou et al. 2000). In chinook salmon, *Oncorhynchus tshawytscha*, and rainbow trout, *Salmo gairdneri*, initial feeding starts after the swim-up stage (Heming et al. 1982; Twongo and MacCrimmon 1976), while in burbot, *Lota lota,* exogenous food needs to be given at 9–10 dph (Palińska-Żarska et al. 2014). The elapsed time from yolk sac depletion to irreversible starvation also differs between fish species (Bisbal and Bengtson 1995; Shan et al. 2008).

Silver perch, *Bidyanus bidyanus* (Mitchell, 1838), endemic to the Murray-Darling river system in New South Wales, is a native Australian freshwater fish, belonging to the Teraponidae family (Merrick and Schmida 1984). Also known as freshwater bream, silver bream and grunter, the silver perch occupies the northern and western rivers and upper reaches of the Murray-Darling river system (Rowland 1995). It is an important freshwater aquaculture species intensively cultured in Australia with production around 500 metric tons per year (Rowland 2009). It has also been cultured in China, Taiwan and Israel (Allan and Rowland 2002). Limited information on its morphogenesis and larval ontogeny originating from wild broodfish has been reported by Lake (1967b), but no detailed information has been published about domesticated broodfish. Since the life cycle of the silver perch can be closed in captivity (Rowland 2004), it is important to assess the ontogeny and morphogenesis of larvae from the domesticated broodfish and compare it with the larvae from the wild broodfish.

This paper delivers early life-history data on the silver perch larvae, obtained from the domesticated broodfish. The research details the information found during the early larval stages with respect to the yolk absorption rates, the onset of feeding, and the mortality response of the unfed larvae.

Materials and methods

Each experiment was accepted by the Animal Ethics Committee of Curtin University and the Australian Code of Practice for the care and use of animals for scientific purposes was also followed.

Fish and experimental setup

The second generation of domesticated silver perch broodfish (6 years old), reared in a farmer's pond at Gingin, Western Australia (31°20″0.5.53″S, 115°46′20.51″E), was used in this experiment. Eggs were obtained after human chorionic gonadotropin (hCG) hormone was injected into both male and female broodfish at a dosage of 200 IU/kg (Levavi-Sivan 2004; Rowland 1984, 2009). Following hormone injection, both males and females were kept together in a 3-tonne fiberglass tank filled with two tons of freshwater until they spawned. The spawning tank was placed in a hatchery hall at a room temperature of 20–25 °C, equipped with gentle aeration that was placed in the mid-bottom of the tank.

After spawning, the eggs were transported from Gingin to the Curtin Aquaculture Research Laboratory (CARL) for incubation and for the production of the fry used in the experiment. On arrival, about 5000 eggs were placed into two 200-L, conical-shaped, fiberglass incubator tanks filled with freshwater, with a flow-through system that had a flow rate of 500 mL/min. With this system, egg shells were retained in the incubation tank while larvae from the two incubation tanks were moved into the sump tank. Larvae from the sump tank were then transferred to the holding tank (200 L volume) at a stocking density of 50 larvae/L. The holding tank was filled with 6 ppt saline water and gently aerated. The water temperature in the holding tank was preserved at 20 ± 0.5 °C by a thermostatically controlled heater (Sonpar, Model-Ha200). The newly hatched larvae were then haphazardly taken to perform morphometric measurements, age at first feeding, and mortality experiments. The rest of the larvae were kept unfed in this tank until the end of the experiment. The swimming behaviour of the larvae was observed simultaneously to identify the initial time of starvation characterized by a slow-moving swimming, hanging the head down into the water column, or almost no reaction to gentle probing.

Morphometric measurements

The hatching time was appointed at which 90% of the viable eggs were hatching (Shan et al. 2008). Meanwhile, morphometric measurements were performed at 6 h intervals, from hatching until 10 days post hatch (dph), using 7–53 larvae (average = 25) per sample. Yolk utilization (yolk sac and oil globule) was dignified based on the reduction in the volume of the yolk-sac and oil globule over time. The yolk utilization of the unfed larvae was studied using a stereoscopic microscope (Olympus-SZH) while morphometric characteristics were monitored using photographs taken with a digital camera (Olympus-SC30) attached to the microscope. On the photographs, five morphometric characteristics (TL: total length; YsH: yolk sac height; YsL: yolk sac length; OgD: oil globule diameter) were quantified to the nearest 0.001 mm (using Photoshop image software), while yolk sac volume (YsV) and oil globule volume (OgV) were estimated. The YsV was estimated using the formula introduced by Bagarinao (1986) for a prolate spheroid, $V = \pi/6\ l\ h^2$, where l is the length of yolk sac and h is the height of yolk sac. The volume of the oil globule in the larvae was computed from the formula $V = \pi/6\ d^3$, where d is the oil globule diameter. The specific growth rate (SGR) of the TL at different periods of their larval stage was calculated by the formula: SGR = (e.g. −1) × 100% (Árnason et al. 2009; Hopkins 1992), where $g = [\ln(l_2)-\ln(l_1)]/(t_2-t_1)$, and l_2 and l_1 are the mean TL on day t_2 and t_1, respectively. All measurements were performed parallel to the body-length axis, and all depths and diameters were measured perpendicular to this axis (Fig. 1). The measurements for the diameter of ova, water-hardened egg, yolk sac, and oil globule, and the TL of each larva at hatching from domesticated broodfish in this study were then compared to the corresponding characteristic for larval performance from wild broodfish as reported by Lake (1967b). The data presented by Lake (1967b) were based on the wild broodfish of silver perch stocked in a pond and spawned naturally before eggs were collected for examination in accordance with the egg and larvae parameters.

Age at first feeding

To determine the age at first feeding, 10 one-day-old larvae were introduced into a beaker glass containing 1 L of water with rotifers (*Brachionus calyciflorus*) at a density of 10 ind/mL. Beakers were placed in the incubator tank, each equipped with a thermostat (Thermomix, B. Braun Biotech International) in order to maintain the water temperature at approximately 20 ± 0.1 °C. To facilitate prey contrast, the trough was covered with black plastic on the sides and bottom, and illuminated from above by a 36 W fluorescent light. The larvae were netted out after 45 min, lightly anaesthetized with AQUI-S (0.025 mg/L) and preserved in 8% buffered formalin for subsequent gut content observation. The incidence of feeding was expressed as the percentage of test fish (n = 10) with food (at least one food item) in their gut. The feeding trials were repeated every 24 h with unfed silver perch until they were 144 h or 6 days' old.

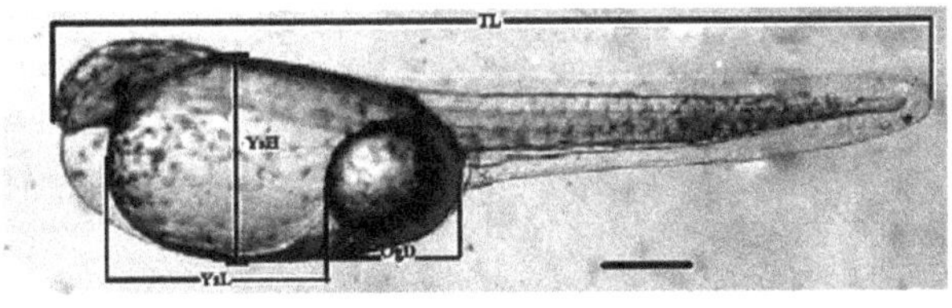

Fig. 1 Newly hatched silver perch, *Bidyanus bidyanus* larvae (eyes, jaws and gut not formed) showing morphometric measurements of total length (TL), yolk sac length (YsL), yolk sac height (YsH), and oil globule diameter (OgD), (*bar* = 250 μm)

Mortality

The experiment was conducted using three 2-L, conical-shaped glass containers placed in an incubator tank to determine the mortality of unfed larvae during their yolk stage. Fifty 1-dph larvae from the holding tank were transferred to the experimental glass container. Mild aeration was provided, along with illumination using 40 W fluorescent light regulated at about 200 lx. Dead fish were counted and removed along with bottom sediments before changing the water about 30% from the rearing volume every 24 h. Cumulative mortality was then calculated daily. Water temperature was maintained at 20 °C.

Data analysis

Statistical differences in egg and larva parameters (the TL, diameter of ova, diameter of water hardened egg, Ysh, and the Ogd) between domesticated and wild broodfish were analyzed by one sample t test. The comparison between TL and SGR at different stages were analyzed by an independent-sample t test between two periods of development i.e. from 0 to 96 hph and from 96 to 240 hph. Mortality rates at a different age of larvae were analyzed by one-way analysis of variance (ANOVA) followed by a post hoc test (Tukey's) with the level of significance below 5% ($P < 0.05$). The relationship between the mortality rate and the time after hatching was studied using regression analysis. Statistical analysis was performed using IBM SPSS Statistic-22 and regression analysis using Microsoft Excel 2013.

Results

Larval size, yolk reserves, and early growth

The eggs obtained from the pond-reared broodfish hatched about 32 h after fertilization, at 20 ± 1.0 °C. The newly-hatched silver perch larvae were transparent with a large yolk sac prolonging from the tip of the snout to about half of the ventrolateral of the body (Fig. 1) with an initial mean length of 1.35 ± 0.03 mm and a volume of 0.71 ± 0.07 mm^3. An oil globule (OG) of 0.64 ± 0.00 mm in diameter and 0.02 ± 0.00 mm^3 mean volume was present at the caudal tip of the yolk sac. The frontal part of the body was bent about the yolk sac and the border of finfold formed surrounded the body from the dorsal and ventral part of the body to the caudal part. The intestine was not fully formed, with anus and mouth sealed. Eye pigmentation was observed 12 h post-hatching (hph) and completed after 3 dph. The intestinal canal started formation at 2 dph, and a simple, straight, tubular gut was observed. Further development of the alimentary system was the formation of the intestinal loop at 3 dph and at the same time the lower jaw was completely formed.

At hatching, the silver perch larvae each measured 2.52 ± 0.05 mm in total length (TL). They grew to 4.12 ± 0.13 mm TL in about 96 h, by which time there was no more yolk sac left and about 45% of the oil globule remained (Fig. 2). After this stage, the TL growth subsequently slowed down and was significantly ($P < 0.05$) lower than the first 96-h period.

Yolk utilization rate

The yolk-sac utilization rates were much higher than the oil-globule exhaustion rate during the early life history of the silver perch larvae. Nearly 34 and 0.5% of the initial yolk sac and oil globule reserves, respectively, were utilized within 6 h. More than 50% of the silver perch larvae exhausted their yolk sac completely by the fourth day at 96 hph, whereas the oil globule traces (<1%) could still be seen up to 240 hph (Fig. 2). At the same time, the TL increased significantly ($P < 0.05$) as the SGR reached approximately 12% per day during the first 96 h, followed by a slow growth rate of 4.8% only from 96 to 240 hph.

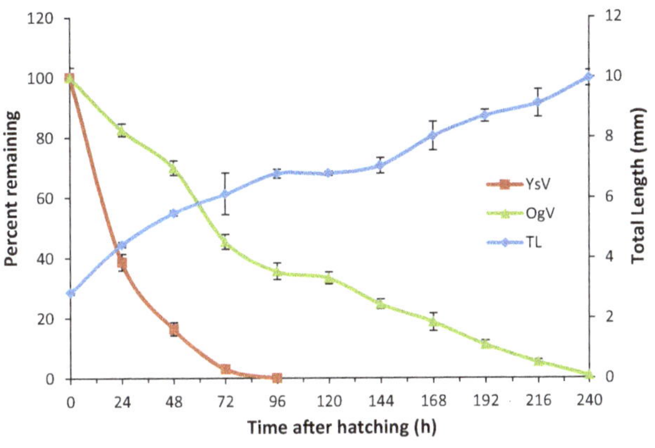

Fig. 2 Yolk sac volume (YsV) and oil globule volume (OgV) utilization and the total length (TL) development in relation to the time after hatching in hours (h) of the unfed silver perch, *Bidyanus bidyanus*, larvae. Data points are mean ± S.E. of 7–20 larvae per sample

Age at first feeding

Even before the yolk exhausted completely, the larvae of the silver perch have demonstrated a sign of food searching behaviour, comprising prey search, pursuit, and attempts to capture prey. The initial timing of first feeding was noticed at 5 dph. However, only about 5% of larvae could successfully initiate feeding instantly after the yolk sac was totally absorbed. The incidence of 100% of larvae feeding was observed at 6 dph; that is, one day after the yolk sac was fully absorbed. When silver perch larvae initiated first feeding, they already had functional mouths and active swimming behaviour, which indicated that most of the larvae needed some time to establish active feeding before and during the yolk exhaustion period. The symptoms of starvation of unfed larvae were observed at 5 dph.

Mortality

All silver perch larvae died no later than 240 hph. The daily cumulative mortality during the early life history of silver perch larvae increased with age (Fig. 3). The daily mortality rate underwent a rapid increase ($P < 0.05$) after the exhaustion of the yolk sac at 96 hph onward and 50% mortality was reached at 175 hph (Fig. 3). The correlation analysis between mortality and yolk sac-volume showed that mortality rate was significantly ($P < 0.05$) affected by the availability of the yolk sac.

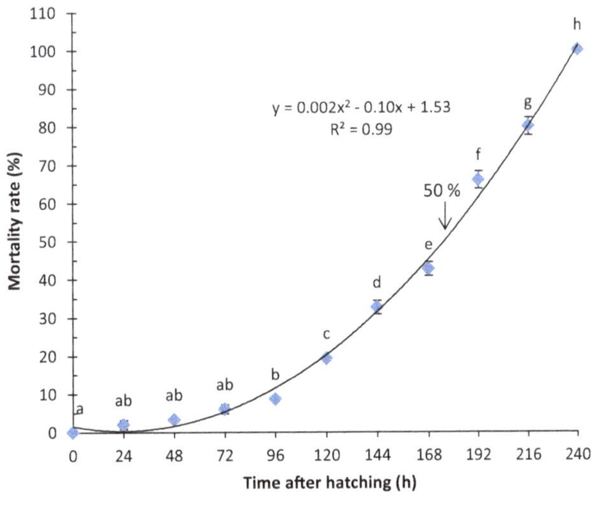

Fig. 3 The effect of starvation on mortality in silver perch, *Bidyanus bidyanus* larvae. Data points are mean ± S.E. of three replications. The *marker* with the same label indicates not significantly different ($P < 0.05$)

Discussion

Earlier research has shown that the silver perch would not spawn at a temperature below 24 °C in the natural environment and at a temperature below 23.3 °C under captive pond conditions (Lake 1967a). The domesticated broodfish of the silver perch does not spawn naturally under the Gingin (Western Australia) pond conditions in temperatures ranging between 11 and 34 °C. However, after a single injection of HCG, the domesticated broodfish in this experiment spawned at 21 °C. This is in the range of temperatures (21–24 °C) within which silver perch spawned in Israel (Levavi-Sivan 2004), which are both lower than the spawning temperature of 25 °C as reported by (Rowland 1984). Hence, no evidence of spawning deterioration was observed in this experiment. This may indicate that the domesticated silver perch could have adapted to spawn under different ranges of temperature after long time acclimation in pond conditions. This phenomenon has also been stated by Lam (1983) for the goldfish, *Cyprinus auratus*, where spawning temperature changed after acclimation to the new environmental condition.

Another interesting information here is the relationship between the incubation temperature and the hatching time (Table 1). The difference in incubation temperature did not make much different in hatching time. This may possible that the incubation period does not only depending on the temperature, but also depending on the broodstock and eggs quality (Mylonas et al. 2010). Therefore, the internal factor related to the eggs development stage at hatching may also affect the hatching time. On the other hand, the hCG injection before spawning may also contribute to influence the period of incubation. However, there is no detailed information to support this argument yet.

The measured TL of the newly-hatched larvae, egg's parameters and larvae morphometric of the domesticated silver perch were significantly lower than for the parameters derived from wild broodfish as reported by Lake (1967b), except for the OgD (Table 1). The reason may be due to differences in parent nutrition resulting in different biochemical composition of eggs. Eggs obtained from naturally foraging broodfish should contain a biochemical composition that is suitable for producing larger eggs and larvae with higher viability of

Table 1 Egg and larval development of silver perch, *Bidyanus bidyanus* from wild and domesticated broodfish

Factor	After Lake (1967b)	This experiment (mean ± S.E.)
Broodfish source	Wild	Domesticated
Broodfish size (kg)	0.7–3.2	1.5–3.5
Temperature at spawning (°C)	23.3	21
Nature of egg	Pelagic	Pelagic
The diameter of ova (mm): averages in parentheses	0.7–1.3 (1.2)^{b}	1.08 ± 0.01^{a} (n = 600)
The diameter of water-hardened egg (mm): averages in parentheses	2.75–2.80 (2.78)^{b}	2.17 ± 0.01^{a} (n = 600)
The diameter of yolk (mm): averages in parentheses	1.00–1.25 (1.14)^{b}	0.81 ± 0.01^{a} (n = 70)
Yolk length at hatch (mm)	–	1.35 ± 0.03 (n = 70)
Yolk volume at hatch (mm^3)	–	0.58 ± 0.06 (n = 70)
The diameter of oil globule (mm): averages in parentheses	0.45–0.59 (0.54)^{a}	0.62 ± 0.01^{b} (n = 70)
Oil globule volume at hatch (mm^3)	–	0.14 ± 0.01 (n = 70)
Number of oil globules and position in yolk sac (minor ones neglected)	1 Posterior	1 Posterior
Egg colour	Colourless	Colourless
Hatching time (h) and temperature range (°C) in parentheses	30–31 (26–27)	30–32 (21–22) (n = 3)
TL of larvae at hatching (mm): averages in parentheses	3.5–3.7 (3.6)^{b}	2.85 ± 0.06^{a} (n = 70)
Maximal larval TL attained on the yolk sac reserves (mm)	–	11.97
Maximal larval TL attained on the oil globule reserves (mm)	–	14.31
Time from hatching to attainment of maximal TL on yolk sac (h)	–	96
Time from hatching to attainment of maximal TL on oil globule (d)	–	10
Age at first feeding (days)	6	5
Age at free swimming (days)	5	4

Data in the same row having the same superscript letters ($^{a, b}$) indicate not significantly different at α level of 0.05

offspring (Kjørsvik et al. 1989). On the other hand, the biochemical composition of eggs resulting from domesticated broodfish can be altered through changes in diet (Bromage 1998; Kjørsvik et al. 1990; Sargent 1995). However, until now, the quality of eggs produced from domesticated silver perch has not been compared to those of wild broodfish. Regardless of the comparison with larvae resulting from wild broodfish, the observed TL in the current study (2.85 ± 0.06) is still comparable to the Japanese flounder, *Paralichthys olivaceus* (2.62 ± 0.03 mm) (Dou et al. 2002). However, it is smaller than the Thai pangas, *Pangasius sutchi,* which has a TL of 2.98–3.10 mm (Islam 2005), the rohu, *Labeo rohita,* which has a TL of 4.58 ± 0.04 mm, and singhi, *Heteropneustes fossilis,* which has a TL of 3.32 ± 0.054 mm (Mookerji and Rao 1999). Even though the initial size of a silver perch larva is smaller, its initial yolk sac volume is relatively larger in terms of the YsL-TL ratio (0.54) than for Thai pangas (0.42) (Islam 2005), illustrating that the larvae is better equipped, in terms of its endogenous reserve as an energy source, to challenge the possibility of delayed exogenous feeding. It also demonstrates that the increase in total length is a biologically important aspect of the yolk-sac larval stages (Peterson et al. 1996) as the TL is directly related to the mouth opening and the size of prey that the larvae are able to ingest (Klimogianni et al. 2011).

During the first 96 hph the TL increased significantly ($P < 0.05$) compared to the next 144 h, whereas at the same time, the yolk sac volume was reduced significantly ($P < 0.05$) due to their absorption. Conversely, the volume of the oil globule was not intensely absorbed. The oil globule consumption during the yolk-sac stage of the silver perch seems lower than the marine species common pandora, *Pagellus erythrinus,* reared at comparable temperature at 18–21 °C (Klimogianni 2004). A probable cause is a delay in oil globule adherence to the larval body (Klimogianni et al. 2011), which occurred mostly at 120–240 hph (Fig. 2). This is much later compared to another marine species, the sharp snout sea bream, *Diplodus puntazzo,* which occurred at 4–6 hph. This could be advantageous to silver perch larvae as they get the opportunity to utilize both endogenous yolk and exogenous food (termed as the 'mixed feeding period') (Chai et al. 2011) for a longer time (6 days). This is much longer than the Chinese sturgeon, *Acipenser sinensis,* whose mixed feeding period only lasts for 2 days (Chai et al. 2011). Larvae have to start their capability to feed on an exogenous diet during the time when both endogenous and exogenous sourçes of diet are available; if not, they will hurt from continuous food shortage (Blaxter and Hempel 1963). An extended period of the mixed-feeding time is advantageous for larvae to gather enough feeding skills to develop their exogenous feeding capability, to escape from food shortages, and hence increase their probability of survival rates.

Since the onset of feeding occurred at 5 dph and the yolk-sac stage of the silver perch larvae lasted after 10 dph, it indicates that the mixed feeding period occurs for 6 days (from 5 to 10 dph). Consequently, the existence of appropriate prey organisms at this stage is of particularly significance for their endurance and the recruitment achievement of the silver perch. The same pattern is demonstrated by the Chinese sturgeon, where exogenous food deprivation can be maintained for 7 days (Chai et al. 2011).

The yolk-utilization rate of the silver perch during the first 24 h was about 61%, which is faster compared to 40% for singhi, *Heteropneustes fossilis*, another freshwater fish with a nearly equal activity pattern, illustrated by the intermittent bursts of rapid movement followed by relatively longer periods of rest during the yolk-sac stage (Mookerji and Rao 1999). However, silver perch has a longer period of mixed feeding (6 days) compared to only 2 days for singhi.

Only about 5% larvae initiated feeding at 5 dph, when <5% of the initial yolk reserves (yolk-sac + oil globule) were remained, lower than for singhi where nearly 40% of the larvae initiate feeding when about 20% of the yolk sac reserves remained at about 26 °C. Furthermore, the rate of oil-globule utilization was slightly lower and about 33% still remained when the yolk sac totally disappeared at 5 dph. These trends were also observed by Mookerji and Rao (1999) in rohu and singhi. Exogenous feeding generally commences before the yolk is fully absorbed (Ghelichi et al. 2010; Gisbert et al. 2000, 2004; Haga et al. 2014; Islam 2005; Kailasam et al. 2007; Klimogianni et al. 2011; Mookerji and Rao 1999; Peña and Dumas 2005; Shan et al. 2008; Wang 2010; Yang 2007), as was also observed in the present study.

The size at first feeding of silver perch must be taken into account for successful rearing of larvae, with the use of suitable live feed or formulated diet. The temperature and level of activity are two important factors influencing the rate of depletion of yolk reserves in newly hatched larvae (Quantz 1985; Rana 1990; Wang et al. 1987). Since, in the current study, the temperature was set at a constant temperature, internal factors related to the natural activity of the silver perch larvae is the only main cause of yolk sac absorption. In an aquaculture context, however, the effect of temperature may be sufficiently relevant to regulate it under controlled nursery conditions to allow larval yolk reserves to last longer by setting a lower rearing temperature (Bagarinao 1986).

In general, nearly all freshwater fish species are demersal spawners, providing their eggs and larvae with a relatively more hospitable and predictable environment. Consequently, relative to marine fishes, freshwater fishes are generally less fecund and produce larvae which, at hatching, are larger and suffer less pre-metamorphosis mortality (Houde 1974; Mookerji and Rao 1999). The mortality of unfed silver perch during the early life history of larvae was significantly affected by the availability of the yolk sac. After the exhaustion of the yolk sac, the larvae may suffer from a nutrient deficiency due to delayed first feeding, and result in significantly increased mortality at 5 dph onward, reaching 50% at 7 dph. This suggests that the availability of the oil globule alone cannot maintain a high survival rate until the end of the yolk-sac stage, but supports fish during the 'mixed feeding period'. At this phase, exogenous feedstuff would help to satisfy the nutritive necessities for larval growth.

In conclusion, at hatching, silver perch larvae from the captive broodfish measured around 2.5 mm in TL and had a large yolk sac which is completely resorbed at 90 hph, whereas the oil globule remained until 240 hph. Even if the onset of feeding occurred at 5 dph, 100% feeding occurred later, at 6 dph. This indicates that silver perch larvae have about 5 days (from 6 to 10 dph) to perform exogenous feeding before the endogenous deposits are totally absorbed. The unfed larvae experienced low mortality (<10%) during the first 96 hph, before it increased significantly (more than double) in the next 24 h. No larvae survived when initial feeding was delayed beyond 240 hph. In addition, although most of the egg and larva parameters of domesticated broodfish were inferior compared to the eggs and larvae produced from wild broodfish, their oil glubule was larger and therefore they may have better viabilty.

Acknowledgements The authors would like to acknowledge the Research Institute for Coastal Aquaculture (RICA), Maros, for administrative support, and the South Sulawesi provincial Government of Indonesia who granted the scholarship. Special thanks to Trevor Blinco, who kindly prepared the broodfish, and to Simon Longbottom, who kindly provided technical assistance during the laboratory work.

References

Allan GL, Rowland SJ (2002) Silver Perch, *Bidyanus bidyanus*. In: Webster CD, Lim C (eds) Nutrient requirements and feeding of finfish for aquaculture. CABI Publishing, Wallingford, pp 358–373

Árnason T, Björnsson B, Steinarsson A, Oddgeirsson M (2009) Effects of temperature and body weight on growth rate and feed conversion ratio in turbot (*Scophthalmus maximus*). Aquaculture 295:218–225

Bagarinao T (1986) Yolk resorption, onset of feeding and survival potential of larvae of three tropical marine fish species reared in the hatchery. Mar Biol 91:449–459

Bisbal GA, Bengtson DA (1995) Effects of delayed feeding on survival and growth of summer flounder, *Paralichthys dentatus* larvae. Mar Ecol Prog Ser 121:301–306

Blaxter J (1974) The early life history of fish. Springer, New York

Blaxter JHS, Hempel G (1963) The influence of egg size on herring larvae (*Clupea harengus* L.). J Cons Int Explor Mer 28:211–240

Bromage N (1998) Broodstock management and the optimisation of seed supplies. Aquacult Sci 46:395–401

Chai Y, Xie C, Wei QW (2011) Yolk-sac absorption and point of no return in Chinese sturgeon, *Acipenser sinensis* larvae. J Appl Ichthyol 27:687–689

Dou S, Seikai T, Tsukamoto K (2000) Cannibalism in Japanese flounder juveniles, *Paralichthys olivaceus*, reared under controlled conditions. Aquaculture 182:149–159

Dou S, Masuda R, Tanaka M, Tsukamoto K (2002) Feeding resumption, morphological changes and mortality during starvation in Japanese flounder larvae. J Fish Biol 60:1363–1380

Dou SZ, Masuda R, Tanaka M, Tsukamoto K (2005) Effects of temperature and delayed initial feeding on the survival and growth of Japanese flounder larvae. J Fish Biol 66:362–377

Ghelichi A, Makhdoomi N, Jorjani S, Taheri A (2010) Effect of water temperature on the timing of initial feeding of Persian sturgeon *Acipenser persicus* larvae. Int Aquat Res 2:113–119

Gisbert E, Williot P (1997) Larval behaviour and effect of the timing of initial feeding on growth and survival of Siberian sturgeon (*Acipenser baeri*) larvae under small scale hatchery production. Aquaculture 156:63–76

Gisbert E, Williot P, Castelló-Orvay F (2000) Influence of egg size on growth and survival of early stages of Siberian sturgeon (*Acipenser baeri*) under small scale hatchery conditions. Aquaculture 183:83–94

Gisbert E, Conklin DB, Piedrahita RH (2004) Effects of delayed first feeding on the nutritional condition and mortality of California halibut larvae. J Fish Biol 64:116–132

Gwak WS, Tanaka M (2001) Developmental change in RNA: DNA ratios of fed and starved laboratory-reared Japanese flounder larvae and juveniles, and its application to assessment of nutritional condition for wild fish. J Fish Biol 59:902–915

Haga Y, Naiki T, Tazaki Y, Takebe T, Kumon K, Tanaka Y, Shiozawa S, Nakamura T, Ishida S, Ide K, Masuma S, Takeuchi T (2014) Effect of feeding microdiet and yolk-sac larvae of spangled emperor *Lethrinus nebulosus* at different ages on survival and growth of pacific bluefin tuna *Thunnus orientalis* larvae. Aquacult Sci 58:491–499

Heming TA, Mclnerney JE, Alderdice DF (1982) Effect of temperature on initial feeding in alevins of chinook salmon (*Oncorhynchus tshawytscha*). Can J Fish Aquat Sci 39:1554–1562

Hopkins K (1992) Reporting fish growth: a review of the basics. J World Aquacult Soc 23:173–179

Houde ED (1974) Effects of temperature and delayed feeding on growth and survival of larvae of three species of subtropical marine fishes. Mar Biol 26:271–285

Islam A (2005) Embryonic and larval development of Thai Pangas (*Pangasius sutchi* Fowler, 1937). Dev Growth Differ 47:1–6

Jug-Dujaković J, Dulčić J, Katavić I (1995) Embryonic and yolk-sac larval development of the sparid Dentex (Dentex) dentex (Linnaeus, 1758). Fish Res 24:91–97

Kailasam M, Thirunavukkarasu AR, Selvaraj S, Stalin P (2007) Effect of delayed initial feeding on growth and survival of Asian sea bass *Lates calcarifer* (Bloch) larvae. Aquaculture 271:298–306

Klimogianni A, Koumoundouros G, Kaspiris P, Kentouri M (2004) Effect of temperature on the egg and yolk-sac larval development of common pandora, *Pagellus erythrinus*. Mar Biol 145:1015–1022

Kjorsvik E, Mangor-Jensen A, Holmefjord I (1989) Egg quality in marine fishes. Adv MarBiol 26:71–113

Kjørsvik E, Mangor-Jensen AT, Holmefjord I (1990) Egg quality in fishes. Adv Mar Biol 6:71–113

Kjørsvik E, van der Meeren T, Kryvi H, Arnfinnson J, Kvenseth PG (1991) Early development of the digestive tract of cod larvae, *Gadus morhua* L., during start-feeding and starvation. J Fish Biol 38:1–15

Klimogianni A, Kalanji M, Pyrenis G, Zouloti A, Trakos G (2011) Ontogeny of embryonic and yolk-sac larval stage of the sparid sharpsnout sea bream (*Diplodus puntazzo* Cetti, 1977). J Fish Aquat Sci 6:63–73

Kohno H, Ordonio-Aguilar RS, Ohno A, Taki Y (1997) Why is grouper larval rearing difficult?: an approach from the development of the feeding apparatus in early stage larvae of the grouper, *Epinephelus coioides*. Ichthyol Res 44:267–274

Koumoundouros G, Divanach P, Anezaki L, Kentouri M (2001) Temperature-induced ontogenetic plasticity in sea bass (*Dicentrarchus labrax*). Mar Biol 139:817–830

Lake JS (1967a) Rearing experiments with five species of Australian freshwater fishes. I. Inducement to spawning. Aust J Mar Freshw Res 18:137–154

Lake JS (1967b) Rearing experiments with five species of Australian freshwater fishes. II. Morphogenesis and ontogeny. Aust J Mar Freshw Res 18:155–176

Lam TJ (1983) Environmental influences on gonadal activity in Fish. In: Hoar WS, Randall DJ, Donaldson EM (eds) Fish physiology, vol 9, Part B. pp 65–116

Levavi-Sivan B (2004) Spawning induction and hormonal levels during final oocyte maturation in the silver perch (*Bidyanus bidyanus*). Aquaculture 229:419–431

Merrick JR, Schmida GE (1984) Australian freshwater fishes: biology and management. John R, Merrick, North Ryde

Mookerji N, Rao TR (1999) Rates of yolk utilization and effects of delayed initial feeding in the larvae of the freshwater fishes rohu and singhi. Aquac Int 7:45–56

Mylonas CC, Fostier A, Zanuy S (2010) Broodstock management and hormonal manipulations of fish reproduction. Gen Comp Endocrinol 165:516–534

Palińska-Żarska K, Żarski D, Krejszeff S, Nowosad J, Biłas M, Trejchel K, Kucharczyk D (2014) Dynamics of yolk sac and oil droplet utilization and behavioural aspects of swim bladder inflation in burbot, *Lota lota* L., larvae during the first days of life, under laboratory conditions. Aquac Int 22:13–27

Paperna I (1978) Swimbladder and skeletal deformations in hatchery bred *Spams aurata*. J Fish Biol 12:109–114

Peña R, Dumas S (2005) Effect of delayed first feeding on development and feeding ability of *Paralabrax maculatofasciatus* larvae. J Fish Biol 67:640–651

Peterson RH, Martin-Robichaud DJ, Berge O (1996) Influence of temperature and salinity on length and yolk utilization of striped bass larvae. Aquac Int 4:89–103

Polo A, Yúfera M, Pascual E (1991) Effects of temperature on egg and larval development of *Sparus aurata* L. Aquaculture 92:367–375

Sargent JR (1995) Origin and functions of eggs lipid: nutritional implications. In: Bromage NR, Roberts RJ (eds) Broodstock management and egg and larval quality. Blackwell, London, pp 353–372

Qin J, Fast AW, DeAnda D, Weidenbach RP (1997) Growth and survival of larval snakehead (*Channa striatus*) fed different diets. Aquaculture 148:105–113

Quantz G (1985) Use of endogenous energy sources by larval turbot *Scophthalmus maximus*. Trans Am Fish Soc 114:558–563

Rana KJ (1990) Influence of incubation temperature on *Oreochromis niloticus* (L.) eggs and fry: II. Survival, growth and feeding of fry developing solely on their yolk reserves. Aquaculture 87:183–195

Rowland SJ (1984) The hormone-induced spawning of silver perch, *Bidyanus bidyanus* (Mitchell) (Teraponidae). Aquaculture 42:83–86

Rowland SJ (1995) High density pond culture of silver perch, *Bidyanus bidyanus*. Asian Fish Sci 8:73–79

Rowland SJ (2004) Domestication of silver perch, *Bidyanus bidyanus*, broodfish. J Appl Aquacult 16:75–83

Rowland SJ (2009) Review of aquaculture research and development of the Australian freshwater fish silver perch, *Bidyanus bidyanus*. J World Aquacult Soc 40:291–324

Shan X, Quan H, Dou S (2008) Effects of delayed first feeding on growth and survival of rock bream *Oplegnathus fasciatus* larvae. Aquaculture 277:14–23

Twongo TK, MacCrimmon HR (1976) Significance of the timing of initial feeding in hatchery rainbow trout (*Salmo gairdneri*). J Fish Res Board Can 33:1914–1921

Wang Y (2010) Effects of the timing of initial feeding on growth and survival of loach (*Misgurnus anguillicaudatus*) larvae. Aquac Int 18:135–148

Wang YL, Buodington RK, Doroshov SI (1987) Influence of temperature on yolk utilization by the white sturgeon, *Acipenser transmontanus*. J Fish Biol 30:263–271

Yang Z (2007) Effect of timing of first feeding on survival and growth of obscure puffer (*Takifugu obscurus*) larvae. J Freshw Ecol 22:387–392

17

Capability of some agricultural wastes for removing some heavy metals from polluted water stocked in combination with Nile tilapia, *Oreochromis niloticus* (L.)

Mohsen Abdel-Tawwab · Gamal O. El-Sayed · Sherien H. H. Shady

Abstract Heavy metal (HM) pollution is one of the major problems that adversely affect the aquatic ecosystem and inhabiting biota. Heavy metals adsorption by low-cost adsorbents is one of the techniques used for HM removing from polluted water. In the present study, agricultural wastes (AW), i.e., rice straw, sugarcane bagasse, and maize stalks, were washed with distilled water, dried in a dry-oven, cut into small pieces (<0.5 cm long), and immersed at 1.0 g/L in aquaria containing synthetic mixture of lead (Pb), cadmium (Cd), copper (Cu), and zinc (Zn). Nile tilapia, *Oreochromis niloticus* (L.), fingerlings (25.2 ± 0.88 g) were stocked at a density of ten fish per 100-L aquarium for 72 h, during which fish were fed on a fish diet containing 25% crude protein ad libitum twice daily. Samples of water, AW, and fish were collected at different times to determine HM concentrations. The HM removal from polluted water was depending on the type of the metal ions, AW, and the contact time. However, HM concentrations in aquaria waters of all AW treatments decreased significantly by increasing contact time up to 24 h after which their concentrations were almost the same. Concentrations of waterborne Pb, Cd, Cu, and Zn in AW-containing aquaria were significantly lower than those of AW-free aquaria. The presence of any AW reduced significantly HM concentrations. In AW-free aquaria, HM-exposed fish accumulated more HM in their body than those reared in AW-containing aquaria. The results of this experiment showed that all AW had the capability to remove HM levels from the polluted water and reduce their bioaccumulation in fish body. However, rice straw was the more efficient adsorbent for all metals.

Keywords Nile tilapia · Agricultural wastes · Rice straw · Maize stalks · Sugarcane bagasse · Heavy metals · Bioaccumulation

Introduction

Heavy metal (HM) pollution is spreading throughout the world due to the expansion of industrial activities and they easily reach the aquatic ecosystem. Unfortunately, HM are non-biodegradable and could deteriorate fish growth and health (Fazio et al. 2014; Savorelli et al. 2016; Abdel-Tawwab 2016; Abdel-Tawwab et al.

M. Abdel-Tawwab (✉) · S. H. H. Shady
Department of Fish Biology and Ecology, Central Laboratory for Aquaculture Research, Abbassa, Abo-Hammad, Sharqia, Egypt
e-mail: Mohsentawwab@gmail.com; Mohsen_tawwab@yahoo.com

G. O. El-Sayed
Chemistry Department, Faculty of Science, Benha University, Benha, Egypt

2013, 2016). Thus, scientific methods for HM removal and/or detoxification are essential to improve the health of the fish inhabiting HM-stressed environment. One of these methods is the use of feed supplements that could adsorb HM and/or facilitate their release out of fish body and subsequently could improve fish health (Abdel-Tawwab 2015; Abdel-Tawwab et al. 2015, 2017a, 2017b). Many other treatment technologies such as adsorption have been developed to remove HM from polluted water.

Adsorption is generally known to be one of the most effective techniques for HM removal from wastewater (Benhima et al. 2008; Zafar et al. 2007; Gupta and Babu 2009; Mosa et al. 2011; Chiban et al. 2012; Witek-Krowiak and Reddy 2013). Considerable attention has been paid to the development of effective and low-cost adsorbents. Thus, studies have been done on the use of AW as alternative biosorbants for HM removal from the aqueous effluents (Tarley and Arruda 2004; Mohan and Pittman 2006; Karnitz Júnior et al. 2010; Dos Santos et al. 2011; Chiban et al. 2012).

Nile tilapia, *Oreochromis niloticus* (L.), is one of the most known freshwater fish species in Egypt and worldwide due to its good growth and high-value market. This fish inhabits different aquatic ecosystems, which may be HM-polluted. On the other hand, some countries use wastewater in fish farming and polluted water may enter the fish farms (Easa et al. 1995; Shereif and Mancy 1995; Shereif et al. 1995). The HM pollution definitely deteriorates fish health and subsequently human health if the fish are consumed. Hence, a low-cost method for HM removal is essential to improve the health of aquatic ecosystem and subsequently the health of the farmed fish. Therefore, the present study was conducted to investigate the capability of rice straw, sugarcane bagasse, and maize stalks to remove HM from the polluted water and reduce their bioaccumulation in Nile tilapia.

Materials and methods

Preparation of the adsorbents

Rice straw, sugarcane bagasse, and maize stalks were collected from a local area (Abbassa, Abo-Hammad, Sharqia, Egypt). They were washed with distilled water many times, dried in a dry-oven (GCA, model 18EM, Precision Scientific group, Chicago, Illinois, USA) at 85 °C until constant weight, and cut into small pieces (<0.5 cm long). Afterward, these wastes were used without any pre-treatment as adsorbents for HM.

Preparation of synthetic mixture of heavy metals

Several wastewater samples were collected randomly from Saft-El-Henna drainage (10 km North-East Zagazig city, Sharqia, Egypt), stored in polyethylene bottles, decanted, and filtered on Whatman paper (0.45 mm porosity). Concentrations of lead (Pb), cadmium (Cd), copper (Cu), and zinc (Zn) were determined as described. Synthetic mixtures of HM, containing the same concentrations as in the drainage samples, were prepared using lead nitrate ($Pb(NO_3)_2$), cadmium chloride ($CdCl_2{\cdot}2.5H_2O$), zinc sulfate ($ZnSO_4$), copper sulfate ($CuSO_4{\cdot}5H_2O$; Merck & Co Inc., Kenilworth, NJ, USA).

Fish rearing and adsorption study

This study was based on a 4 × 6 factorial design with four adsorbents (control, rice straw, sugarcane bagasse, and maize stalks) and six contact times (0, 3, 6, 24, 48, and 72 h) in triplicates. Nile tilapia, *O. niloticus* (L.), fingerlings were obtained from the nursery ponds of the Central Laboratory for Aquaculture Research (CLAR), Abbassa, Abo-Hammad, Sharqia, Egypt. Fish were transferred to the wet laboratory and kept in an indoor fiberglass tank for 2 weeks for adaptation to the laboratory conditions where light–dark photoperiod cycle was maintained at 12–12 h using fluorescent tubes as a light source. During the adaptation period and the experimental running, fish were fed on a fish diet containing 25% crude protein as used by Abdel-Tawwab (2016) ad libitum twice daily at 9:00 and 14:00 h. The synthetic HM mixture was added to each 100-L aquarium and vigorously stirred. Then, fish (25.2 ± 0.88 g) were distributed into aquaria at a density of 10 fish per aquarium. Each aquarium was supplied with compressed air via air-stones using aquarium's air pump.

Each AW was kept in a net (<0.5-mm pore) and inserted in each corresponding aquarium at 1.0 g/L. At each contact time, samples from aquarium's water, AW, and whole-fish body were taken for HM determination.

Water quality parameters

Water samples were collected at each contact time from each aquarium to monitor different water quality parameters. Water temperature and dissolved oxygen were measured in site using a portable oxygen meter (Jenway, London, UK). The pH was measured using a pH-meter (Digital Mini-pH Meter, model 55, Fisher Scientific, Denver, CO, USA). The unionized ammonia (NH_3) was measured using Multi-parameters Ion Analyzer (HANNA Instruments, Rhode Island, USA). Total alkalinity and total hardness were determined by titration according to Boyd (1984).

Heavy metals residue

For measuring HM concentrations in water, 1-L water sample was filtered via 0.8 μm Millipore acetylcellulose filter paper (Millipore, Bedford, MA, USA), digested with 10 ml concentrated H_2SO_4 on a hot plate at 70 °C, concentrated to 50 ml, and transferred to a volumetric flask. Samples were adjusted up to 100 ml with redistilled water.

For measuring HM concentrations in AW and whole-fish body, a 1.0-g sample was oven-dried at 85 °C until constant weight and ashed in a muffle furnace (Thermolyne Corporation, Dubuque, Iowa, USA) at 550 °C for 6 h. Ash was digested with 5 ml concentrated H_2SO_4 and gradually kept at 130 °C on a hot plate until complete dryness. Then, the digests were diluted with 2 N HCl to a constant volume. Heavy metal concentrations were determined by using an atomic absorption spectrophotometer (Thermo 6600, Thermo Electron Corporation, Cambridge, UK), which was calibrated using standard solutions.

Statistical analysis

The obtained data were subjected to two-way ANOVA to evaluate effects of agricultural wastes and contact time as factors. Differences between means were tested at the 5% probability level using Duncan test. All the statistical analyses were performed using SPSS program version 20 (SPSS, Richmond, VA, USA) as described by Dytham (1999).

Results and discussion

Water quality parameters (Table 1) did not show significant differences among the different treatments ($P > 0.05$) and they were within the acceptable ranges for fish culture (Boyd 1984). The metal concentrations used in the present study were 58.4 μg/L, 1.31 mg/L, 22.15 μg/L, and 2.72 mg/L for Pb, Cd, Cu, and Zn, respectively (Table 2). Only, Cd concentration exceeded the permissible limits (USEPA 1986; WHO 1989).

Metal concentrations in aquaria's water were significantly affected by AW, contact time, and their interaction ($P < 0.05$; Table 2). In addition, concentrations of waterborne Pb, Cd, Cu, and Zn in AW-containing

Table 1 Changes in physico-chemical parameters of aquaria's water stocked with different agricultural wastes in combination with Nile tilapia exposed to synthetic heavy metals mixture for different contact times

	3 h	6 h	24 h	48 h	72 h	*P* value
Water temperature (°C)	26.4 ± 0.22	26.8 ± 0.09	26.6 ± 0.06	26.7 ± 0.22	26.6 ± 0.13	0.192
Dissolved oxygen (mg/L)	5.8 ± 0.11	5.9 ± 0.12	5.7 ± 0.12	5.9 ± 0.09	5.8 ± 0.14	0.530
pH	7.65 ± 0.022	7.68 ± 0.031	7.68 ± 0.054	7.68 ± 0.031	7.67 ± 0.021	0.081
Unionized ammonia (mg/L)	0.55 ± 0.022	0.54 ± 0.016	0.56 ± 0.025	0.53 ± 0.025	0.59 ± 0.015	0.074
Total alkalinity (mg/L)	158.3 ± 1.67	160.0 ± 2.58	160.0 ± 2.01	160.0 ± 2.88	155.8 ± 1.54	0.371
Total hardness (mg/L)	244.2 ± 4.91	244.2 ± 4.91	253.3 ± 3.08	250.0 ± 4.83	249.2 ± 2.71	0.482

Table 2 Concentrations of lead (Pb), cadmium (Cd), copper (Cu), and zinc (Zn) in aquaria's water stocked with different agricultural wastes in combination with Nile tilapia exposed to synthetic heavy metals mixture for different contact times

Exposure time (h)	Wastes	Lead (μg/L)	Cadmium (mg/L)	Copper (μg/L)	Zinc (mg/L)
	Control				
0		56.5a	1.12a	21.45a	2.22a
3		38.9b	0.59c	15.66b	0.70cd
6		33.7cd	0.51cd	12.18cd	0.61de
24		29.6e	0.45def	12.00de	0.51efgh
48		27.3ef	0.41defg	10.95de	0.49efgh
72		24.2f	0.39efgh	10.58de	0.47fghi
	Rice straw				
0		56.5a	1.12a	21.45a	2.22a
3		37.1bc	0.51cd	10.16def	1.06b
6		27.0ef	0.43def	9.11efgh	0.77c
24		25.1f	0.39efgh	7.19ghi	0.50efgh
48		24.6f	0.38efgh	7.17ghi	0.43ghi
72		24.1f	0.34gh	6.50hi	0.40hij
	Sugarcane bagasse				
0		56.5a	1.12a	21.45a	2.22a
3		30.2de	0.58c	10.58de	0.60def
6		17.7g	0.58c	6.54hi	0.55efg
24		15.2gh	0.50cd	6.51hi	0.52efgh
48		15.1gh	0.48cde	5.92i	0.51efgh
72		15.1gh	0.48cde	5.48i	0.50efgh
	Maize stalks				
0		56.5a	1.12a	21.45a	2.22a
3		26.2ef	0.69b	14.78bc	0.78c
6		16.6g	0.58c	9.62defg	0.36ijk
24		14.2gh	0.45def	7.38fghi	0.28jkl
48		12.1h	0.35fgh	6.32hi	0.26kl
72		11.9h	0.29h	4.74i	0.231
Pooled SE		1.769	0.031	0.668	0.079
Two-way ANOVA		*P* value			
Agricultural wastes (AW)		0.0001	0.0001	0.0001	0.0001
Contact time		0.0001	0.0001	0.0001	0.0001
AW × contact time		0.0001	0.008	0.013	0.0001

Means followed by the same letter are not significantly different

aquaria were significantly lower than those of AW-free aquaria. Moreover, metal concentrations in AW were significantly affected by AW, contact time, and their interaction ($P < 0.05$; Table 3). Concentrations of all tested metals in AW increased significantly by time up to 24–48 h after that their concentrations were approximately the same (Table 3). These results could be attributed to the availability of adsorbing site (surface functional groups) on the adsorbent surface. Additionally, AW may be fermented within days and subsequently their adsorption capacity was low. Similar results with rice straw were obtained by Swelam et al. (2016) who found that the Cu adsorption by straw rice increased significantly with increasing contact time and became almost constant for ~3.5 h.

All AW adsorbed metals, but rice straw, were the most efficient adsorbent for all metals. It is clear that metal removal was depending on the type of the metal ions, the adsorbent, and the contact time. Each AW contains many compounds, which have a variety of functional groups present in the binding process, for example carboxyl, amino, alcohol, and esters (Gupta and Ali 2000). These functional groups have the ability

Table 3 Concentrations of lead, cadmium, copper, and zinc (mg/g dry weight) in different agricultural wastes stocked in combination with Nile tilapia exposed to synthetic heavy metals mixture for different contact times

Exposure time (h)	Wastes	Lead	Cadmium	Copper	Zinc
	Rice straw				
0		0.23j	0.65g	0.28f	1.32h
3		1.50e	2.89f	0.74e	2.71g
6		2.41c	6.91e	1.28d	5.36f
24		3.04b	14.72d	1.64b	10.39c
48		3.48a	35.49a	1.98a	14.85a
72		3.64a	35.68a	2.07a	15.31a
	Sugarcane bagasse				
0		0.27j	0.72g	0.26f	1.37h
3		0.87g	2.11f	0.61e	2.59g
6		0.94fg	6.73e	1.15d	5.45f
24		1.94d	13.23	1.43c	6.14ef
48		2.48c	21.66c	1.48c	7.02de
72		2.52c	23.99c	1.74b	8.10d
	Maize stalks				
0		0.15j	0.37g	0.22f	1.28h
3		0.42i	2.30f	0.68e	2.98g
6		0.68h	6.07e	1.19d	5.28f
24		0.82g	13.66d	1.48c	11.14c
48		0.93fg	26.22b	1.74b	13.78bc
72		0.97fg	29.61b	1.96a	14.18ab
Pooled SE		0.162	0.225	0.191	0.960
Two-way ANOVA		*P* value			
Agricultural wastes (AW)		0.0001	0.0001	0.0001	0.0001
Contact time		0.0001	0.0001	0.0001	0.0001
AW × contact time		0.0001	0.001	0.0001	0.0001

Means followed by the same letter are not significantly different

to bind metal ions in solution or by donation of an electron pair from these groups to form metal complexes. Due to the differences in chemical compositions and functional groups of each AW used in the present study, the adsorption capability differs. Many researchers found a relationship between the presence of various functional groups and their complexation with metals during biosorption process (Tarley and Arruda 2004). The AW capacity for metal removal depends on many factors such as metallic elements, pretreatments of AW, and operating conditions; however, AW tended to prefer some metals more than others (Nguyen et al. 2013). Benaissa (2006) investigated the capacity of four inexpensive materials which are peels of peas, broad bean, medlar and fig leaves, to remove Cd from aqueous solutions; it was noted that the broad bean peel has the maximum adsorption capacity for Cd.

In the present study, rice straw was more effective adsorbent than sugarcane bagasse and maize stalks. The higher adsorption capacity of rice straw is probably due to the presence of silanol (SiOH) groups in its structure and more surface area (Fatemeh et al. 2008). In this regard, Osman et al. (2010) reported that rice hull showed highest efficiency in sequestering Zn, Cd, and iron among biosorbents investigated and the removal efficiencies by rice hull, sawdust, sugarcane bagasse, and wheat straw were 98.2, 96.9, 93.0, and 91.2%, respectively. Additionally, the difference in AW capacity to adsorb metals could be attributed to the abundant availability of binding sites, which enhances the retention of metals onto AW surface (Marin-Rangel et al. 2012; Jiménez-Cedillo et al. 2013; Nguyen et al. 2013). Kelly-Vargas et al. (2012) reported that lemon peel and orange peel demonstrated adsorption capacities for Cu and Pb were 48 and 15% higher than banana peel, respectively, and the Cd uptake by banana peel was higher than that of lemon peel and orange peel 82 and 57%, respectively. On the other hand, Mosa et al. (2011) reported that the removal efficiency of metals decreased in the order of cotton stalks, maize stalks, and rice straw. They attributed highest removal by cotton stalks to its highest concentration of cellulose, hemicellulose, and lignin as compared to other crop-residues.

All AW adsorbents reduced significantly HM concentrations in whole-fish body ($P < 0.05$; Table 4). In AW-free aquaria, HM-exposed fish accumulated more HM than those reared in AW-containing aquaria. These results suggest that AW immersion in aquaria stocked with Nile tilapia could reduce metal bioaccumulation in their body. It also noticed that metal concentrations in fish body were higher than those in polluted water. Concentration of Pb, Cd, Cu, and Zn in whole-body of Nile tilapia 72-h post exposure in AW-free aquaria were 0.714, 5.47, 1.309, and 7.37 mg/g dry weight, respectively, (Table 4). However, Pb concentrations (0.267, 0.286, and 0.290 mg/g dry weight), Cd concentrations (3.98, 3.52, and 3.58 mg/g dry weight), Cu concentrations (0.284, 0.262, and 0.283 mg/g dry weight), and Zn concentration (3.27, 3.64, and 3.13 mg/g dry weight) in fish body were approximately the same when fish were stocked with rice straw, sugarcane bagasse, and maize stalks for 72 h, respectively.

Table 4 Concentrations of lead, cadmium, copper, and zinc (mg/g dry weight) in whole-body of Nile tilapia exposed to synthetic heavy metals mixture in combination with different agricultural wastes for different contact times

Exposure time (h)	Wastes	Lead	Cadmium	Copper	Zinc
	Control				
0.0		0.031g	1.67e	0.082h	1.42f
3		0.372d	3.41c	0.412d	4.23c
6		0.449c	4.23b	0.617c	5.62b
24		0.575b	4.84ab	0.975b	6.39ab
48		0.656ab	5.22a	1.249a	6.91a
72		0.714a	5.47a	1.309a	7.37a
	Rice straw				
0.0		0.031g	1.67e	0.082h	1.42f
3		0.112f	1.98de	0.102g	1.83ef
6		0.175ef	2.23d	0.161fg	2.28e
24		0.218ef	2.84d	0.216ef	2.73de
48		0.251ef	3.52c	0.252cf	3.13d
72		0.267ef	3.98c	0.284e	3.27d
	Sugarcane bagasse				
0.0		0.031g	1.67e	0.082h	1.42f
3		0.135f	1.72e	0.125g	1.83ef
6		0.190ef	1.83de	0.174fg	2.24e
24		0.231ef	1.97de	0.213ef	2.83de
48		0.263ef	2.38de	0.243ef	3.18d
72		0.286e	2.86d	0.262ef	3.64d
	Maize stalks				
0.0		0.031g	1.67e	0.082h	1.42f
3		0.113f	1.81de	0.132g	1.78ef
6		0.175ef	2.31d	0.198fg	2.10e
24		0.228ef	2.92d	0.237ef	2.97de
48		0.262ef	3.46c	0.261ef	3.05d
72		0.290e	3.58c	0.283e	3.13d
Pooled SE		0.854	0.449	0.659	1.519
Two-way ANOVA		*P* value			
Agricultural wastes (AW)		0.0001	0.0001	0.0001	0.0001
Contact time		0.0001	0.0001	0.0001	0.0001
AW × contact time		0.0001	0.0001	0.0001	0.0001

Means followed by the same letter are not significantly different

Metal concentrations in whole-fish body were higher than those of HM-polluted water. This is because fish accumulated more metals in their bodies. Similar results were obtained by Abdel-Tawwab et al. (2013, 2016). Khalil and Hussein (1997) reared Nile tilapia for 9 months in secondary effluents with 1.0 Pb mg/L and found that the concentration of Pb in fish flesh was tenfold higher than the food standard. It is also noticed that HM accumulation in fish body increased as contact time increased. Easa et al. (1995), Shereif et al. (1995) and Shereif and Mancy (1995) studied Pb, Cu, Zn, and Cd in Nile tilapia reared in fishponds received wastewater. They found that HM concentrations in the fish increased by increasing the culture period.

Conclusions

Rice straw, sugarcane bagasse, and maize stalks are effective and cheap adsorbents for removal of Pb, Cd, Cu, and Zn from polluted water. They also reduced HM bioaccumulation in fish body. Rice straw showed higher adsorption efficiency than the others.

References

Abdel-Tawwab M (2015) The use of American ginseng (*Panax quinquefolium*) in practical diets for Nile tilapia (*Oreochromis niloticus*): resistance to waterborne copper toxicity. Aquac Res 46:1001–1006

Abdel-Tawwab M (2016) Effect of feed availability on susceptibility of Nile tilapia, *Oreochromis niloticus* (L.) to environmental zinc toxicity: growth performance, biochemical response, and zinc bioaccumulation. Aquaculture 464:309–315

Abdel-Tawwab M, Mousaad MNM, Sharafeldin KM, Ismaiel NEM (2013) Changes in growth and biochemical status of common carp, *Cyprinus carpio* L. exposed to water-born zinc toxicity for different periods. Internat. Aquat Res 5:11 **(1–9)**

Abdel-Tawwab M, Sharafeldin KM, Mousaad MNM, Ismaiel NEM (2015) Coffee bean in common carp, *Cyprinus carpio* L. diets: effect on growth performance, biochemical status, and resistance to waterborne zinc toxicity. Aquaculture 448:207–213

Abdel-Tawwab M, El-Sayed GO, Shady SH (2016) Growth, biochemical variables, and zinc bioaccumulation in Nile tilapia, *Oreochromis niloticus* (L.) as affected by waterborne zinc toxicity and exposure period. Int Aquat Res 8:197–206

Abdel-Tawwab M, El-Sayed GO, Monier MN, Shady SH (2017a) Dietary EDTA supplementation improved growth performance, biochemical variables, antioxidant response, and resistance of Nile tilapia, *Oreochromis niloticus* (L.) to environmental heavy metals exposure. Aquaculture 473:478–486

Abdel-Tawwab M, Sharafeldin KM, Ismaiel NEM (2017b) Interactive effects of coffee bean supplementation and waterborne zinc toxicity on growth performance, biochemical variables, antioxidant activity and zinc bioaccumulation in whole body of common carp, *Cyprinus carpio* L. Aquac Nutr 2017:1–8. doi:10.1111/anu.12540

Benaissa H (2006) Screening of new sorbent materials for cadmium removal from aqueous solutions. Hazard Mater 132:189–195

Benhima H, Chiban M, Sinan F, Seta P, Persin M (2008) Removal of lead and cadmium ions from aqueous solution by adsorption onto micro-particles of dry plants. Colloids Surf. B: Biointerfaces 61:10–16

Boyd CE (1984) Water quality in warm fishponds. Auburn University Agriculture Experimental Station, Auburn

Chiban M, Soudani A, Sinan F, Persin M (2012) Wastewater treatment by batch adsorption method onto micro-particles of dried *Withania frutescens* plant as a new adsorbent. J Environ Manage 95:61–65

dos Santos WNL, Cavalcante DD, da Silva EGP, das Virgens CF, Dias FdS (2011) Biosorption of Pb(II) and Cd(II) ions by *Agave sisalana* (sisal fiber). Microchemical J 97:269–273

Dytham C (1999) Choosing and using statistics: a biologist's guide. Blackwell Science Ltd, London

Easa M, El-S Shereif MM, Shaaban AI, Mancy KH (1995) Public health implications of wastewater reuse for fish production. Water Sci Technol 32:145–152

Fatemeh A, Shariatmadari H, Mirghaffari N (2008) Modification of rice hull and sawdust sorptive characteristics for remove heavy metals from synthetic solutions and wastewater. J Hazard Mater 154:451–458

Fazio F, Piccione G, Tribulato K, Ferrantelli V, Giangrosso G, Arfuso F, Faggio C (2014) Bioaccumulation of heavy metals in blood and tissue of striped mullet in two Italian lakes. J Aquat Anim Health 26:278–284

Gupta VK, Ali I (2000) Utilization of bagasse fly ash (a sugar industry waste) for the removal of copper and zinc from wastewater. Sep Puri Technol 18:131–140

Gupta S, Babu BV (2009) Utilization of waste product (tamarind seeds) for the removal of Cr(VI) from aqueous solutions: equilibrium, kinetics, and regeneration studies. J Environ Manage 90:3013–3022

Jiménez-Cedillo MJ, Olguín MT, Fall C, Colin-Cruz A (2013) As(III) and As(V) sorption on iron-modified non-pyrolyzed and pyrolyzed biomass from *Petroselinum crispum* (parsley). J Environ Manage 117:242–252

Karnitz Júnior O, Gurgel LVA, Gil LF (2010) Removal of Ca(II) and Mg(II) from aqueous single metal solutions by mercerized cellulose and mercerized sugarcane bagasse grafted with EDTA dianhydride (EDTAD). Carbohyd Polym 79:184–191

Kelly-Vargas K, Cerro-Lopez M, Reyna-Tellez S, Bandala ER, Sanchez-Salas JL (2012) Biosorption of heavy metals in polluted water, using different waste fruit cortex. Phys Chem Earth 37–39:26–29

Khalil MT, Hussein HA (1997) Use of waste water for aquaculture: an experimental field study at a sewage-treatment plant, Egypt. Aquacult Res 28:859–865

Marin-Rangel VM, Cortes-Martines R, Villanueva RAC, Garnica-Romo MG, Martinez-Flores HE (2012) As(V) biosorption in an aqueous solution using chemically treated lemon (*Citrus aurantifolia* swingle) residues. J Food Sci 71:10–14

Mohan D, Pittman CU Jr (2006) Activated carbons and low cost adsorbents for remediation of tri- and hexavalent chromium from water. J Hazard Mater 137(2):762–811

Mosa AA, El-Ghamry A, Trüby P (2011) Chemically modified crop residues as a low-cost technique for the removal of heavy metal ions from wastewater. Water Air Soil Pollut 217:637–647

Nguyen TAH, Ngo HH, Guo WS, Zhang J, Liang S, Yue QY, Li Q, Nguyen TV (2013) Applicability of agricultural waste and by-products for adsorptive removal of heavy metals from wastewater. Biores Technol 148:574–585

Osman HE, Badwy RK, Ahmad HF (2010) Usage of some agricultural byproducts in the removal of some heavy metals from industrial wastewater. J Phytol 2:51–62

Savorelli F, Manfra L, Croppo M, Tornambè A, Palazzi D, Canepa S, Trentini PL, Cicero AM, Faggio C (2016) Fitness evaluation of *Ruditapes philippinarum* exposed to Ni. Biol Trace Elem Res. doi:10.1007/s12011-016-0885-y

Shereif M, Mancy K (1995) Organochlorine pesticides and heavy metals in fish reared in treated sewage effluents and fish grown in farms using polluted surface waters in Egypt. Water Sci Technol 32:153–161

Shereif M, El-S Easa M, El-Samra M, Mancy K (1995) A demonstration of wastewater treatment for reuse applications in fish production and irrigation in Suez. Egypt. Water Sci Technol 32:137–144

Swelam AA, Awad MB, Salem AMA, El-Feky AS (2016) Removal of copper ions from synthetic waste water using some agricultural wastes as a low-cost adsorbent. AOP J Environ Waste Manag 1:103–117

Tarley CRT, Arruda MAZ (2004) Biosorption of heavy metals using rice milling byproducts. Characterization and application for removal of metals from aqueous effluents. Chemosphere 54:987–995

USEPA (United States Environmental Protection Agency) (1986). Quality Criteria for Water. EPA 440/5-86-001

WHO (World Health Organization) (1989) Guidelines for drinking water quality, vol 1. Recommendations WHO, Geneva, p 19

Witek-Krowiak A, Reddy DHK (2013) Removal of microelemental Cr(III) and Cu(II) by using soybean meal waste: unusual isotherms and insights of binding mechanism. Bioresour Technol 127:350–357

Zafar MN, Nadeem R, Hanif MA (2007) Biosorption of nickel from protonated rice bran. J Hazard Mater 143:478–485

18

Diversity of denitrifying bacteria in the greenwater system of coastal aquaculture

V. Kathiravan · K. K. Krishnani

Abstract Greenwater technology is based on integrating finfishes inside pens for zero water exchange system of shrimp aquaculture. Nitrogen transformation could be improved by rearing euryhaline finfishes like grey mullet, milkfish which have a broad diet and tolerate poor water quality. The abundance of four denitrifying functional genes coding for *nitrate reductase* (*narG* and *napA*), *nitrite reductase* (*nirS*), *nitric oxide reductase* (*qnorB*) and *nitrous oxide reductase* (*nosZ*) has been examined in the greenwater system through a metagenomic approach. Phylogeny revealed homology of *narG* clones with uncultured environmental clones, whereas *napA* clone sequences were found to have homology with cultured (*Stappia aggregata*) and uncultured microorganisms. The *nirS* clones show uniqueness with *Marinobacter hydrocarbonoclasticus, Aromatoleum aromaticum,* and *Ruegeria pomeroyi*. The *qnorB* gene has been reported for the first time from culture systems along the Indian coast and clone exhibited 84–87 % identity with different uncultured bacteria. The *nosZ* clones are closely affiliated with *S. aggregate* and alpha bacterium. This study revealed denitrifying diversity from a greenwater system which could eventually be used in planning the future strategy for comprehending nitrogen fluxes, greenhouse gases and their mitigation in coastal aquaculture systems.

Keywords Greenwater · Coastal aquaculture · Denitrification · Nitrogen flux · Functional gene · Diversity

Introduction

Nitrates are a by-product of excess feed and decayed organic wastes from animals in aquaculture ponds and can be toxic to animals at higher concentrations. Denitrification is a reductive process which converts nitrate (NO_3) into atmospheric nitrogen (N_2) through nitrite (NO_2), nitric oxide (NO), and nitrous oxide (N_2O). These intermediate products can be used as alternative electron acceptors for energy production in anaerobic condition (Stres and Murovec 2008). Nitric oxide and nitrous oxide attract the modern scientist as they are 310

V. Kathiravan · K. K. Krishnani
Central Institute of Brackishwater Aquaculture, 75, Santhome High Road, R.A. Puram, Chennai 600028, India

V. Kathiravan
Department of Botany and Research Center, The Madura College (Autonomous), Madurai 625011, India

K. K. Krishnani (✉)
School of Edaphic Stress Management, National Institute of Abiotic Stress Management, Baramati (Dist. Pune) 413 115, India
e-mail: krishnanik@hotmail.com

times more potent greenhouse gases than carbon dioxide (Fernandez et al. 2008). In the denitrification pathway, different functional genes are involved that encode for enzyme to use nitrogenous metabolites by targeting conserved regions. Complete denitrification needs the sequential action of four enzymes: *nitrate reductase* (*narG* and *napA*), *nitrite reductase* (*nirK/nirS*), *nitric oxide reductase* (*nor*), and *nitrous oxide reductase* (*nosZ*) (Chen et al. 2011). They can be used to remove excess nitrogen from wastewater treatment plants (Throback et al. 2004). Cultivation-independent functional gene-based approaches have an important role to play in examining the entire diversity (Palumbo et al. 2004). The use of functional genes involved in denitrification is a more direct method of detecting or identifying denitrifying bacteria (Wolsing and Prieme 2004). Analysis of functional diversity and its dynamics in the environment could help to understand the microbial ecology and biogeochemistry of aquatic systems (Taroncher-Oldenburg et al. 2003).

Previous studies have focused mainly on the composition and activities of soil microbiota in aquaculture. A few studies have employed molecular tools to understand the diversity of archaeal and bacterial community structures in coastal aquaculture. The composition of nitrate reducing community has been explored from various habitats. Studies conducted by Sanford et al. (2012) have expanded the current understanding of the nitrogen cycle. Denitrification in aquatic environment leads to nitrate removal from the aquatic system; therefore, it is necessary to understand functional diversity of denitrifier communities in the system (Huang et al. 2011).

Herbivores finfish mainly the grey mullet (*Mugil cephalus*) and milkfish (*Chanos chanos*) are propagated as bioremediators in fish cages in shrimp growing ponds. This is an innovative bioaugmentation technique called as greenwater technology, which is proven most functional among all others for environmental management in coastal aquaculture (Baliao et al. 1999; Baliao 2000). We have previously examined the diversity of sulfur oxidizing bacteria, nitrogen fixing bacteria and nitrite oxidizing bacteria in greenwater system (Krishnani et al. 2010; Kathiravan and Krishnani 2012; Kathiravan et al. 2012). In the present study, functional genes: *narG, napA*, *nirS*, *qnorB,* and *nosZ* are used as molecular markers to identify denitrifying population in the greenwater system of coastal aquaculture.

Methods

Sampling and analysis

The greenwater technology experiment with milkfish-*C. chanos* was conducted in a 0.5 ha shrimp pond at Karapadagai (10°60'84"N, 79°79'19") village in Nagapattinam Dist., Tamil Nadu, India. 100-g soil and 200-ml water samples each were collected from four corners and a center of the pond and pooled in order to prepare a composite soil sample (0.5 kg) and water sample (1 l). Soil pH (Cyberscan, Canada), electrical conductivity (Elico, India), available phosphorous, total nitrogen (APHA 1989) and organic carbon were determined using the chromic acid digestion method (Walkley and Black 1934).

Metagenomic DNA isolation

Composite water samples (1 l) collected from the pond were transferred to an ultra-filtration unit to filter the bacteria onto a membrane (0.2 μm). The filter membrane containing bacteria was used for metagenomic DNA isolation using modified CTAB-phenol chloroform method (Krishnani et al. 2009a). The DNA was extracted from composite soil using the PowerSoil DNA isolation kit (PowerSoil, Mobio Laboratories Inc., CA, USA), according to the manufacturer's instructions. Metagenomic DNA from composite soil and water samples were pooled to obtain the absolute diversity of the pond. Purity was determined by measuring the 260/280 nm absorbance ratio. The DNA was stored at −20 °C for further molecular analyses.

PCR amplification of functional genes

Polymerase chain reaction was performed with a negative control of 40-μl reaction mixture using the mastercycler (Eppendorf Mastercycler gradient, Germany). The denitrifying functional genes were amplified using the primer pairs shown in Table 1. Two microliters of pooled metagenomic DNA were added to a 40-μl

Table 1 Oligonucleotides used for PCR amplification and sequencing of functional genes

Genes	Primer	References
narG	narG1960f-5′TAYGTSGGSCARGARAA 3′	Philippot et al. (2002)
	narG2650r-5′TTYTCRTACCABGTBGC 3′	
napA	V67F-5′TAYTTYYTNHSNAARATHATGTAYGG 3′	Flanagan et al. (1999)
	V67R-5′DATNGGRTGCATYTCNGCCATRTT 3′	
nirS	NirS1F-5′CCTAYTGGCCGCCRCART 3′	Braker et al. (1998)
	NirS6R-5′CGTTGAACTTRCCGGT 3′	
qnorB	qnorB2F-5′ GGNCAYCARGGNTAYGA 3′	Braker and Tiedje (2003)
	qnorB7R-5′ GGNGGRTTDATCADGAANCC 3′	
nosZ	Nos1527F-5′CGCTGTTCHTCGACAGYCA 3′	Scala and Kerkhof (1998)
	NosZ-F-5′CG(CT)TGTTCMTCGACAGCCAG 3′	Kloos et al. (2001)
	NosZR-5′CATGTGCAGNGCRTGGCAGAA 3′	

PCR mixture containing 1X PCR buffer (INVITROGEN, USA), 1.5 mM $MgCl_2$, 0.25 mM of each dNTP, 30 pM of each primer, and 1 unit of Taq polymerase (INVITROGEN, USA). The reaction was initially denatured at 94 °C for 2 min; followed by 34 cycles of 94 °C for 40 s, 57 °C for 40 s, and 72 °C for 40 s; followed by a final extension step at 72 °C for 8 min. The PCR products were analyzed on 1.5 % (w/v) ethidium bromide-stained agarose gel to ensure the correct size fragment were amplified.

Cloning and sequence analysis

The amplified functional genes were purified with a gel extraction kit (Himedia, India) according to manufacturer's instructions. The purified PCR products were ligated using the pDK101 as recommended by the manufacturer and were transformed into high efficiency competent cells. Clones were confirmed using *NcoI* restriction endonuclease. Sequencing was carried out in an ABI 3100 Genetic Analyzer (Applied Biosystems, USA). Nucleotide sequences were translated into protein sequences using the Translate tool on the ExPASy (Expert Protein Analysis System) proteomics server of the Swiss Institute of Bioinformatics (http://us.expasy.org/tools/dna.html). The nucleotide and protein sequences were compared with the entries in GenBank using BlastN and BlastP, respectively (Altschul et al. 1990). The phylogenetic tree based on the alignment of predicted amino acids was generated using the Maximum Composite Likelihood from MEGA 4 (Tamura et al. 2007). Bootstrap analyses of amino acid sequences were also performed using this program. Signature motif of *qnorB* was identified by using PROSTIE (http://www.prosite.expasy.org/). The richness of denitrifying bacterial community in greenwater system has been analyzed by EcoSim (Gotelli and Entsminger 2012).

Results and discussions

Soil characteristics of greenwater system

Soil quality parameters such as pH, electrical conductivity (EC), available phosphorous, total nitrogen, and organic carbon were found to be in the range of 8.29–8.40, 10.86–15.92 dS/m, 0.0288–0.0682 ppm, 8.4–10.12 ppm and 0.19–0.31 %, respectively, which are well within the safe levels prescribed for finfish and shellfish aquaculture.

Detection of denitrifying functional genes

Denitrifying bacterial community was detected in greenwater-bioaugmentation system using PCR amplification of functional genes such as *narG napA*, *nirS*, *qnorB* and *nosZ* gene. The primers used in this study amplified well and yielded PCR products of the expected 650 bp (Fig. 1a), 414 bp (Fig. 1b), 890 bp (Fig. 1c), 660 bp (Fig. 1d) and 700 bp (Fig. 1e) for *narG, napA, nirS, qnorB,* and *nosZ,* respectively.

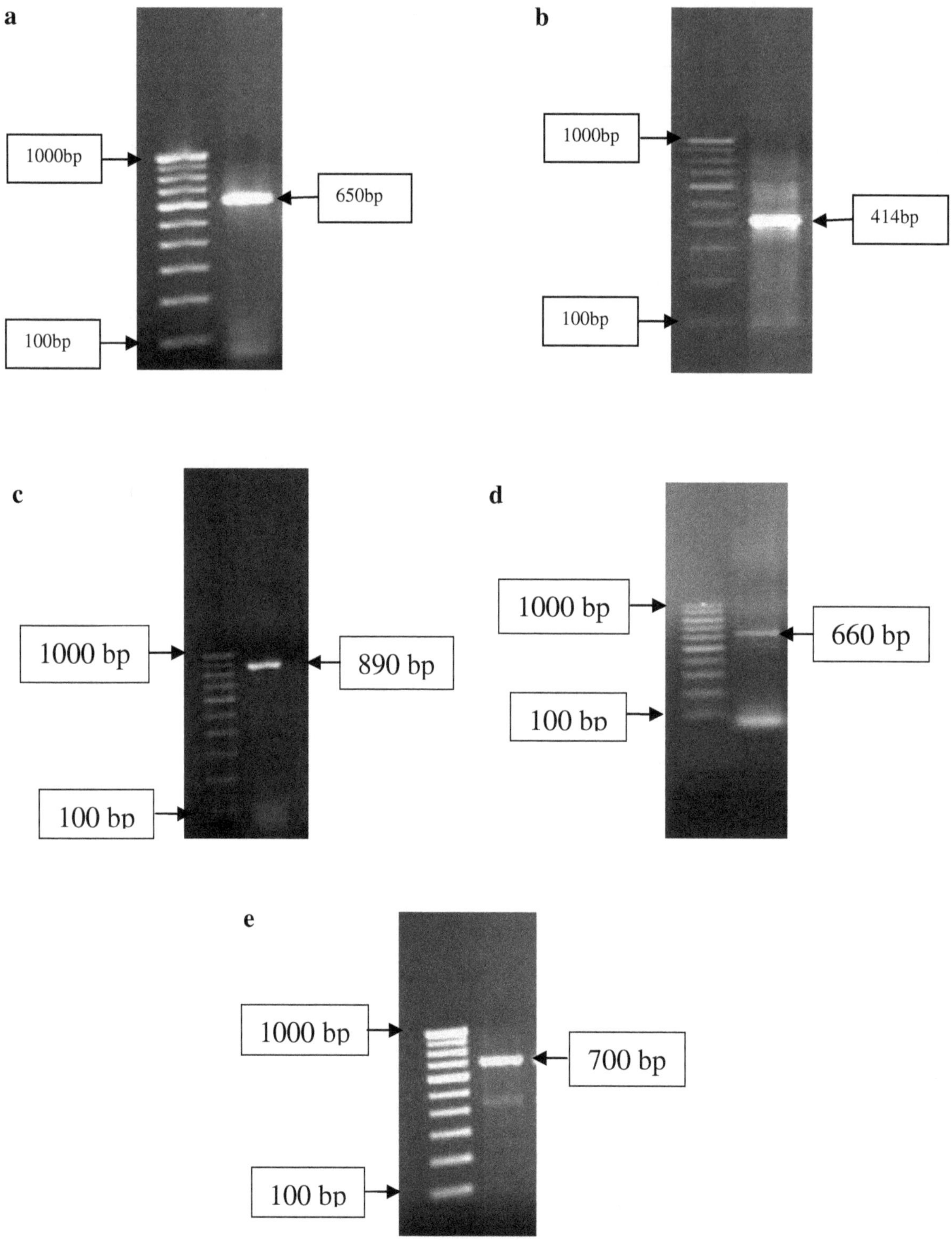

Fig. 1 PCR amplification of functional genes. **a** 650 bp fragment of narG gene; **b** 414 bp fragment of napA gene; **c** 890 bp fragment of nirS gene; **d** 660 bp fragment of qnorB gene; **e** 700 bp fragment of nosZ gene

Nucleotide sequence accession numbers

The nucleotide sequences of denitrifying functional genes determined in this study have been deposited in the GenBank database. The accession numbers for the *narG, napA, nirS* and *nosZ* gene sequences are FJ905006 to

Table 2 Percent similarity of amino acid predicted from denitrifying functional genes

Environmental clone	Accession number	Homology on amino acid levels
narG gene		
V18DBT21	FJ905013-ACQ65795	u.bacterium(ABF20959) 80 %
V18DBT17	FJ905012-ACQ65794	u.bacterium (ACH90505) 81 %
V18DBT14	FJ905011-ACQ65793	u.bacterium(CAL92046) 78 %
V18DBT9	FJ905010-ACQ65792	u.bacterium (ACH90505) 81 %
V18DBT24	FJ905009-ACQ65791	u.bacterium(CAL92058) 97 %
V18DBT23	FJ905008-ACQ65790	u.bacterium(CAL92058) 96 %
V18DBT18b	FJ905007-ACQ65789	u.bacterium(CAL92058) 97 %
V18DBT4	FJ905006-ACQ65788	u.bacterium(CAL63724) 95 %
napA gene		
GSV13DBT14	FJ905004-ACQ65787	*S.aggregata*(ZP_01549680) 99 %
GSV13DBT4	FJ905003-ACQ65786	u.bacterium(CAL63635) 85 %
GSV13DBT11	FJ905002-ACQ65785	u.bacterium(CAL63627) 87 %
GSV13DBT20	FJ905001-ACQ65784	u.bacterium(CAL63659) 89 %
GSV13DBT21	FJ905000-ACQ65783	u.bacterium(CAL63637) 93 %
GSV13DBT24	FJ904999-ACQ65782	u.bacterium(ABW17415) 88 %
GSV13DBT23	FJ904998-ACQ65781	u.bacterium(ABW17415) 88 %
GSV13DBT19	FJ904997-ACQ65780	u.bacterium(CAL63637) 93 %
GSV13DBT12	FJ904996-ACQ65779	*S.aggregata*(ZP_01549680) 99 %
GSV13DBT7	FJ904995-ACQ65778	u.bacterium(ABW17415) 86 %
GSV13DBT3	FJ904994-ACQ65777	u.bacterium(ABW17382) 85 %
GSV13DBT2	FJ904993-ACQ65776	u.bacterium(CAL63637) 93 %
nirS gene		
DBT26	GU553355-ADD54535	*M.hydrocarbonoclasticus*(ACN97382) 99 %
DBT2GW17	GU553357-ADD54537	*R.pomeroyi*_(YP_165049) 86 %
DBT9GW17	GU553358-ADD54538	*A.aromaticum*(YP_157499) 85 %
DBT13GW17	GU553359-ADD54539	*A.aromaticum*(YP_157499) 88 %
qnorB gene		
DBT20AA10	GU553356-ADD54536	u.bacterium (AEH50055) 87 %
		u.bacterium (AEH49944) 86 %
		u.bacterium (CAD45402) 84 %
		u.bacterium (AEH49905) 86 %
		u.bacterium (AEH50007) 86 %
		u.bacterium (AEH50064) 86 %
		u.bacterium (AEH49888) 84 %
		u.bacterium (CAD45392) 85 %
		u.bacterium (CAD45394) 85 %
nosZ gene		
DBTNZ6	GU122973-ACY92298	*S. aggregata* (ZP_01549649) 99 %
DBTNZ7	GU122967-ACY92292	u.bacterium(ACJ02316) 83 %
DBTNZ51	GU122968-ACY92293	*Rhodobacterales*_sp (ZP_05078392) 98 %
DBTNZ52	GU122969-ACY92294	a.proteobacterium(AFC41135) 100 %
DBTNZ56	GU122970-ACY92295	*S. aggregata* (ZP_01549649) 100 %
DBTNZ58	GU122972-ACY92297	a.proteobacterium(AFC41135) 99 %
DBTNZ67	GU122971-ACY92296	*S. aggregata* (ZP_01549649) 99 %

FJ905013, FJ904993 to FJ905004, GU553355 to GU553359, and GU122967 to GU122973, respectively. The accession number of the *qnorB* gene sequence is GU553356.

Diversity of denitrifying functional genes

Determination of the physiological characteristics of unknown groups of denitrifiers and the development of approaches for identifying active denitrifiers are crucial (Philippot 2005). Predicted amino acid sequences of functional genes were compared with sequences available in the GenBank. Percent similarity of amino acids predicted from gene sequences is depicted in Table 2. Phylogenetic tree based on predicted amino acids encoded by *narG* with bootstrap support of 100 % is presented in Fig. 2. Eight distinct clones were yielded from 28 clones. Five clones (V1DBT4, V1DBT14, V1DBT18b, V1DBT23 and V1DBT24) were most closely related to sequences from estuarine sediments (GenBank: CAL63724, CAL92046, CAL92071, CAL92086, CAL92058, and CAL92131) (Smith et al. 2007) with the identity ranging 78–97 %. The Clone V1DBT21 isolate was closely related to *narG* sequence (GenBank: ABF20959). A further two clones (V1DBT9 and V1DBT17) were related to *narG* sequence (GenBank: ACH90505) derived from sediments of the river (Reyna et al. 2010).

Phylogenetic tree based on predicted amino acids encoded by *napA* was shown in Fig. 3. Comparison with the GenBank database using BLASTN search revealed that sequences showed homology to *napA* sequences from uncultured environmental clones. The Clones GSV13DBT3 and GSV13DBT7 showed 85–86 % similarity at the amino acid level with the clones GenBank: ABW17382 and ABW1741 isolated from coral reef sediments. Clones GSV13DBT12 and GSV13DBT14 have 99 % amino acid homology with *nitrate reductase* gene of *Stappia aggregata* IAM 12614 strain (King 2006). Clones GSV13DBT23 and GSV13DBT24 exhibited 88 % similarity with clone GenBank: ABW17415. The remaining six *napA* clones (GSV13DBT2, GSV13DBT4, GSV13DBT11, GSV13DBT19, GSV13DBT20, and GSV13DBT21) were closely related with the clones from estuarine sediment (Smith et al. 2007).

Phylogenetic tree based on predicted amino acids encoded by *nirS* is presented in Fig. 4. Clone DBT26 matched 99 % with *Marinobacter hydrocarbonoclasticus* strain DSM 8798 (GenBank: ACN97382) at amino

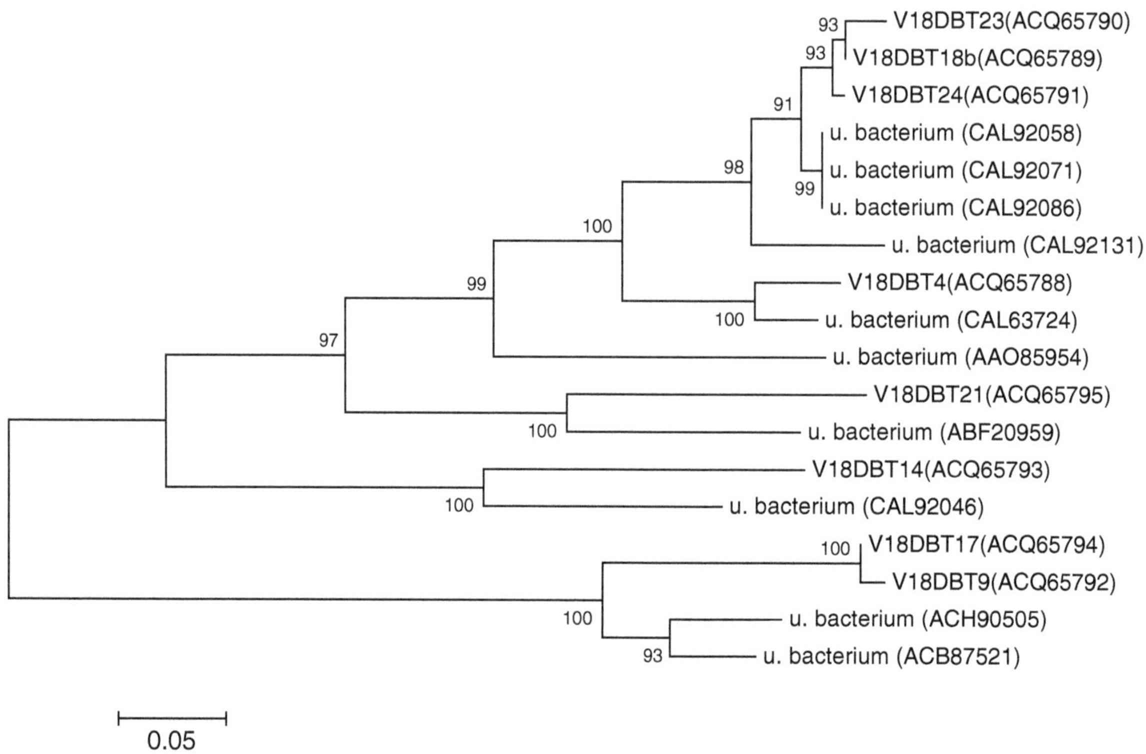

Fig. 2 Phylogenetic tree based on predicted amino acids encoded by narG as determined by maximum likelihood method

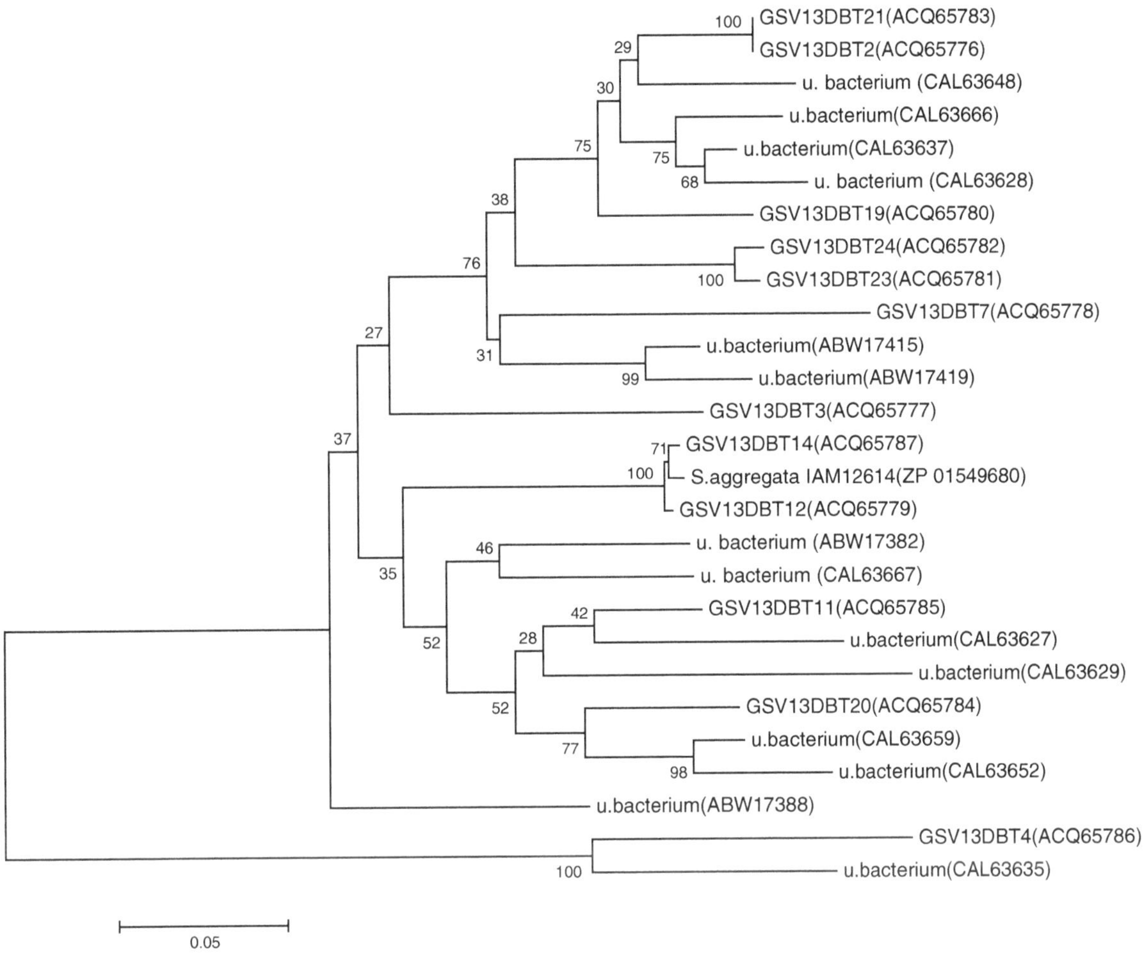

Fig. 3 Phylogenetic tree based on predicted amino acids encoded by napA as determined by maximum likelihood method

acid level (Gonzalez-Domenech et al. 2010). Tiedje (1988) has reported *M. hydrocarbonoclasticus,* which is a true denitrifier as it can reduce nitrate into nitric oxide (N_2O) to the extent of 90 % in the presence of acetylene. Clones DBT13GW17 and DBT9GW17 show 88 and 87 % similarity at amino acid levels with *nirS* gene from *Aromatoleum aromaticum* EbN1 (GenBank: YP_157499) (Kuhner et al. 2005) and *Azoarcus*_sp.KH32C (GenBank: BAL25833), respectively. Clone DBT2GW17 exhibits 86 % similarity with *Ruegeria pomeroyi* DSS-3 (GenBank: YP_165049) (Moran et al. 2007). Prieme et al. (2002) amplified *nirS* genes from marsh soil and most of the clones were not found in cultivated denitrifiers. Krishnani (2010) has reported a *nirS* clone in coastal soil showing similarity with uncultured bacteria from coastal aquaculture.

The *qnorB* clone DBT20AA10 is not correlated to any known cultivable denitrifying isolate. It has a maximum similarity with uncultured bacterial clones retrieved from the Taoyuan Agro-ecosystem Research Station of the Chinese Academy of Sciences, China (Chen et al. 2011). The homology level of clone with uncultured clones ranged 84–87 % at the amino acid level (Table 2). This shows that the retrieved clone is diverse from other *nitric oxide reductase* present in the GenBank. Phylogenetic tree based on predicted amino acids encoded by *qnorB* is presented in Fig. 5. One signature motif such as [YWG]-[LIVFYWTA](2)-[VGS]-H-[LNP]-x-V-x(44,47)-H-H was found in clone sequence. This confirmed that the clone belongs to *nitric oxide reductase*.

Phylogenetic tree based on predicted amino acids encoded by *nosZ* is presented in Fig. 6. The biodiversity in greenwater-bioaugmentation system of coastal aquaculture is reflected in the finding that 23 *nosZ* clones from greenwater yielded seven different clones. The Clones DBTZ6, DBTNZ56, and DBTZ67 show 99–100 % similarity at amino acid level with *nosZ* gene from *S. aggregata* IAM 12614 (GenBank: ZP_01549649) (King 2006). The Clones DBTNZ7, DBTNZ52, DBTNZ58 have 99–100 % sequence

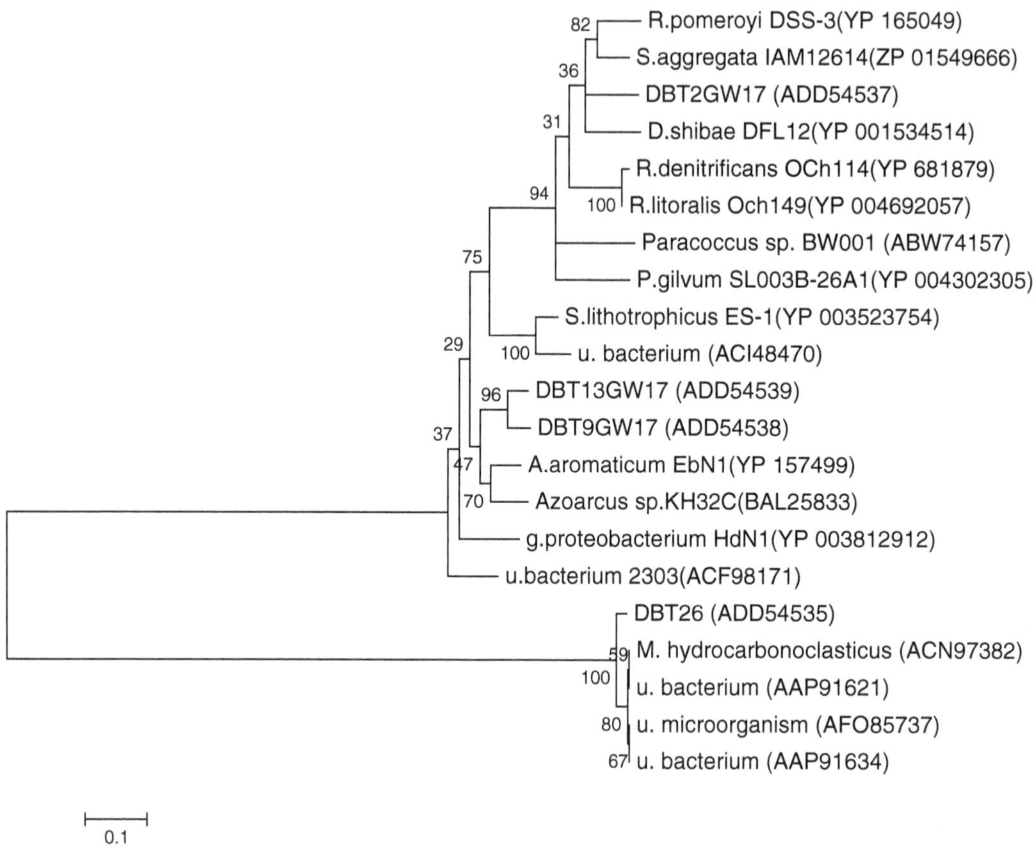

Fig. 4 Phylogenetic tree based on predicted amino acids encoded by nirS as determined by maximum likelihood method

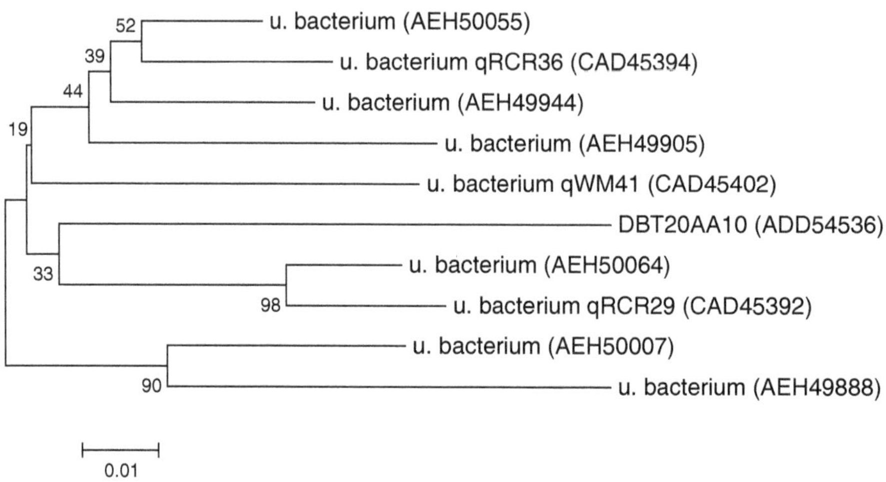

Fig. 5 Phylogenetic tree based on predicted amino acids encoded by qnorB as determined by maximum likelihood method

similarity with *nitrous oxide reductase* genes of Alpha proteobacterium (GenBank: AFC41135). The Clone DBTNZ51 has 93–98 % similarity with *nosZ* sequences from *Rhodobacterales* bacterium Y4I (GenBank: ZP_05078392), *Ruegeria pomeroyi* DSS-3 (GenBank: YP_164881) and *Roseobacter* sp. SK209-2-6 (GenBank: ZP_01754658). Krishnani (2010) has reported three *nosZ* clones having 82 % similarity with *Marinobacter* sp.

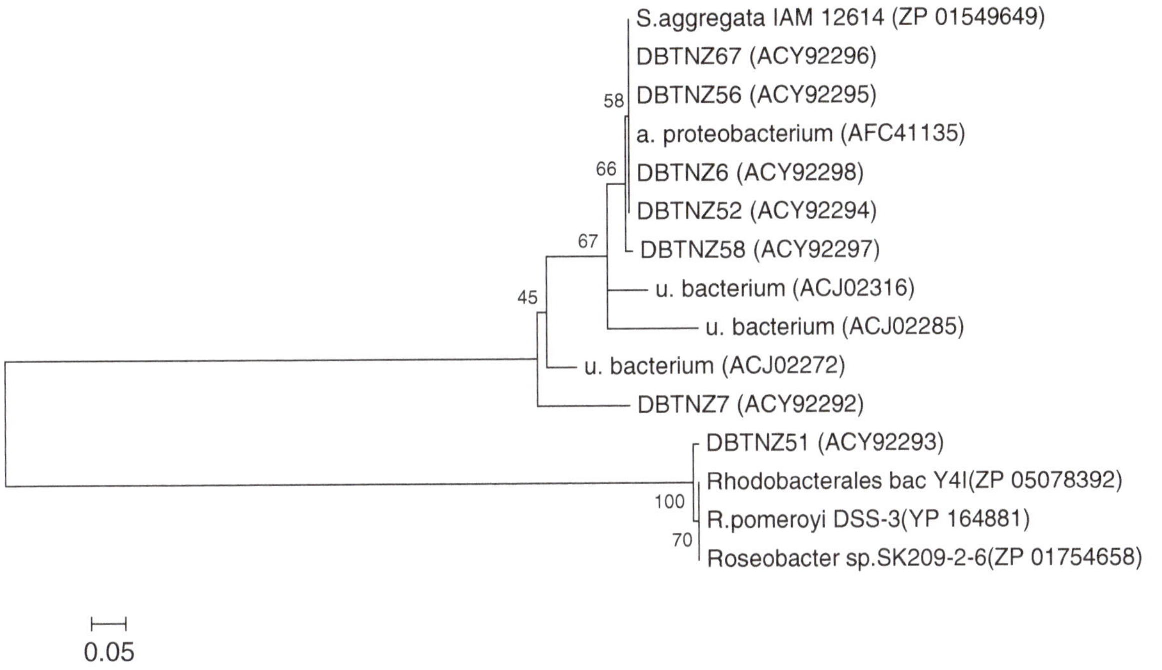

Fig. 6 Phylogenetic tree based on predicted amino acids encoded by nosZ as determined by maximum likelihood method

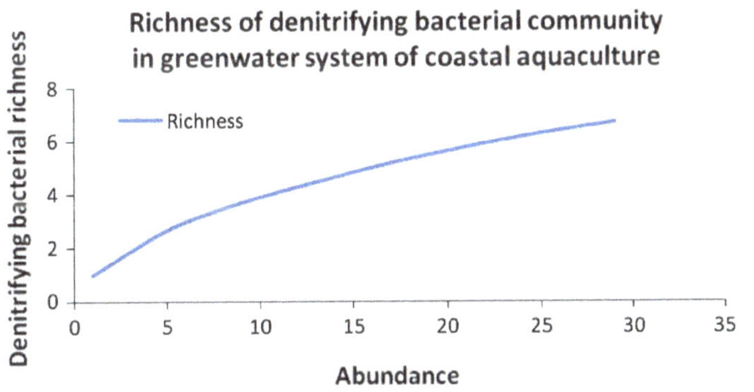

Fig. 7 Richness of denitrifying bacterial community in greenwater system of coastal aquaculture

The richness of denitrifying bacteria is presented in Fig. 7. The results revealed higher denitrifying bacterial diversity in the greenwater system of coastal aquaculture as compared to earlier studies on denitrifying functional genes in Indian coastal aquaculture, which retrieved less denitrifying bacterial diversity with the same primers (Krishnani et al. 2009a, b; Krishnani 2010).

Conclusions

In the present study, the diversity of genes (*narG, napA, nirS, qnorB,* and *nosZ*) encoding complete denitrification in Indian coastal aquaculture has been investigated. To our knowledge, this is the first report for identification of complete denitrification genes especially *qnorB* gene in Indian coastal aquaculture. Maximum clones have similarity with uncultured bacterial clones which revealed that many bacteria yet to be cultured in future to utilize their metabolites. To achieve this objective, studies must be performed in a way that successfully relates microbial diversity and activity. The results confirmed that metagenomic library creation is specific, sensitive and can be used for analysis of uncultivable denitrifying bacterial community from aquatic environment. In conclusion, the denitrifying communities from the greenwater system of coastal aquaculture are phylogenetically diverse from other environments. This work could be useful in planning the future strategy for understanding of denitrifying communities, nitrogen fluxes, greenhouse gases and their mitigation in coastal aquaculture systems.

Acknowledgments The authors are grateful to Dr. A.G. Ponniah, Director, Central Institute of Brackishwater Aquaculture, Chennai for providing facilities to carry out this work. Financial assistance from the Department of Biotechnology, Ministry of Science and Technology is gratefully acknowledged. The authors are thankful to Dr G Gopikrishna, Principal Scientist, CIBA, Chennai for critically reviewing the manuscript.

Conflict of interest The authors declare that they have no competing interests.

Authors' contributions VK and KKK defined the research theme, designed methods and experiments. VK carried out the laboratory experiments. VK and KKK analyzed the data and interpreted the results. VK and KKK wrote the manuscript. Both authors read and approved the final manuscript.

References

Altschul SF, Gish W, Miller W, Myers EW, Lipman DJ (1990) Basic local alignment search tool. J Mol Biol 215:403–410

APHA, AWWA, WEF (1989) Standard methods for the examination of water and wastewater, 17th edn., American Public Health Association, Washington, DC, USA

Baliao DD (2000) Environment-friendly schemes in intensive shrimp farming. State of the art Series. Ilo-ilo: SEAFDEC

Baliao DD, De Los Santo MA, Franco NM (1999) Milkfish pond culture. Aquaculture Extension manual no. 25. Ilo-ilo: SEAFDEC

Braker G, Tiedje JM (2003) Nitric oxide reductase (*norB*) genes from pure cultures and environmental samples. Appl Environ Microbiol 69(6):3476–3483

Braker G, Fesefeldt A, Witzel KP (1998) Development of PCR primer systems for amplification of nitrite reductase genes (nirK and nirS) to detect denitrifying bacteria in environmental samples. Appl Environ Microbiol 64(10):3769–3775

Chen Z, Liu J, Wu M, Xie X, Wu J, Wei W (2011) Differentiated response of denitrifying communities to fertilization regime in paddy soil. Microb Ecol 63(2):446–459

Fernandez LA, Perotti EB, Sagardoy MA, Gomez MA (2008) Denitrification activity of *Bradyrhizobium* sp. isolated from Argentine soybean cultivated soils. World J Microbiol Biotechnol 24(11):2577–2585

Flanagan DA, Gregory LG, Carter JP, Karakas-Sen A, Richardson DJ, Spiro S (1999) Detection of genes for periplasmic nitrate reductase in nitrate respiring bacteria and in community DNA. FEMS Microbiol Lett 177(2):263–270

Gonzalez-Domenech CM, Martínez-Checa F, Béjar V, Quesada E (2010) Denitrification as an important taxonomic marker with in the genus *Halomonas*. Syst Appl Microbiol 33(2):85–93

Gotelli NJ, Entsminger GL (2012) EcoSim 7.72. Acquired Intelligence, Inc. http://www.uvm.edu/~ngotelli/EcoSim/EcoSim.html

Huang S, Chen C, Yang1 X, Wu1 Q, Zhang R (2011) Distribution of typical denitrifying functional genes and diversity of the nirS-encoding bacterial community related to environmental characteristics of river sediments. Biogeosci 8: 3041–3051

Kathiravan V, Krishnani KK (2012) Diversity of diazotrophic bacteria in greenwater system of coastal aquaculture. J Environ Appl Biores 1(1):5–11

Kathiravan V, Krishnani KK, Azad IS, Kailasam M, Ponniah AG (2012) Diversity of nitrite oxidizers in greenwater system of coastal aquaculture by metagenomic approach targeting *nxrb* genes. Biotechnol Indian J 6(11):359–366

King GM (2006) Nitrate-dependent anaerobic carbon monoxide oxidation by aerobic CO-oxidizing bacteria. FEMS Microbiol Ecol 56(1):1–7

Kloos K, Mergel A, Rösch C, Bothe H (2001) Denitrification within the genus *Azospirillum* and other associative bacteria. Aust J Plant Physiol 28(9):991–998

Krishnani KK (2010) Detection and diversity of nitrifying and denitrifying bacteria in coastal aquaculture. Aquaculture 302(1–2):57–70

Krishnani KK, Shekhar MS, Gopikrishna G, Gupta BP (2009a) Molecular biological characterization and biostimulation of ammonia-oxidizing bacteria in brackishwater aquaculture. J Environ Sci Health A 44(14):1598–1608

Krishnani KK, Shekhar MS, Gopikrishna G, Gupta BP (2009b) Sequence Similarity based identification of nitrifying bacteria in coastal aquaculture for bioremediation predictability. Asian Fish Sci 22(1):41–49

Krishnani KK, Kathiravan V, Natarajan M, Kailasam M, Pillai SM (2010) Diversity of sulfur-oxidizing bacteria in Greenwater System of coastal aquaculture. Appl Biochem Biotechnol 162(5):1225–1237

Kuhner S, Wöhlbrand L, Fritz I, Wruck W, Hultschig C, Hufnagel P, Kube M, Reinhardt R, Rabus R (2005) Substrate-dependent regulation of anaerobic degradation pathways for toluene and ethylbenzene in a denitrifying bacterium, strain EbN1. J Bacteriol 187(4):1493–1503

Moran MA, Belas R, Schell MA, González JM, Sun F, Sun S, Binder BJ, Edmonds J, Ye W, Orcutt B, Howard EC, Meile C, Palefsky W, Goesmann A, Ren Q, Paulsen I, Ulrich LE, Thompson LS, Saunders E, Buchan A (2007) Ecological genomics of marine Roseobacters. Appl Environ Microbiol 73(14):4559–4569

Palumbo AV, Schryver JC, Fields MW, Bagwell CE, Zhou JZ, Yan T, Liu X, Brandt CC (2004) Coupling of functional gene diversity and geochemical data from environmental samples. Appl Environ Microbiol 70(11):6525–6534

Philippot L (2005) Tracking nitrate reducers and denitrifiers in the environment. Biochem Soc Trans 33(1):200–204

Philippot L, Piutti S, Martin-Laurent F, Hallet S, Germon JC (2002) Molecular analysis of the nitrate-reducing community from unplanted and maize-planted soils. Appl Environ Microbiol 68(12):6121–6128

Prieme A, Braker G, Tiedje JM (2002) Diversity of nitrite reductase (*nirK* and *nirS*) gene fragments in forested upland and wetland soils. Appl Environ Microbiol 68(4):1893–1900

Reyna L, Wunderlin DA, Genti-Raimondi S (2010) Identification and quantification of a novel nitrate-reducing community in sediments of Suquía River basin along a nitrate gradient. Environ Pollut 158(5):1608–1614

Sanford RA, Wagner DD, Wu Q, Chee-Sanford JC, Thomas SH, Cruz-García C, Rodríguez G, Massol-Deyá A, Krishnani KK, Ritalahti KM, Nissen S, Konstantinidis KT, Löffler FE (2012) Unexpected nondenitrifier nitrous oxide reductase gene diversity and abundance in soils. Proc Natl Acad Sci 109(48):19709–19714

Scala DJ, Kerkhof LJ (1998) Nitrous oxide reductase (*nosZ*) gene-specific PCR primers for detection of denitrifiers and three *nosZ* genes from marine sediments. FEMS Microbiol Lett 162(1):61–68

Smith CJ, Nedwell DB, Dong LF, Osborn AM (2007) Diversity and abundance of nitrate reductase genes (narG and napA), nitrite reductase genes (nirS and nrfA), and their transcripts in estuarine sediments. Appl Environ Microbiol 73(11):3612–3622

Stres B, Murovec B (2008) Differences in melting temperatures of degenerated oligonucleotides targetting nitrous oxide reductase (*nosz*) genes. Acta agriculturae Slovenica 92(1):75–82

Tamura K, Dudley J, Nei M, Kumar S (2007) MEGA4: Molecular Evolutionary Genetics Analysis (MEGA) software version 4.0. Mol Biol Evol 24(8):1596–1599

Taroncher-Oldenburg G, Griner EM, Francis CA, Ward BB (2003) Oligonucleotide microarray for the study of functional gene diversity in the nitrogen cycle in the environment. Appl Environ Microbiol 69(2):1159–1171

Throback IN, Enwall K, Jarvis A, Hallin S (2004) Reassessing PCR primers targeting *nirS*, *nirK* and *nosZ* genes for community surveys of denitrifying bacteria with DGGE. FEMS Microbiol Ecol 49(3):401–417

Tiedje JM (1988) Ecology of denitrification and dissimilatory nitrate reduction to ammonium. In: Zehnder AJB (ed) Biology of anaerobic microorganisms, tome I. Wiley, New York

Walkley A, Black CA (1934) An estimation of the Degtijareff method for determining soil organic matter and a proposed modification of the chromic acid titration method. Soil Sci 37:93–101

Wolsing M, Prieme A (2004) Observation of high seasonal variation in community structure of denitrifying bacteria in arable soil receiving artificial fertilizer and cattle manure by determining T-RFLP of nir gene fragments. FEMS Microbiol Ecol 48(2):261–271

19

Acute toxicity and risk assessment of three commercial detergents using the polychaete *Capitella* sp. C from Chetumal Bay, Quintana Roo, Mexico

R. G. Uc-Peraza · V. H. Delgado-Blas

Abstract The acute toxicity of three formulations of commercial detergent (ROMA®, FOCA® and BLANCA NIEVES®) was evaluated using the polychaete *Capitella* sp. C in static bioassays over a 48-h exposure period. The probit method was used to determine the median lethal concentration (LC_{50}) of each formulation as a whole as well as the LC_{50} of the active ingredient, linear alkylbenzene sulfonate (LAS), using 95 % confidence intervals. The formulations and LAS showed LC_{50} values of between 70.79 and 147.91 ppm and 15.48 and 22.38 ppm, respectively, at 48 h. FOCA® was the most toxic detergent, followed by BLANCA NIEVES® and finally ROMA®. The variation in the toxicity of the three detergents could have been caused both by differences in the relative concentrations of the anionic surfactant LAS contained in each formulation and the presence of other ingredients (enzymes, sodium silicate, sodium tripolyphosphate, bleachers and perfumes) which can also increase formulation toxicity. Correlation analysis revealed that percent mortality of *Capitella* sp. C increased with increase in the concentrations of the detergent over the 48-h exposure period. The risk quotient was greater than one for all three evaluated detergents, indicating that there is a high risk that they adversely affect the aquatic biota, particularly sediment-dwelling organisms such as the test species.

Keywords Polychaeta · *Capitella* · Bioassays · Acute toxicity · Detergents

Introduction

The contamination of aquatic ecosystems is increasing at an alarming rate as a consequence of the discharge of untreated sewage of urban and industrial origin into coastal zones, rivers, streams and lakes. Among the contaminants found in these sewage effluents are organic pollutants such as detergents, which can cause toxicity problems for the aquatic biota found in the receiving water bodies (Uc-Peraza and Delgado-Blas 2012). Furthermore, the phosphorus component of detergents contributes to eutrophication which results in an increase in the cell numbers of some algal species leading to decreased species diversity (Tkachenko and Kutsyn 2002; Markina and Aizdaicher 2007).

Detergents and surfactants are among the extensively used chemicals at home as well as in industry (Azizullah et al. 2011). Currently, the world production of detergent amounts to 10 millions of tons per years (IMRG 2012), and the worldwide total annual production of surfactants is estimated to exceed 15 million tons

---

To the memory of Jonatán Delgado Martínez, poet, composer, musician and rock drummer.

R. G. Uc-Peraza (✉) · V. H. Delgado-Blas
División de Ciencias e Ingeniería, Universidad de Quintana Roo, 77019 Chetumal, Quintana Roo, Mexico
e-mail: russ_net42@hotmail.com; russell.peraza@gmail.com

(Van Bogaert et al. 2007; Henkel et al. 2012, 2014). As a result, large quantities of detergents and their components enter the environment (Warne and Schifko 1999; Liwarska-Bizukojc et al. 2005). McAvoy et al. (1993) estimated that 5 % of all linear alkylbenzene sulfonate (LAS) produced in the United States reach the aquatic environment. In Mexico, 463,965 tons of detergents are manufactured annually (Sobrino-Figueroa 2013) most of which eventually find their way into freshwater and marine ecosystems (Cserháti et al. 2002).

The toxic effect of a detergent depends on its mode of action, the toxicity of the active ingredient (surfactant) in relation to its chemical and physical structure and the response of the test organism (Ofojekwu et al. 1999). LAS is the most common anionic surfactant used in the formulation of domestic and industrial detergents, and is considered an important potential pollutant (Ainsworth 1992). However, despite the fact that several studies have been undertaken to determine the toxicity of LAS in aquatic organisms, little is known about the toxic effects of detergent as a whole (Markina and Aizdaicher 2007).

The detergents we use in everyday life are complex mixtures of various compounds. A typical detergent consists of surfactants, builders, anti-redeposition agents, zeolite, alkaline agents, corrosion inhibitors, processing aids, colorants, fragrances, oxygen bleach, suds control agents, opacifiers, bleaching agents, enzymes and other minor constituents (Bajpai and Tyagi 2007). These components can interact either antagonistically, additively or synergistically to increase or diminish the toxicity of the detergent in the aquatic environment (Warne and Schifko 1999). Thus, evaluating the individual effects of surfactants or other ingredients does not reflect either the actual or net influence of a detergent on the aquatic environment, it is thus necessary to estimate the biological effects of detergent formulations as a whole (Markina and Aizdaicher 2007; Azizullah et al. 2011).

Polychaetes are generally the most abundant taxon in benthic communities and are commonly utilized as indicator species of environmental conditions in marine ecosystems (Dean 2008). *Capitella capitata* (Fabricius 1780), a species of Capitellidae Grube, 1862, is an opportunistic polychaete that has often been used as an indicator of organic pollution in marine sediments (Reish 1959; Bellan 1967; Pearson and Rosenberg 1978). It has also been extensively employed as a test organism in toxicological bioassays (Reish and LeMay 1991). The *C. capitata* species complex includes at least 50 cryptic siblings that differ mainly in their enzyme and general protein patterns, ecophysiological characteristics and reproduction modes (Méndez et al. 2000). The feeding characteristics of the adults and juveniles mean that they play an important role in the recycling and elimination of toxic substances associated with sediments. Ecotoxicological studies performed with metals (Reish 1988; Reish and LeMay 1991; Méndez and Green-Ruiz 2005, 2006) and other pollutants (Méndez 2005, Méndez et al. 2008, 2013) have registered diverse adverse effects on this species complex.

The introduction of toxic chemical substances, such as detergents, to the aquatic environment could thus negatively impact on the population dynamics of this species in their natural ecosystems. However, in Mexico, ecotoxicological studies of commercial detergent formulations are limited. The aim of this study was to determine the mean lethal concentration (LC_{50}) of three commercial detergents and their ecological risk in sediments, using the polychaete *Capitella* sp. C as the test species. We hope that the information gained will add to the available knowledge about toxicity levels of these detergents in aquatic environments in Mexico. Such knowledge is crucial for the development and implementation of policies associated with the protection of aquatic life in Chetumal Bay, Quintana Roo, Mexico.

Materials and methods

Worms

Adult *Capitella* sp. C individuals were collected from Chetumal Bay, Quintana Roo, Mexico (18°30′47.89″N, 88°16′30.87″W) in a zone free from wastewater discharges. The worms were collected with a PVC corer (0.018 m^2). Deep cores 20 cm deep were taken and then sieved using a 0.5-mm open mesh. The polychaetes were transported to the laboratory in vials containing water from the sampling site. In addition, sediment from the site was collected and transported in sealed polyethylene bags, with the purpose of using it during the adaptation period of the test organisms as well as in the bioassays. In the laboratory, undamaged organisms, 20–25 mm long were selected (polychaetes that were fragmented, pale in color or with reduced motility were discarded). Specimens were identified as *Capitella* sp. C with the aid of a stereoscopic microscope and an optical microscope using the keys given by Garza-García (2009). Individual *Capitella* sp. C were then placed in 3 L, constantly aired aquariums, together with water and sediment from the sampling site and maintained at room temperature (25 ± 1 °C) in a natural light–dark cycle for 2 days prior to the bioassays to allow them to adapt to the conditions (APHA et al. 1992). No food was provided either during acclimation or over the ecotoxicological test period.

Chemicals

Three commercial detergents labeled as biodegradable were evaluated, all of which are among the most commonly used domestic detergents in Mexico: ROMA®, FOCA® and BLANCA NIEVES®. In all cases, the active ingredient was the anionic surfactant LAS. Their usage and other ingredients that make up the detergents are listed in Table 1. It is important to note that neither the LAS concentrations nor those of the other components are stated on the product container: this datum is confidential and inaccessible to the consumer. Nevertheless, the proportions of some of the main constituents of detergents are generally: surfactant 15 %, poly-phosphate + silicate 30 %, sodium perborate 20 %, fluorescent pigment 0.1 %, sodium sulfate 20 % and enzymes 0.5 % (Sobrino-Figueroa 2013).

A stock solution was prepared for each detergent at 0.3 % (3 g) in 1 L of distilled water. This was then used to obtain serial dilutions (uniformly in a 0.5 logarithmic scale) as follows: 15.62, 31.25, 62.5, 125, 250 ppm (test concentrations). The LAS concentration was then calculated for each dilution of each detergent using the analysis method established by the Official Mexican Guideline NMX-AA-039-SCFI-2001 (Secretaria de Económica 2001).

Sediment

The sediment used for the ecotoxicological tests was previously analyzed (very fine sand with 0.81 % organic matter) and sieved with a 0.5-mm mesh to extract any organisms possibly present in the sample. The sediment samples were then refrigerated at 4 °C for 48 h before testing to avoid potential oxidation–reduction reactions and to eliminate any other undesired organisms.

Toxicity tests

Bioassays were of the static type without renewal of the test solution and for each assay a control (without detergent, 0 ppm) and five concentrations were used, each with three replicates. Tank dimensions were 10 × 15 × 22 cm. A total of 100 g of previously treated sediment were placed in each tank (replicate). Saline water was then added and 10 *Capitella* sp. C individuals were randomly selected from a total of 180 organisms for each bioassay chamber and placed in each tank. After waiting 45 min to allow the organisms to bury into the sediment (5 mm thick), the toxic solution was added to give a total volume of 2 L per replicate.

For each bioassay chamber, mortality readings were performed at 1, 2, 4, 8, 18, 24, 36 and 48 h of exposure. Dead organisms were immediately removed after examination. Individuals were considered dead when they fulfilled the following criteria: pale color; swollen and showing no movement on the surface of the sediment (APHA et al. 1992; Uc-Peraza and Delgado-Blas 2012); unresponsive to physical stimulation. Tests were rejected if the survival rate of the control group was 90 % or less. The following physico-chemical parameters were measured before and after the bioassays: temperature, pH, salinity and dissolved oxygen.

Table 1 Summary of the use and ingredients of the LAS-containing commercial detergents tested

Detergents	Detergent use	Ingredients
ROMA®	Multi-propose	Phosphates, sodium silicate, anti-redeposition agents, bleachers and perfume
FOCA®	Laundry	Proteolytic enzyme, phosphates, sodium silicate, anti-redeposition agents, bleachers and perfume
BLANCA NIEVES®	Multi-propose	Sodium silicate, carbonate, sulfate, sodium tripolyphosphate, anti-redeposition agents and perfume

Statistical analysis

The LC_{50} was calculated using the probit method (APHA et al. 1992). The results were then graphed as regression curves with probit units against the logarithm base 10 of detergent concentration using the software SigmaPlot 12.0, with 95 % confidence intervals. The values obtained from the acute toxicity tests were then subjected to a descriptive analysis to assess the general behavior of the data. A one-way analysis of variance (ANOVA, alpha = 0.05) was then performed to analyze differences in percent mortality of *Capitella* sp. C at 48 h among the concentrations of the detergents tested and among replicates. Pearsońs product–moment correlation analyses were then done to assess the relationship between detergent concentration and percent mortality.

Characterization of ecological risk

According to USEPA guidelines, ecological risk assessment (ERA) is defined as a process that evaluates the likelihood of adverse ecological effects on ecosystems exposed to one or more stressors (US EPA 1998; Gao et al. 2013). Risk assessment is thus, essentially, a simple comparison of predicted environmental concentration (PEC) with predicted no-effect concentration (PNEC), normally expressed as the "risk quotient" (RQ = PEC/PNEC) (Jensen et al. 2001), which is then used to determine ecological risk (Cristale et al. 2013; Gao et al. 2013). The principle is that when the risk quotient is greater than or equal to one there is a stronger likelihood of an impact. In contrast, a risk quotient of less than one indicates that the likelihood of an effect is low, and thus of less concern (Lam and Gray 2001). In this study, we examined the ERA by taking the PEC to be equivalent to the LAS value in sediments of 5.3 ppm (Sánchez-Peinado 2007). The PNEC was then calculated using the following equation: $PNEC = LC_{50}/AF$, where LC_{50} is the value obtained from the acute toxicity tests and AF is the assessment factor (100). For data interpretation, an RQ value ≥ 1 indicates a potentially high risk, and an RQ value <1 a nonsignificant (low) risk.

Results

The average values of the physico-chemical parameters of the water used in the bioassays showed little variation (Table 2). However, salinity was the only parameter that remained constant at 10 ‰.

The LC_{50} values at 48 h for the three detergents tested ranged from 70.79 to 147.91 ppm (Figs. 1, 2, 3), giving the following acute toxicity sequence: FOCA® > BLANCA NIEVES® > ROMA®. Table 3 shows the LC_{50} for the active ingredient (LAS) in each detergent, with their corresponding 95 % confidence intervals. These ranged from 15.48 and 22.38 ppm. It can be appreciated that FOCA® produced the lowest LC_{50} values, both for the formulation as a whole and the active ingredient.

The percentage mortality of *Capitella* sp. C exposed to different concentrations of the three formulations of commercial detergents at 48 h is presented in Table 4. It can be seen that no mortality was registered throughout the bioassays in any of the controls used (0 ppm). However, 3–27 % mortality was recorded for the most dilute detergent concentrations: 15.62 and 31.25 ppm, and between 20 and 57 % mortality was registered for 62.5 and 125 ppm. The solutions with the highest detergent concentration (250 ppm) produced the highest mortality percentages for all three formulations tested as follows: ROMA (70 %), BLANCA NIEVES® (87 %) and FOCA®(93 %). The ANOVA detected significant differences ($p < 0.05$) in percent mortality among the concentrations of each detergent. However, there were no significant differences ($p > 0.05$) between replicates or formulations. The correlation coefficient between detergent concentration and percent

Table 2 Means and standard errors for the physico-chemical parameters measured in the bioassays

Detergents	Temperature (°C)	Dissolved oxygen (mg/L)	pH	Salinity (‰)
ROMA®	24.97 ± 0.04	4.81 ± 0.05	8.15 ± 0.04	10
FOCA®	25.00 ± 0.07	3.98 ± 0.11	7.54 ± 0.06	10
BLANCA NIEVES®	25.02 ± 0.02	3.56 ± 0.09	7.88 ± 0.09	10

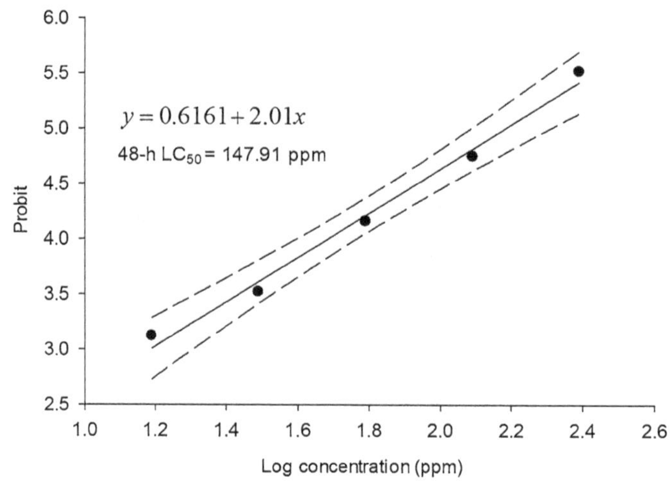

Fig. 1 Probit for the percent mortality of *Capitella sp.* C exposed to varying concentrations of ROMA® for 48 h under laboratory conditions

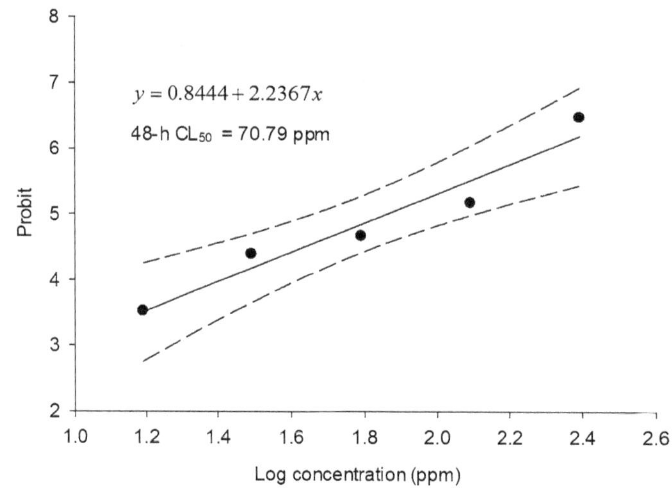

Fig. 2 Probit for the percent mortality of *Capitella sp.* C exposed to varying concentrations of FOCA® for 48 h under laboratory conditions

mortality was statistically significant ($p < 0.05$) in all cases. This means that percent mortality of *Capitella* sp. C increased with increase in the concentrations of the detergents tested over the assay period.

The RQ values calculated for the three commercial detergents were greater than one in all cases (Table 5). The highest RQ value was 7.48 (FOCA®) and the lowest 3.58 (ROMA®) with an average of 5.37. There is thus a high risk that these detergents will cause damage to aquatic ecosystems, particularly sediments.

Discussion

The three commercial detergents tested contain different toxic chemical compounds that may adversely affect aquatic ecosystems. It is known that detergents can be toxic to aquatic invertebrates, fish and plants (Warne 1995). Ecotoxicological studies with fish (Okoli-Anunobi et al. 2002; Omotoso and Fagbenro 2005; Sobrino-Figueroa 2013), algae (Aizdaicher and Markina 2006; Markina and Aizdaicher 2007; Azizullah et al. 2012), crustaceans and bacteria (Pedrazzani et al. 2012) have shown that these organisms are affected morphologically, physiologically and biochemically when exposed to detergents. It has been documented that the use of household washing detergents and softeners can be toxic for aquatic organisms in concentrations from 0.07 up

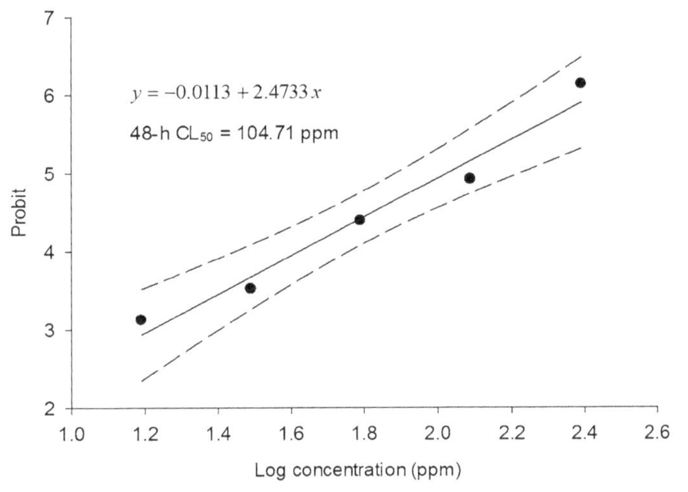

Fig. 3 Probit for the percent mortality of *Capitella sp.* C exposed to varying concentrations of BLANCA NIEVES® for 48 h under laboratory conditions

Table 3 LAS equivalent LC_{50} values at 48 h of *Capitella* sp. C. exposed to varying concentrations of ROMA®, FOCA® and BLANCA NIEVES® for 48 h under laboratory conditions

Detergents	LC_{50} (ppm)	Confidence intervals (95 %)
ROMA®	22.38	±13.42
FOCA®	15.48	±10.32
BLANCA NIEVES®	19.05	±12.55

Table 4 Percent mortality of *Capitella* sp. C. exposed to varying concentrations of ROMA®, FOCA® and BLANCA NIEVES® for 48 h under laboratory conditions

Concentration (ppm)	Mortality (%)		
	ROMA®	FOCA®	BLANCA NIEVES®
250	70	93	87
125	40	57	47
62.5	20	37	27
31.25	7	27	7
15.62	3	7	3
Control	0	0	0

Table 5 Ecological risk to the sediment of the detergents tested

Detergents	PNEC	RQ	Risk level
ROMA®	1.47	3.58	High
FOCA®	0.70	7.48	High
BLANCA NIEVES®	1.04	5.06	High

to 35.4 ppm (Ankley and Burkhard 1992; Pettersson et al. 2000). According to the literature, aquatic organisms are negatively affected by the presence of active ingredients in detergents, such as anionic and non-ionic surfactants, at concentrations ranging from 0.0025 to 300 ppm, and 0.3 to 200 ppm, respectively (Liwarska-Bizukojc et al. 2005). This may be more or less severe depending on the organism tested. In this study, the three detergents evaluated gave LC_{50} values of 70.79–147.91 ppm (formulation as a whole) and

15.48–22.38 ppm (active ingredient only) against *Capitella* sp. C after 48 h exposure, and the percent mortalities registered for the different concentrations of the detergents ranged from 3 to 93 %.

FOCA®, with LC_{50} values of 70.79 (whole formulation) and 15.48 ppm (active ingredient), was the most toxic detergent tested and also produced the highest percent mortality (93 %). These results are similar to those reported by Uc-Peraza and Delgado-Blas (2012) who examined the acute toxicity of these same commercial brands of detergents against other polychaete species. It is known that the toxicity of the detergents is due to the action the surfactants (Markina and Aizdaicher 2007) together with the other ingredients contained in the formulation (Sobrino-Figueroa 2013). It has been demonstrated that surfactants present in detergents can negatively affect living cells in different ways by damaging cell membranes (Mikolajczyk and Diehn 1978; Chawla et al. 1987), attaching to proteins, and affecting cell physiological and biochemical processes (Markina and Aizdaicher 2007; Azizullah et al. 2012). At low concentrations, detergents produce changes in cell membrane permeability, thus altering the function of the respiratory system in aquatic species (Sobrino-Figueroa 2013). In particular, anionic surfactants can solubilize and denature proteins and alter enzyme activity by binding or disrupting enzyme structure (Argese et al. 1994). In polychaetes, the anionic surfactant LAS can cause morphological damages (Conti 1987). Regarding other ingredients, it is known that enzymes (protease and lipase), sodium silicate, perfumes and bleachers (sodium perborate) can contribute to an increase in detergent toxicity for a wide range of aquatic organisms (Warne 1995; Warne and Schifko 1999; Markina and Aizdaicher 2007; Pedrazzani et al. 2012; Sobrino-Figueroa 2013).

BLANCA NIEVES® was less toxic to *Capitella sp.* C than FOCA®, possibly because this formulation does not contain enzymes and bleachers. However, it does contain other ingredients such as sodium silicate and perfumes, which may have contributed to the (lower) toxicity of the detergent. Furthermore, it cannot be ruled out that the other ingredients in the formulation, such as carbonates, sulfates, sodium and anti-redeposition agents may have contributed to its toxicity. Finally, ROMA® was found to be the least toxic of the three detergents tested.

When evaluating detergent toxicity, studies have often focused on surfactants as these are considered to be among the most toxic compounds contained in these materials, and also represent a high percentage, 15–40 %, of the total number of ingredients in the formulations (Scheibel 2004). Warne and Schifko (1999) found that surfactants contributed between 10.4 and 98.8 % of the toxicity of detergents they analyzed with a mean contribution of 40.7 %. Thus, the low toxicity of ROMA® might be due to the low concentrations of surfactant it contains, as well as the absence of other toxic ingredients (enzymes and sodium tripolyphosphate). In conclusion, the variations in toxicity between the detergents tested could be caused both by the concentration of the anionic surfactant LAS in the formulation and the presence of other ingredients (enzymes, sodium silicate, sodium tripolyphosphate, bleachers and perfumes).

The LAS LC_{50} values reported by this study with *Capitella sp.* C are similar to those registered by Uc-Peraza and Delgado-Blas (2012). These authors reported LAS toxicity (LC_{50}) values of between 12.88 and 14.12 ppm against *Laeonereis culveri* at 48 h. Similar values were also reported by Conti (1987) who determined the toxicity of two anionic surfactants: sodium dodecyl sulfate (SDS) and LAS in another polychaete species, *Arenicola marina* (SDS: LC_{50} 15.2 ppm; LAS: LC_{50} 12.5 ppm, at 48 h). This shows that *Capitella sp.* C is less sensitive to anionic surfactants than *A. marina* and *L. culveri* in acute toxicity tests.

Comparing our results with those for other groups of aquatic organisms, we can observe that *Capitella* sp. C is more resistant to detergents than crustaceans and fish. For example, Warne (1995) examined the toxicity of 24 detergents to the cladoceran *Ceriodaphnia* cf. *dubia* and found that toxicity levels (measured as EC_{50} at 48 h, immobilization) varied between 1.6 and 70.3 ppm. In another study, Pettersson et al. (2000) evaluated the toxicity of 26 commercial detergents using *Daphnia magna* as the test organism, reporting values between 4 and 85 ppm (EC_{50} at 48 h) for 25 of them. In fish, Okoli-Anunobi et al. (2002) calculated the acute toxicity of a commercial detergent to *Oreochromis niloticus* finding an LC_{50} of 9.77 ppm at 96 h. Omotoso and Fagbenro (2005) determined the toxicity of three commercial detergents with this same species registering LC_{50} values of between 12.04 and 41.88 ppm at 96 h. However, if we look at studies performed with mollusk species as test organisms, it can be observed that *Capitella* sp. C is more sensitive. For example, Iannacone and Alvariño (2002) undertook an ecotoxicological evaluation of three commercial detergents against different species of mollusks and reported LC_{50} values of 201.07 ppm (*Melanoides tuberculata*), 82.93 ppm (*Physa venustula*) and 71.41 ppm (*Heleobia cumingii*) at 48 h. The differences in sensitivity between polychaetes and mollusks may be explained by the way in which the detergent acts on the organism. In polychaetes

(*Arenicola marina*), anionic detergents (LAS) cause severe damage mainly at the level of the epidermis and in the region of the gills (Conti 1987). This is because, in the majority of cases, the soft bodies of the polychaetes are exposed to the environment and thus come into direct contact with pollutants. In contrast, the hard resistant shells of mollusks afford some protection from contaminants making them more tolerant. It is important to mention that the comparisons made here are based on a limited number of experiments, and the response of different species within taxa may vary with their biochemistry and physiology, the age of the test organisms, the detergent evaluated and the conditions under which the bioassays are conducted. More toxicity tests performed under similar conditions with a wide variety of aquatic species are thus needed.

Our results of the correlation analysis show that *Capitella* sp. C percent mortality increased significantly ($p < 0.05$) with increase in the concentrations of the detergents during the 48-h exposure period, coinciding with the results of Uc-Peraza and Delgado-Blas (2012). This observation also agrees with Okwuosa and Omoregie (1995) and Ogundele et al. (2005) who determined the acute toxicity of anionic surfactants in fish. In another study, Aizdaicher and Markina (2006) evaluated the toxicity of two detergents and one anionic surfactant to an algal species (*Plagioselmis prolonga*) and also observed that the effects on the test organisms increased with increasing concentrations of these substances.

The risk quotient value of the three LAS-containing detergents indicates that they are high-risk substances likely to cause damage to aquatic biota, particularly sediment-dwelling organisms such as *Capitella* sp. C. This is because the motility and feeding characteristics of these organisms make them more vulnerable to the impact of detergents. It has been demonstrated that 20–30 % of surfactants entering the sea environment easily accumulate in sediments and can affect bottom-dwelling biota (Marín et al. 1991; Marin et al. 1994; Markina and Aizdaicher 2007). The effects are particularly acute in those zones where there are outfalls of raw sewage or sewage effluent, as is the case at several sites within the urban shoreline zone at Chetumal, Quintana Roo, where clandestine discharges of soapy water have been detected. Surfactants and other ingredients contained in detergents are common components of the domestic and municipal residual output which eventually reaches the natural environment causing diverse toxic effects to the aquatic organisms present (Ankley and Burkhard 1992; Pettersson et al. 2000, Azizullah et al. 2011). Warne and Schifko (1999) mention that surfactants and sodium silicates are the main contributors to detergent toxicity, and were components of the three detergents evaluated in this study. Surfactants are used as cleaning agents while sodium silicates act as water softeners (Table 1). Specifically, the toxic effects of the anionic surfactant LAS have been examined in several experiments with different groups of aquatic organisms providing evidence of its toxicity (Marín et al. 1991; Bao-Quey and Dar-Yi 1994; Jorgensen and Christoffersen 2000; Rosety et al. 2001; Hampel et al. 2001; Christoffersen et al. 2003; Stefanoni and Abessa 2008; Coelho and Rocha 2010). However, it has been determined that the ecological risk of LAS to aquatic organisms is low (Fendinger et al. 1994; Van de Plassche et al. 1999; Versteeg et al. 1999; Temara et al. 2002). Despite this, both this study and that undertaken by Uc-Peraza and Delgado-Blas (2012) show that when the effects of the surfactant and the other ingredients contained in detergents are considered as a whole, there is a high risk that the aquatic biota will be adversely affected. Commercial detergents are complex mixtures of different compounds which can interact either antagonistically, additively or synergistically, thus modifying their toxicity to the aquatic environment (Warne and Schifko 1999). The observed effects can thus be considered as the net impact of the detergent as a whole (Azizullah et al. 2011). Studies undertaken using different aquatic organisms (Aizdaicher and Markina 2006; Sobrino-Figueroa 2013) have demonstrated that detergent formulations taken as a whole may be more toxic than the active ingredients they contain, with significant ecological consequences for shore ecosystems both at high (10 ppm) and low (0.1–1 ppm) concentrations (Aizdaicher and Markina 2006). Nevertheless, given that the results obtained in this study with *Capitella* sp. C cannot be generalized with other taxa, we recommend that ecotoxicological bioassays at sub-lethal doses to test for chronic toxicity are performed, both with polychaetes and other groups of aquatic organisms representative of the aquatic biota in Chetumal Bay, Quintana Roo, Mexico.

Conclusions

Three formulations of commercial detergents tested showed LC_{50} values (48 h exposure) for *Capitella* sp. C between 70.79–147.91 ppm and 15.48–22.38 ppm corresponding to the formulation and the active ingredient, respectively. The percentage of mortality among the concentrations of the three detergents was 3–93 %. FOCA® was the most toxic detergent of all. The variation in the toxicity of the three detergents could have been caused both by differences in the relative concentrations of LAS contained in each formulation and the presence of other ingredients (enzymes, sodium silicate, sodium tripolyphosphate, bleachers and perfumes) which can also increase formulation toxicity. Finally, the ecological risk assessment for the three detergents indicates that there is a high risk to the aquatic biota, especially sediment-dwelling organisms such as the test species.

Acknowledgments This study was done with the support of the División de Ciencias e Ingeniería of the Universidad of Quintana Roo, México. We would like to thank two anonymous reviewers whose comments and observations have substantially improved this manuscript.

References

Ainsworth SJ (1992) Soaps and detergents. Chem Eng News 70(3):27–37

Aizdaicher NA, Markina ZV (2006) Toxicity effects of detergents on the alga *Plagioselmis prolonga* (Cryptophyta). Russ J Mar Biol 32:45–49

Ankley GT, Burkhard P (1992) Identification of surfactants as toxicants in a primary effluent. Environ Toxicol Chem 11(9):1235–1248

APHA, Awwa, WPCF (American Public Health Association, American Water Works Association and Water Pollution Control Federation) (1992) Métodos normalizados para el análisis de aguas potables y residuals. Ediciones Díaz de Santos, Madrid

Argese E, Marcomini A, Miana P, Bettiol C, Perin G (1994) Submitochondrial particle response to linear alkylbenzene sulfonates, nonylphenol polyethoxylates and their biodegradation derivatives. Environ Toxicol Chem 13(5):737–742

Azizullah A, Richter P, Häder DP (2011) Toxicity assessment of a common laundry detergent using the freshwater flagellate *Euglena gracilis*. Chemosphere 84(10):1392–1400

Azizullah A, Richter P, Jamil M, Häder DP (2012) Chronic toxicity of a laundry detergent to the freshwater flagellate *Euglena gracilis*. Ecotoxicology 21(7):1957–1964

Bajpai D, Tyagi VK (2007) Laundry detergents: an overview. J Oleo Sci 56(7):327–340

Bao-Quey H, Dar-Yi W (1994) Effects of linear alkylbenzene sulfonate (LAS) on the respiratory functions of Tigerperch (*Terapon jarbua*). Zool Stud 33(3):205–210

Bellan G (1967) Pollution et peuplement benthiques sur substrat meuble dans la region de Marseille. 1 Partie. Le secteur de Cortiu. Rev Int Océanogr Méd 6–7:53–87

Chawla G, Viswanathan PN, Devi S (1987) Biochemical studies on the toxicity of linear alkylbenzene sulphonate to *Scenedesmus quadricauda* in culture. Environ Exp Bot 27:311–319

Christoffersen K, Hansen BW, Johansson LS, Krog E (2003) Influence of LAS on marine calanoid copepod population dynamics and potential reproduction. Aquat Toxicol 63(4):405–416

Coelho KS, Rocha O (2010) Assessment of the potential toxicity of a linear alkylbenzene sulfonate (LAS) to freshwater animal life by means of cladoceran bioassays. Ecotoxicology 19(4):812–818

Conti E (1987) Acute toxicity of three detergents and two insecticides in the lugworm, *Arenicola marina* (L.): a histological and a scanning electron microscopic study. Aquat Toxicol 10:325–334

Cristale J, Katsoyiannis A, Sweetman AJ, Jones KC, Lacorte S (2013) Occurrence and risk assessment of organophosphorus and brominated flame retardants in the River Aire (UK). Environ Pollut 179:194–200

Cserháti T, Forgacs E, Oros G (2002) Biological activity and environmental impact of anionic surfactants. Environ Inter 28(5):337–348

Dean HK (2008) The use of polychaetes (Annelida) as indicator species of marine pollution: a review. Rev Biol Trop 56(4):11–38

Fendinger NJ, Versteeg DJ, Weeg E, Dyer S, Rapaport RA (1994) Environmental behavior and fate of anionic surfactants. In: Baker L (ed) Environmental chemistry of lakes and reservoirs. American Chemical Society, Washington, pp 527–557

Gao P, Li Z, Gibson M, Gao H (2013) Ecological risk assessment of nonylphenol in coastal waters of China based on species sensitivity distribution model. Chemosphere 104:113–119

Garza-García ME (2009) Capitellidae Grube, 1862. In: de León-González JA, Bastida-Zabala JR, Carrera-Parra LF, García-Garza ME, Peña-Rivera A, Salazar-Vallejo SI, Solís-Weiss V (eds) Poliquetos (Annelida: Polychaeta) de México y América Tropical. Monterrey, pp 101–114

Hampel M, Moreno-Garrido I, Sobrino C, Lubian LM, Blasco J (2001) Acute toxicity of LAS in marine microalgae: esterase activity and inhibition growth as endpoints of toxicity. Ecotoxicol Environ Saf 48(3):287–292

Henkel M, Müller MM, Kügler JH, Lovaglio RB, Contiero J, Syldatk C, Huasmann R (2012) Rhamnolipids as biosurfactants from renewable resources: concepts for next-generation rhamnolipid production. Process Biochem 47(8):1207–1219

Henkel M, Sydatk C, Hausmann R (2014) The prospects for the production of Rhamnolipids on renewable resources. In: Kosaric N, Sukan FV (eds) Biosurfactants: production and utilization-processes technologies, and economics, 1st edn. CRC Press, Florida, pp 83–99

Iannacone J, Alvariño L (2002) Efecto del detergente doméstico alquil aril sulfonato de sodio lineal (LAS) sobre la mortalidad de tres caracoles dulceacuícolas en el Perú. Ecol Aplic 1:81–87

Info Mine Research Group (2012) Synthetic detergents and cleaning products in the CIS and Baltic Countries: production, market and forecast. Info Mine Market Research Group. http://www.infomine.ru/files/catalog/166/file_166_eng.pdf. Accessed 24 August 2015

Jensen J, Løkke H, Holmstrup M, Krogh PH, Elsgaard L (2001) Effects and risk assessment of linear alkylbenzene sulfonates in agricultural soil. 5. Probabilistic risk assessment of linear alkylbenzene sulfonates in sludge-amended soil. Environ Toxicol Chem 20(8):1690–1697

Jorgensen E, Christoffersen K (2000) Short-term effects of linear alkylbenzene sulfonate on freshwater plankton studied under field conditions. Environ Toxicol Chem 19(4):904–911

Lam PK, Gray JS (2001) Predicting effects of toxic chemicals in the marine environment. Mar Pollut Bull 42(3):169–173

Liwarska-Bizukojc E, Miksch K, Malachowska-Jutsz A, Kalka J (2005) Acute toxicity and genotoxicity of five selected anionic and nonionic surfactants. Chemosphere 58(9):1249–1253

Marin MG, Pivotti L, Campesan G, Turchetto M, Tallandini L (1994) Effects and fate of sediment-sorbed linear alkylbenzene sulphonate (LAS) on the bivalve mollusc *Mytilus galloprovincialis* Lmk. Wat Res 28:85–90

Marín MG, Bressan M, Brunetti R (1991) Effects of linear alkylbenzene sulphonate (LAS) on two marine benthic organisms. Aquat Toxicol 19(3):241–248

Markina ZV, Aizdaicher NA (2007) Influence of laundry detergents on the abundance dynamics and physiological state of the benthic microalga *Attheya ussurensis* (Bacillariophyta) in laboratory culture. Russ J Mar Biol 33(6):391–398

McAvoy DC, Eckhoff WS, Rapaport RA (1993) Fate of linear alkylbenzene sulfonate in the environment. Environ Toxicol Chem 12(6):977–987

Méndez N (2005) Effects of teflubenzuron on larvae and juveniles of the polychaete Capitella sp. B from Barcelona, Spain. Water Air Soil Pollut 160:259–269

Méndez N, Green-Ruiz C (2005) Preliminary observations of cadmium and copper effects on juveniles of the polychaete *Capitella* sp. Y (Annelida: Polychaeta) from Estero del Yugo, Mazatlán, México. Rev Chil Hist Nat 78:701–710

Méndez N, Green-Ruiz C (2006) Cadmium and copper effects on larval development and mortality of the polychaete *Capitella* sp. Y from Estero del Yugo, Mazatlán, México. Water Air Soil Pollut 171:291–299

Méndez N, Linke-Gamenick INEZ, Forbes VE (2000) Variability in reproductive mode and larval development within the *Capitella capitata* species complex. Invert Reprod Develop 38(2):131–142

Méndez N, Anguas-Cabrera DN, García-de la Parra LM (2008) Effects of methamidophos on sediment processing and body mass of *Capitella sp.* Y from Estero del Yugo, Mazatlán, México. J Exp Mar Biol Ecol 361:92–97

Méndez N, Lacorte S, Barata C (2013) Effects of the pharmaceutical fluoxetine in spiked-sediments on feeding activity and growth of the polychaete *Capitella teleta*. Mar Environ Res 89:76–82

Mikolajczyk E, Diehn B (1978) Morphological alteration in Euglena gracilis induced by treatment with CTAB (Cetyltrimethylammonium bromide) and Triton X-100: correlations with effects on photophobic behavior response. J Protozool 25:461–470

Ofojekwu PC, Okwuosa VN, Omoregie E (1999) Effect of alkylbenzene sulphonate (ABS) detergent on the weight gain of the toothed carp, *Aphyosemion gairdneri* (L.). J Environ Sci 3(2):234–238

Ogundele O, Ihuahi JA, Omojowo FS, Bitrus P (2005) Toxicity of linear alkylbenzene sulphonate (LAS) detergent to *Clarias gariepinus* fingerlings. http://aquaticcommons.org/4036/1/273.pdf. Accessed 24 January 2015

Okoli-Anunobi CAIN, Ufodike EBC, Chude LA (2002) Lethal effects of the detergent, Elephant Blue® on the Nile Tilapia, *Oreochromis niloticus* (L.). J Aquat Sci 17(2):95–97

Okwuosa VN, Omoregie EO (1995) Acute toxicity of alkylbenzene sulphonate (ABS) detergent to the toothed carp, *Aphyosemion gairdneri* (L.). Aquacult Res 26(10):755–758

Omotoso FO, Fagbenro OA (2005) A comparative study on the toxicity of three commercial detergents on the survival of the Nile Tilapia, *Oreochromis niloticus*. J Agric Res Develop 4(2):139–147

Pearson TH, Rosenberg R (1978) Macrobenthic succession in relation to organic enrichment and pollution of the marine environment. Oceanogr Mar Biol Ann Rev 16:229–311

Pedrazzani R, Ceretti E, Zerbini I, Casale R, Gozio E, Bertanza G, Gelatti U, Donato F, Feretti D (2012) Biodegradability, toxicity and mutagenicity of detergents: integrated experimental evaluations. Ecotoxicol Environ Saf 84:274–281

Pettersson A, Adamsson M, Dave G (2000) Toxicity and detoxification of Swedish detergents and softener products. Chemosphere 41(10):1611–1620

Reish DJ (1959) An ecological study of pollution in Los Angeles-Long Beach Harbors, California. Occas Papers Allan Hancock Found Publ 22:1–117

Reish DJ (1988) The use of toxicity testing in marine environmental research. In: Soule DF, Kleppel GS (eds) Marine organisms as indicators. Springer, New York, pp 231–245

Reish DJ, LeMay JA (1991) Toxicity and bioconcentration of metals and organic compounds by polychaeta. Ophelia Suppl 5:653–660

Rosety M, Ordóñez FJ, Rosety-Rodríguez M, Rosety JM, Rosety I, Carrasco C, Ribelles A (2001) Acute toxicity of anionic surfactants sodium dodecyl sulphonate (SDS) and linear alkylbenzene sulphonate (LAS) on the fertilizing capability of gilthead (*Sparus aurata* L.) sperm. Histol Histopathol 16(3):839–843

Sánchez-Peinado MDM (2007) Efectos biológicos de los sulfonatos de alquilbenceno lineales (LAS) en suelo agrícola: biotransformación y estudios de biodiversidad. http://hdl.handle.net/10481/1555. Accessed 13 March 2015

Scheibel J (2004) The evolution of anionic surfactant technology to meet the requirements of the laundry detergent industry. J Surfact Deterg 7(4):319–328

Secretaria de Económica (2001) Análisis de aguas–Determinación de sustancias activas al azul de metileno (SAAM) en aguas naturales, potables, residuales y residuales tratadas—Método de Prueba. http://www.conagua.gob.mx/CONAGUA07/Noticias/NMX-AA-039-SCFI-2001.pdf. Accessed 22 November 2014

Sobrino-Figueroa AS (2013) Evaluation of oxidative stress and genetic damage cause by detergents in the zebrafish *Dario rerio* (Cyprinidae). Comp Biochem Physiol Part A Mol Integr Physiol 165(4):528–532

Stefanoni MF, Abessa DMS (2008) Lysossomal membrane stability of the brown mussel *Perna perna* (Linnaeus) (Mollusca, Bivalvia) exposed to the anionic surfactant linear alkylbenzene sulphonate (LAS). Panam J Aquat Sci 3(1):6–9

Temara A, Carr G, Webb S, Versteeg D, Feijtel T (2002) Marine risk assessment: linear alkylbenzene sulponates (LAS) in the North Sea. Mar Pollut Bull 42(8):635–642

Tkachenko FP, Kutsyn EB (2002) Detergent effect upon amino-acid composition of protein in the green alga *Cladophora vagabunda* (L.) Hoek. Gidrobiologicheskii Zh 38(3):94–98

Uc-Peraza RG, Delgado-Blas VH (2012) Determinación de la concentración letal media (CL_{50}) de cuatro detergentes domésticos biodegradables en *Laeonereis culveri* (Webster, 1879) (Polychaeta, Annelida). Rev Int Contam Ambient 28(2):137–144

US EPA (US Environmental Protection Agency) (1998) Guidelines for ecological risk assessment. US Environmental Protection Agency. EPA/630/R095/002F Washington, DC

Van Bogaert IN, Saerens K, De Muynck C, Develter D, Soetaer W, Vandamme EJ (2007) Microbial production and application of sophorolipids. Appl Microbiol Biotechnol 76:23–24

van de Plassche EJ, de Bruijn JHM, Stephenson RR, Marshall SJ, Feijtel TCJ, Belanger SE (1999) Predicted no-effect concentrations and risk characterization of four surfactants: linear alkyl benzene sulfonate, alcohol ethoxylates, alcohol ethoxylated sulfates and soap. Environ Toxicol Chem 18(11):2653–2663

Versteeg LD, Belanger SE, Carr GJ (1999) Understanding single species and model ecosystem sensitivity: a data based comparison. Environ Toxicol Chem 18(6):1329–1346

Warne MSJ (1995) Acute toxicity of laundry detergents to on Australian Cladoceran (*Ceriodaphnia dubia*). Aust J Ecotoxicol 1(2):127–135

Warne MSJ, Schifko AD (1999) Toxicity of laundry detergent components to a freshwater cladoceran and their contribution to detergent toxicity. Ecotoxicol Environ Saf 44(2):196–206

PERMISSIONS

All chapters in this book were first published in IAR, by Springer; hereby published with permission under the Creative Commons Attribution License or equivalent. Every chapter published in this book has been scrutinized by our experts. Their significance has been extensively debated. The topics covered herein carry significant findings which will fuel the growth of the discipline. They may even be implemented as practical applications or may be referred to as a beginning point for another development.

The contributors of this book come from diverse backgrounds, making this book a truly international effort. This book will bring forth new frontiers with its revolutionizing research information and detailed analysis of the nascent developments around the world.

We would like to thank all the contributing authors for lending their expertise to make the book truly unique. They have played a crucial role in the development of this book. Without their invaluable contributions this book wouldn't have been possible. They have made vital efforts to compile up to date information on the varied aspects of this subject to make this book a valuable addition to the collection of many professionals and students.

This book was conceptualized with the vision of imparting up-to-date information and advanced data in this field. To ensure the same, a matchless editorial board was set up. Every individual on the board went through rigorous rounds of assessment to prove their worth. After which they invested a large part of their time researching and compiling the most relevant data for our readers.

The editorial board has been involved in producing this book since its inception. They have spent rigorous hours researching and exploring the diverse topics which have resulted in the successful publishing of this book. They have passed on their knowledge of decades through this book. To expedite this challenging task, the publisher supported the team at every step. A small team of assistant editors was also appointed to further simplify the editing procedure and attain best results for the readers.

Apart from the editorial board, the designing team has also invested a significant amount of their time in understanding the subject and creating the most relevant covers. They scrutinized every image to scout for the most suitable representation of the subject and create an appropriate cover for the book.

The publishing team has been an ardent support to the editorial, designing and production team. Their endless efforts to recruit the best for this project, has resulted in the accomplishment of this book. They are a veteran in the field of academics and their pool of knowledge is as vast as their experience in printing. Their expertise and guidance has proved useful at every step. Their uncompromising quality standards have made this book an exceptional effort. Their encouragement from time to time has been an inspiration for everyone.

The publisher and the editorial board hope that this book will prove to be a valuable piece of knowledge for researchers, students, practitioners and scholars across the globe.

LIST OF CONTRIBUTORS

Mehmet Borga Ergönül
Department of Biology, Faculty of Science, Ankara University, Tandoğan, 06100 Ankara, Turkey

Jan Breine
Instituut voor Natuur-en Bosonderzoek, Dwersbos 28, Linkebeek, 1630 Brussels, Belgium

Erika Van den Bergh
Instituut voor Natuur-en Bosonderzoek, Havenlaan 88 bus 73, 1000 Brussels, Belgium

G. V. Garaffo, M. L. Jaubet, E. N. Llanos, M. A. Saracho Bottero and R. Elías
Departamento de Ciencias Marinas, Facultad de Ciencias Exactas y Naturales, Instituto de Investigaciones Marinas y Costeras (IIMyC), Universidad Nacional de Mar del Plata, Deán Funes 3350, 7602 AYL Mar del Plata, Argentina Consejo Nacional de Investigaciones Científicas y Técnicas (CONICET), Av. Rivadavia 1917, C1033AAJ Buenos Aires, Argentina

Mercedes González-Wangüemert, Maria João Rodrigues, Luísa Barreira and Luísa Custódio
Center of Marine Sciences, Faculty of Sciences and Technology, University of Algarve, Ed. 7, Campus of Gambelas, 8005-139 Faro, Portugal

Christina C. Roggatz
Center of Marine Sciences, Faculty of Sciences and Technology, University of Algarve, Ed. 7, Campus of Gambelas, 8005-139 Faro, Portugal
Department of Chemistry, University of Hull, Cottingham Road, Hull HU6 7RX, UK

Manuela Moreira da Silva
Center of Marine and Environmental Research (CIMA), University of Algarve, Campus de Gambelas, 8005-139 Faro, Portugal

N. Heydari, S. M. R. Fatemi and A. Mashinchian
Department of Natural Resources and Environment, Science and Research Branch, Islamic Azad University, Tehran, Iran

R. Musavi Nadushan
Department of Fishery, Marine Science and Technology Campus, Islamic Azad University, North Branch of Tehran, Zaaferaniyeh, Tehran, Iran

B. Raeisi
Department of Mathematics, Basic Sciences School, Shahed University, Tehran, Iran

Takeo Kurihara, Kazumasa Hashimoto, Shoji Nakano, Yukihiko Matsuyama and Kazumaro Okamura
Seikai National Fisheries Research Institute, Japan Fisheries Research and Education Agency, 1551-8 Tairamachi, Nagasaki 851-2213, Japan

Atsushi Ito and Masaei Kanematsu
National Research Institute of Fisheries and Environment of Inland Sea, Japan Fisheries Research and Education Agency, 1760 Momoshima, Onomichi, Hiroshima 722-0061, Japan

Jaakko Johannes Leppänen
Environmental Change Research Unit (ECRU) and Helsinki Institute of Sustainability Science, Faculty of Biological and Environmental Sciences, University of Helsinki, 00014 Helsinki, Finland

Yaqin Li, Shannon L. Meseck, Mark S. Dixon and Gary H. Wikfors
NOAA, National Marine Fisheries Service, Northeast Fisheries Science Center, Milford Laboratory, Milford, CT 06460, USA

Armin Namayandeh
Taxanama Corp., 25 Mabelle Ave., Toronto, ON, Canada

David V. Beresford
Environmental and Life Sciences-Biology Department, Trent University, 2140 East Bank Drive, Peterborough, ON K9L 1Z8, Canada

Keith M. Somers
Department of Ecology and Evolutionary Biology, University of Toronto, Toronto, Canada

Peter J. Dillon
Environmental and Life Sciences-Chemistry Department, Trent University, 1600 West Bank Drive, Peterborough, ON K9J 7B8, Canada

Eduardo Antonio Ríos-Villamizar and Maria T. F. Piedade
Instituto Nacional de Pesquisas da Amazônia/ Ecologia, Monitoramento e Uso Sustentável de Áreas Úmidas Amazônicas (Grupo MAUA/ CDAM/INPA), Campus I, Av. André Araújo, 2.936, Petrópolis, Caixa Postal 478, CEP: 69067-375 Manaus, Brazil

Wolfgang J. Junk
Instituto Nacional de Ciência e Tecnologia em Áreas Úmidas (INCT-INAU -UFMT), Cuiabá, Brazil

A. V. Waichman
Universidade Federal do Amazonas - UFAM/ICB, Manaus, Brazil

James R. Tweedley, Chris S. Hallett and Stephen J. Beatty
School of Veterinary and Life Sciences, Centre for Fish and Fisheries Research, Murdoch University, 90 South Street, Murdoch, WA 6150, Australia

Jason A. Crisp, James R. Tweedley and Navid R. Moheimani
School of Veterinary and Life Sciences, Murdoch University, Perth, Australia

Gavin J. Partridge
School of Veterinary and Life Sciences, Murdoch University, Perth, Australia
Australian Centre for Applied Aquaculture Research, South Metropolitan TAFE, Fremantle, Australia

Frances M. L. D'Souza
Department of Water, Perth, Australia

Ritindra N. Bhaduri
Department of Biological Sciences, California State University Stanislaus, One University Circle, Turlock, CA 95382, USA

M. Wolf
Oregon Institute of Marine Biology, University of Oregon, Charleston, OR 97420, USA

Ulf Båmstedt and Henrik Larsson
Umeå Marine Sciences Centre, University of Umeå, Norrbyn, 905 71 Hörnefors, Sweden

Vanessa Ochi Agostini
Laboratório de Zooplâncton - Instituto de Oceanografia da Universidade Federal do Rio Grande (FURG), Caixa Postal, 474, CEP: 96203-900 Rio Grande, RS, Brazil
Programa de Pós-graduação em Oceanografia Biológica (PPGOB), Bolsista do Conselho Nacional de Desenvolvimento Científico e Tecnológico (CNPq), Rio Grande, Brazil
Laboratório de Microcontaminantes Orgânicos e Ecotoxicologia Aquática, CAPES Postdoctoral fellow at Programa de Pós- Graduação em Oceanografia Física, Química e Geológica, Instituto de Oceanografia, Universidade Federal do Rio Grande, Rio Grande, Brazil

Erik Muxagata
Laboratório de Zooplâncton - Instituto de Oceanografia da Universidade Federal do Rio Grande (FURG), Caixa Postal, 474, CEP: 96203-900 Rio Grande, RS, Brazil

Alexandre José Macedo
Laboratório de Biofilmes e Diversidade Microbiana - Faculdade de Farmácia e Centro de Biotecnologia da Universidade Federal do Rio Grande do Sul (UFRGS), Av. Ipiranga, 2752, Bairro Azenha, 90610-000 Porto Alegre, RS, Brazil

Sulaeman
Curtin Aquaculture Research Laboratory, Department of Environment and Agriculture, Curtin University, 1 Turner Avenue, Bentley, WA 6102, Australia
Research Institute for Coastal Aquaculture (RICA), Jl. Makmur Dg. Sitakka 129, Maros 90511, Indonesia

Ravi Fotedar
Curtin Aquaculture Research Laboratory, Department of Environment and Agriculture, Curtin University, 1 Turner Avenue, Bentley, WA 6102, Australia

Mohsen Abdel-Tawwab and Sherien H. H. Shady
Department of Fish Biology and Ecology, Central Laboratory for Aquaculture Research, Abbassa, Abo-Hammad, Sharqia, Egypt

Gamal O. El-Sayed
Chemistry Department, Faculty of Science, Benha University, Benha, Egypt

V. Kathiravan
Central Institute of Brackishwater Aquaculture, 75, Santhome High Road, R.A. Puram, Chennai 600028, India
Department of Botany and Research Center, The Madura College (Autonomous), Madurai 625011, India

K. K. Krishnani
School of Edaphic Stress Management, National Institute of Abiotic Stress Management, Baramati (Dist. Pune) 413 115, India
Central Institute of Brackishwater Aquaculture, 75, Santhome High Road, R.A. Puram, Chennai 600028, India

R. G. Uc-Peraza and V. H. Delgado-Blas
Divisio´n de Ciencias e Ingenierı´a, Universidad de Quintana Roo, 77019 Chetumal, Quintana Roo, Mexico

Index

A
Anadara Kagoshimensis, 61-66, 68, 70-71
Aspartic Acid, 37
Atrina Pectinata, 62, 70-71
Atrina Sp., 61-70
Azti Marine Biotic Index, 17, 19

B
Babka Gymnotrachelus, 112
Belostomatidae, 112-113, 121-122
Benthic Invertebrates, 100-102, 107-109
Bioaccumulation, 32, 201-202, 206-207
Biological Traits Analysis, 17-18, 20, 22
Bosmina Spp., 72, 78, 82

C
Calcareous Rhodoliths, 154
Carassius Auratus, 131, 133, 135-136, 139-142
Caulerpa Prolifera, 33
Chaetoceros, 50, 53, 92, 145, 186
Chlorella Sp., 52
Chlorophyta, 45, 48-49, 51-52, 56-57, 160
Chydorus Sphaericus, 72, 75, 83-84
Cladocera, 72-74, 76-78, 82-86
Coastal Waters, 16, 19, 26, 99, 228
Cymodocea Nodosa, 33
Cyprinus Carpio, 207

D
Daphnia Magna, 73, 85, 226
Diplonychus Esakii, 112, 117
Docosahexaenoic Acid, 38

E
Ecological Quality Ratio, 1-2
Euglena, 50, 53, 92-93, 99, 228-229
Euphotic Zone, 87, 89-90, 92, 97, 174-175
Exuviaella Cordata, 51-52, 55, 59

F
Feed Conversion Ratio, 198
Fish-based Index, 1-3, 5, 9, 13, 15-16

G
Gambusia Affinis, 112-114, 118, 120-121, 141
Gambusia Holbrooki, 11, 131, 135-136, 141
Gracilaria Tikvahiae, 97

H
H. Polii, 32-34, 36-41
H. Tubulosa, 32-34, 36-42
Halacarid Mites, 153
Halosaccion Glandiforme, 153-155, 158-160
Headwater Streams, 30, 100-101, 103, 107, 109-110
Heteropneustes Fossilis, 197
Holothuria Mammata, 32-33, 36, 41

I
Index of Biotic Integrity, 1-2, 13, 15-16
Invertebrate Fauna, 153, 155, 157

L
Lepthatherina Wallacei, 131, 135-137
Litopenaeus Vannamei, 150

M
Macrobenthic Functional Structure, 17, 28
Mesocosms, 161-165, 169, 173-174, 176-177
Metapenaeus Dalli, 143, 148-149, 151-152
Molt Remnants, 153, 155-156
Multi-mesh Gill Nets, 4
Mytilus Galloprovincialis Lmk., 229

N
Natural Flow Regime, 1, 4
Nile Tilapia, 201-207, 229
Non-metric Multidimensional Scaling, 103, 106, 134, 137

O
Oreochromis Niloticus, 199, 201-202, 207, 226, 229
Oscillatoria Limosa, 51, 55, 59
Oxidative Stress, 230

P
Palmitoleic Acid, 38
Phyla Diatoms, 45
Phytoplankton Biomass, 98-99
Phytoplankton Diversity, 45, 48, 54
Poecilia Mexicana, 113, 118, 121-122
Precambrian Shield, 100-102, 108, 110

Predator-prey Interaction, 112-113
Pseudogobius Olorum, 131, 135-137
Pycnocline Strength, 161, 164, 174
Pyrrophyta, 45, 48-49, 51-52, 55-57, 59

R
Red Alga, 153-155, 157-160
Ruditapes Philippinarum, 61-68, 70, 208

S
Swan-canning Estuary, 143-145, 150, 152

T
Thalassionema Nitzschiodes, 45, 51, 54-55, 59
Thermocline Strength, 166
Total Deforestation Values, 123-124
Total Suspended Solids, 89-92, 123, 125, 127, 129

U
Unionized Ammonia, 203

W
Water Framework Directive, 1, 13-15, 29-30, 43, 84
Wet Weight Biomass, 35
Whitewater Systems, 123, 129

Z
Zooplankton, 7, 72, 75-76, 83-86, 97, 139, 144, 175, 177-181, 188-189

www.ingramcontent.com/pod-product-compliance
Lightning Source LLC
LaVergne TN
LVHW080735210926
848485LV00003B/563